Feedback Systems

Feedback Systems

An Introduction for Scientists and Engineers

Karl Johan Åström
Richard M. Murray

PRINCETON UNIVERSITY PRESS

PRINCETON AND OXFORD

Published by Princeton University Press
41 William Street, Princeton, New Jersey 08540

In the United Kingdom: Princeton University Press
6 Oxford Street, Woodstock, Oxfordshire OX20 1TW

Library of Congress Cataloging-in-Publication Data
Åström, Karl J. (Karl Johan), 1934-
    Feedback systems : an introduction for scientists and engineers / Karl Johan
Åström and Richard M. Murray
p. cm.
    Includes bibliographical references and index.
    ISBN-13: 978-0-691-13576-2 (alk. paper)
    ISBN-10: 0-691-13576-2 (alk. paper)
    1. Feedback control systems. I. Murray, Richard M., 1963-. II. Title.
TJ216.A78 2008
629.8′3–dc22   2007061033

British Library Cataloging-in-Publication Data is available

This book has been composed in LaTeX

The publisher would like to acknowledge the authors of this volume for providing
the camera-ready copy from which this book was printed.

Printed on acid-free paper. ∞

press.princeton.edu

Printed in the United States of America

10 9 8 7 6 5

Contents

Preface

This book provides an introduction to the basic principles and tools for the design and analysis of feedback systems. It is intended to serve a diverse audience of scientists and engineers who are interested in understanding and utilizing feedback in physical, biological, information and social systems. We have attempted to keep the mathematical prerequisites to a minimum while being careful not to sacrifice rigor in the process. We have also attempted to make use of examples from a variety of disciplines, illustrating the generality of many of the tools while at the same time showing how they can be applied in specific application domains.

A major goal of this book is to present a concise and insightful view of the current knowledge in feedback and control systems. The field of control started by teaching everything that was known at the time and, as new knowledge was acquired, additional courses were developed to cover new techniques. A consequence of this evolution is that introductory courses have remained the same for many years, and it is often necessary to take many individual courses in order to obtain a good perspective on the field. In developing this book, we have attempted to condense the current knowledge by emphasizing fundamental concepts. We believe that it is important to understand why feedback is useful, to know the language and basic mathematics of control and to grasp the key paradigms that have been developed over the past half century. It is also important to be able to solve simple feedback problems using back-of-the-envelope techniques, to recognize fundamental limitations and difficult control problems and to have a feel for available design methods.

This book was originally developed for use in an experimental course at Caltech involving students from a wide set of backgrounds. The course was offered to undergraduates at the junior and senior levels in traditional engineering disciplines, as well as first- and second-year graduate students in engineering and science. This latter group included graduate students in biology, computer science and physics. Over the course of several years, the text has been classroom tested at Caltech and at Lund University, and the feedback from many students and colleagues has been incorporated to help improve the readability and accessibility of the material.

Because of its intended audience, this book is organized in a slightly unusual fashion compared to many other books on feedback and control. In particular, we introduce a number of concepts in the text that are normally reserved for second-year courses on control and hence often not available to students who are not control systems majors. This has been done at the expense of certain traditional topics, which we felt that the astute student could learn independently and are often explored through the exercises. Examples of topics that we have included are non-linear dynamics, Lyapunov stability analysis, the matrix exponential, reachability and observability, and fundamental limits of performance and robustness. Topics that we have deemphasized include root locus techniques, lead/lag compensation and detailed rules for generating Bode and Nyquist plots by hand.

Several features of the book are designed to facilitate its dual function as a basic engineering text and as an introduction for researchers in natural, information and social sciences. The bulk of the material is intended to be used regardless of the audience and covers the core principles and tools in the analysis and design of feedback systems. Advanced sections, marked by the "dangerous bend" symbol shown here, contain material that requires a slightly more technical background, of the sort that would be expected of senior undergraduates in engineering. A few sections are marked by two dangerous bend symbols and are intended for readers with more specialized backgrounds, identified at the beginning of the section. To limit the length of the text, several standard results and extensions are given in the exercises, with appropriate hints toward their solutions.

To further augment the printed material contained here, a companion web site has been developed and is available from the publisher's web page:

http://press.princeton.edu/titles/8701.html

The web site contains a database of frequently asked questions, supplemental examples and exercises, and lecture material for courses based on this text. The material is organized by chapter and includes a summary of the major points in the text as well as links to external resources. The web site also contains the source code for many examples in the book, as well as utilities to implement the techniques described in the text. Most of the code was originally written using MATLAB M-files but was also tested with LabView MathScript to ensure compatibility with both packages. Many files can also be run using other scripting languages such as Octave, SciLab, SysQuake and Xmath.

The first half of the book focuses almost exclusively on state space control systems. We begin in Chapter 2 with a description of modeling of physical, biological and information systems using ordinary differential equations and difference equations. Chapter 3 presents a number of examples in some detail, primarily as a reference for problems that will be used throughout the text. Following this, Chapter 4 looks at the dynamic behavior of models, including definitions of stability and more complicated nonlinear behavior. We provide advanced sections in this chapter on Lyapunov stability analysis because we find that it is useful in a broad array of applications and is frequently a topic that is not introduced until later in one's studies.

The remaining three chapters of the first half of the book focus on linear systems, beginning with a description of input/output behavior in Chapter 5. In Chapter 6, we formally introduce feedback systems by demonstrating how state space control laws can be designed. This is followed in Chapter 7 by material on output feedback and estimators. Chapters 6 and 7 introduce the key concepts of reachability and observability, which give tremendous insight into the choice of actuators and sensors, whether for engineered or natural systems.

The second half of the book presents material that is often considered to be from the field of "classical control." This includes the transfer function, introduced in Chapter 8, which is a fundamental tool for understanding feedback systems. Using transfer functions, one can begin to analyze the stability of feedback systems using frequency domain analysis, including the ability to reason about the closed loop behavior of a system from its open loop characteristics. This is the subject of Chapter 9, which revolves around the Nyquist stability criterion.

In Chapters 10 and 11, we again look at the design problem, focusing first on proportional-integral-derivative (PID) controllers and then on the more general process of loop shaping. PID control is by far the most common design technique in control systems and a useful tool for any student. The chapter on frequency domain design introduces many of the ideas of modern control theory, including the sensitivity function. In Chapter 12, we combine the results from the second half of the book to analyze some of the fundamental trade-offs between robustness and performance. This is also a key chapter illustrating the power of the techniques that have been developed and serving as an introduction for more advanced studies.

The book is designed for use in a 10- to 15-week course in feedback systems that provides many of the key concepts needed in a variety of disciplines. For a 10-week course, Chapters 1–2, 4–6 and 8–11 can each be covered in a week's time, with the omission of some topics from the final chapters. A more leisurely course, spread out over 14–15 weeks, could cover the entire book, with 2 weeks on modeling (Chapters 2 and 3)—particularly for students without much background in ordinary differential equations—and 2 weeks on robust performance (Chapter 12).

The mathematical prerequisites for the book are modest and in keeping with our goal of providing an introduction that serves a broad audience. We assume familiarity with the basic tools of linear algebra, including matrices, vectors and eigenvalues. These are typically covered in a sophomore-level course on the subject, and the textbooks by Apostol [10], Arnold [13] and Strang [187] can serve as good references. Similarly, we assume basic knowledge of differential equations, including the concepts of homogeneous and particular solutions for linear ordinary differential equations in one variable. Apostol [10] and Boyce and DiPrima [42] cover this material well. Finally, we also make use of complex numbers and functions and, in some of the advanced sections, more detailed concepts in complex variables that are typically covered in a junior-level engineering or physics course in mathematical methods. Apostol [9] or Stewart [186] can be used for the basic material, with Ahlfors [6], Marsden and Hoffman [146] or Saff and Snider [172] being good references for the more advanced material. We have chosen not to include appendices summarizing these various topics since there are a number of good books available.

One additional choice that we felt was important was the decision not to rely on a knowledge of Laplace transforms in the book. While their use is by far the most common approach to teaching feedback systems in engineering, many students in the natural and information sciences may lack the necessary mathematical background. Since Laplace transforms are not required in any essential way, we have included them only in an advanced section intended to tie things together for students with that background. Of course, we make tremendous use of *transfer functions*, which we introduce through the notion of response to exponential inputs, an approach we feel is more accessible to a broad array of scientists and engineers. For classes in which students have already had Laplace transforms, it should be quite natural to build on this background in the appropriate sections of the text.

Acknowledgments

The authors would like to thank the many people who helped during the preparation of this book. The idea for writing this book came in part from a report on future directions in control [155] to which Stephen Boyd, Roger Brockett, John Doyle and Gunter Stein were major contributors. Kristi Morgansen and Hideo Mabuchi helped teach early versions of the course at Caltech on which much of the text is based, and Steve Waydo served as the head TA for the course taught at Caltech in 2003–2004 and provided numerous comments and corrections. Charlotta Johnsson and Anton Cervin taught from early versions of the manuscript in Lund in 2003–2007 and gave very useful feedback. Other colleagues and students who provided feedback and advice include Leif Andersson, John Carson, K. Mani Chandy, Michel Charpentier, Domitilla Del Vecchio, Kate Galloway, Per Hagander, Toivo Henningsson Perby, Joseph Hellerstein, George Hines, Tore Hägglund, Cole Lepine, Anders Rantzer, Anders Robertsson Dawn Tilbury and Francisco Zabala. The reviewers for Princeton University Press and Tom Robbins at NI Press also provided valuable comments that significantly improved the organization, layout and focus of the book. Our editor, Vickie Kearn, was a great source of encouragement and help throughout the publishing process. Finally, we would like to thank Caltech, Lund University and the University of California at Santa Barbara for providing many resources, stimulating colleagues and students, and pleasant working environments that greatly aided in the writing of this book.

Karl Johan Åström   Richard M. Murray
Lund, Sweden   Pasadena, California
Santa Barbara, California

Chapter One
Introduction

Feedback is a central feature of life. The process of feedback governs how we grow, respond to stress and challenge, and regulate factors such as body temperature, blood pressure and cholesterol level. The mechanisms operate at every level, from the interaction of proteins in cells to the interaction of organisms in complex ecologies.

M. B. Hoagland and B. Dodson, *The Way Life Works*, 1995 [99].

In this chapter we provide an introduction to the basic concept of *feedback* and the related engineering discipline of *control*. We focus on both historical and current examples, with the intention of providing the context for current tools in feedback and control. Much of the material in this chapter is adapted from [155], and the authors gratefully acknowledge the contributions of Roger Brockett and Gunter Stein to portions of this chapter.

1.1 What Is Feedback?

A *dynamical system* is a system whose behavior changes over time, often in response to external stimulation or forcing. The term *feedback* refers to a situation in which two (or more) dynamical systems are connected together such that each system influences the other and their dynamics are thus strongly coupled. Simple causal reasoning about a feedback system is difficult because the first system influences the second and the second system influences the first, leading to a circular argument. This makes reasoning based on cause and effect tricky, and it is necessary to analyze the system as a whole. A consequence of this is that the behavior of feedback systems is often counterintuitive, and it is therefore necessary to resort to formal methods to understand them.

Figure 1.1 illustrates in block diagram form the idea of feedback. We often use

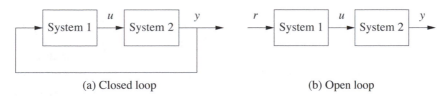

(a) Closed loop   (b) Open loop

Figure 1.1: Open and closed loop systems. (a) The output of system 1 is used as the input of system 2, and the output of system 2 becomes the input of system 1, creating a closed loop system. (b) The interconnection between system 2 and system 1 is removed, and the system is said to be open loop.

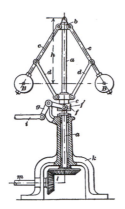
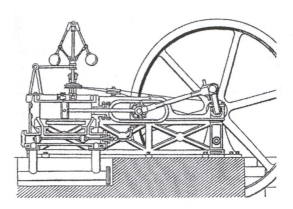

Figure 1.2: The centrifugal governor and the steam engine. The centrifugal governor on the left consists of a set of flyballs that spread apart as the speed of the engine increases. The steam engine on the right uses a centrifugal governor (above and to the left of the flywheel) to regulate its speed. (Credit: Machine a Vapeur Horizontale de Philip Taylor [1828].)

the terms *open loop* and *closed loop* when referring to such systems. A system is said to be a closed loop system if the systems are interconnected in a cycle, as shown in Figure 1.1a. If we break the interconnection, we refer to the configuration as an open loop system, as shown in Figure 1.1b.

As the quote at the beginning of this chapter illustrates, a major source of examples of feedback systems is biology. Biological systems make use of feedback in an extraordinary number of ways, on scales ranging from molecules to cells to organisms to ecosystems. One example is the regulation of glucose in the bloodstream through the production of insulin and glucagon by the pancreas. The body attempts to maintain a constant concentration of glucose, which is used by the body's cells to produce energy. When glucose levels rise (after eating a meal, for example), the hormone insulin is released and causes the body to store excess glucose in the liver. When glucose levels are low, the pancreas secretes the hormone glucagon, which has the opposite effect. Referring to Figure 1.1, we can view the liver as system 1 and the pancreas as system 2. The output from the liver is the glucose concentration in the blood, and the output from the pancreas is the amount of insulin or glucagon produced. The interplay between insulin and glucagon secretions throughout the day helps to keep the blood-glucose concentration constant, at about 90 mg per 100 mL of blood.

An early engineering example of a feedback system is a centrifugal governor, in which the shaft of a steam engine is connected to a flyball mechanism that is itself connected to the throttle of the steam engine, as illustrated in Figure 1.2. The system is designed so that as the speed of the engine increases (perhaps because of a lessening of the load on the engine), the flyballs spread apart and a linkage causes the throttle on the steam engine to be closed. This in turn slows down the engine, which causes the flyballs to come back together. We can model this system as a closed loop system by taking system 1 as the steam engine and system 2 as the governor.

When properly designed, the flyball governor maintains a constant speed of the engine, roughly independent of the loading conditions. The centrifugal governor was an enabler of the successful Watt steam engine, which fueled the industrial revolution.

Feedback has many interesting properties that can be exploited in designing systems. As in the case of glucose regulation or the flyball governor, feedback can make a system resilient toward external influences. It can also be used to create linear behavior out of nonlinear components, a common approach in electronics. More generally, feedback allows a system to be insensitive both to external disturbances and to variations in its individual elements.

Feedback has potential disadvantages as well. It can create dynamic instabilities in a system, causing oscillations or even runaway behavior. Another drawback, especially in engineering systems, is that feedback can introduce unwanted sensor noise into the system, requiring careful filtering of signals. It is for these reasons that a substantial portion of the study of feedback systems is devoted to developing an understanding of dynamics and a mastery of techniques in dynamical systems.

Feedback systems are ubiquitous in both natural and engineered systems. Control systems maintain the environment, lighting and power in our buildings and factories; they regulate the operation of our cars, consumer electronics and manufacturing processes; they enable our transportation and communications systems; and they are critical elements in our military and space systems. For the most part they are hidden from view, buried within the code of embedded microprocessors, executing their functions accurately and reliably. Feedback has also made it possible to increase dramatically the precision of instruments such as atomic force microscopes (AFMs) and telescopes.

In nature, homeostasis in biological systems maintains thermal, chemical and biological conditions through feedback. At the other end of the size scale, global climate dynamics depend on the feedback interactions between the atmosphere, the oceans, the land and the sun. Ecosystems are filled with examples of feedback due to the complex interactions between animal and plant life. Even the dynamics of economies are based on the feedback between individuals and corporations through markets and the exchange of goods and services.

1.2 What Is Control?

The term *control* has many meanings and often varies between communities. In this book, we define control to be the use of algorithms and feedback in engineered systems. Thus, control includes such examples as feedback loops in electronic amplifiers, setpoint controllers in chemical and materials processing, "fly-by-wire" systems on aircraft and even router protocols that control traffic flow on the Internet. Emerging applications include high-confidence software systems, autonomous vehicles and robots, real-time resource management systems and biologically engineered systems. At its core, control is an *information* science and includes the use of information in both analog and digital representations.

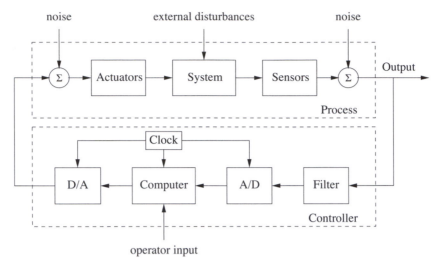

Figure 1.3: Components of a computer-controlled system. The upper dashed box represents the process dynamics, which include the sensors and actuators in addition to the dynamical system being controlled. Noise and external disturbances can perturb the dynamics of the process. The controller is shown in the lower dashed box. It consists of a filter and analog-to-digital (A/D) and digital-to-analog (D/A) converters, as well as a computer that implements the control algorithm. A system clock controls the operation of the controller, synchronizing the A/D, D/A and computing processes. The operator input is also fed to the computer as an external input.

A modern controller senses the operation of a system, compares it against the desired behavior, computes corrective actions based on a model of the system's response to external inputs and actuates the system to effect the desired change. This basic *feedback loop* of sensing, computation and actuation is the central concept in control. The key issues in designing control logic are ensuring that the dynamics of the closed loop system are stable (bounded disturbances give bounded errors) and that they have additional desired behavior (good disturbance attenuation, fast responsiveness to changes in operating point, etc). These properties are established using a variety of modeling and analysis techniques that capture the essential dynamics of the system and permit the exploration of possible behaviors in the presence of uncertainty, noise and component failure.

A typical example of a control system is shown in Figure 1.3. The basic elements of sensing, computation and actuation are clearly seen. In modern control systems, computation is typically implemented on a digital computer, requiring the use of analog-to-digital (A/D) and digital-to-analog (D/A) converters. Uncertainty enters the system through noise in sensing and actuation subsystems, external disturbances that affect the underlying system operation and uncertain dynamics in the system (parameter errors, unmodeled effects, etc). The algorithm that computes the control action as a function of the sensor values is often called a *control law*. The system can be influenced externally by an operator who introduces *command signals* to the system.

Control engineering relies on and shares tools from physics (dynamics and modeling), computer science (information and software) and operations research (optimization, probability theory and game theory), but it is also different from these subjects in both insights and approach.

Perhaps the strongest area of overlap between control and other disciplines is in the modeling of physical systems, which is common across all areas of engineering and science. One of the fundamental differences between control-oriented modeling and modeling in other disciplines is the way in which interactions between subsystems are represented. Control relies on a type of input/output modeling that allows many new insights into the behavior of systems, such as disturbance attenuation and stable interconnection. Model reduction, where a simpler (lower-fidelity) description of the dynamics is derived from a high-fidelity model, is also naturally described in an input/output framework. Perhaps most importantly, modeling in a control context allows the design of *robust* interconnections between subsystems, a feature that is crucial in the operation of all large engineered systems.

Control is also closely associated with computer science since virtually all modern control algorithms for engineering systems are implemented in software. However, control algorithms and software can be very different from traditional computer software because of the central role of the dynamics of the system and the real-time nature of the implementation.

1.3 Feedback Examples

Feedback has many interesting and useful properties. It makes it possible to design precise systems from imprecise components and to make relevant quantities in a system change in a prescribed fashion. An unstable system can be stabilized using feedback, and the effects of external disturbances can be reduced. Feedback also offers new degrees of freedom to a designer by exploiting sensing, actuation and computation. In this section we survey some of the important applications and trends for feedback in the world around us.

Early Technological Examples

The proliferation of control in engineered systems occurred primarily in the latter half of the 20th century. There are some important exceptions, such as the centrifugal governor described earlier and the thermostat (Figure 1.4a), designed at the turn of the century to regulate the temperature of buildings.

The thermostat, in particular, is a simple example of feedback control that everyone is familiar with. The device measures the temperature in a building, compares that temperature to a desired setpoint and uses the *feedback error* between the two to operate the heating plant, e.g., to turn heat on when the temperature is too low and to turn it off when the temperature is too high. This explanation captures the essence of feedback, but it is a bit too simple even for a basic device such as the thermostat. Because lags and delays exist in the heating plant and sensor, a good

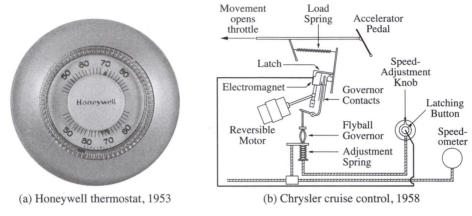

(a) Honeywell thermostat, 1953   (b) Chrysler cruise control, 1958

Figure 1.4: Early control devices. (a) Honeywell T87 thermostat originally introduced in 1953. The thermostat controls whether a heater is turned on by comparing the current temperature in a room to a desired value that is set using a dial. (b) Chrysler cruise control system introduced in the 1958 Chrysler Imperial [170]. A centrifugal governor is used to detect the speed of the vehicle and actuate the throttle. The reference speed is specified through an adjustment spring. (Left figure courtesy of Honeywell International, Inc.)

thermostat does a bit of anticipation, turning the heater off before the error actually changes sign. This avoids excessive temperature swings and cycling of the heating plant. This interplay between the dynamics of the process and the operation of the controller is a key element in modern control systems design.

There are many other control system examples that have developed over the years with progressively increasing levels of sophistication. An early system with broad public exposure was the *cruise control* option introduced on automobiles in 1958 (see Figure 1.4b). Cruise control illustrates the dynamic behavior of closed loop feedback systems in action—the slowdown error as the system climbs a grade, the gradual reduction of that error due to integral action in the controller, the small overshoot at the top of the climb, etc. Later control systems on automobiles such as emission controls and fuel-metering systems have achieved major reductions of pollutants and increases in fuel economy.

Power Generation and Transmission

Access to electrical power has been one of the major drivers of technological progress in modern society. Much of the early development of control was driven by the generation and distribution of electrical power. Control is mission critical for power systems, and there are many control loops in individual power stations. Control is also important for the operation of the whole power network since it is difficult to store energy and it is thus necessary to match production to consumption. Power management is a straightforward regulation problem for a system with one generator and one power consumer, but it is more difficult in a highly distributed system with many generators and long distances between consumption and generation. Power demand can change rapidly in an unpredictable manner and

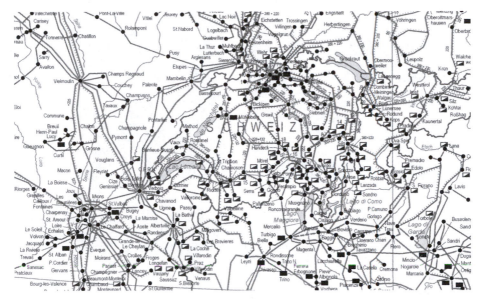

Figure 1.5: A small portion of the European power network. By 2008 European power suppliers will operate a single interconnected network covering a region from the Arctic to the Mediterranean and from the Atlantic to the Urals. In 2004 the installed power was more than 700 GW (7×10^{11} W). (Source: UCTE [www.ucte.org])

combining generators and consumers into large networks makes it possible to share loads among many suppliers and to average consumption among many customers. Large transcontinental and transnational power systems have therefore been built, such as the one show in Figure 1.5.

Most electricity is distributed by alternating current (AC) because the transmission voltage can be changed with small power losses using transformers. Alternating current generators can deliver power only if the generators are synchronized to the voltage variations in the network. This means that the rotors of all generators in a network must be synchronized. To achieve this with local decentralized controllers and a small amount of interaction is a challenging problem. Sporadic low-frequency oscillations between distant regions have been observed when regional power grids have been interconnected [134].

Safety and reliability are major concerns in power systems. There may be disturbances due to trees falling down on power lines, lightning or equipment failures. There are sophisticated control systems that attempt to keep the system operating even when there are large disturbances. The control actions can be to reduce voltage, to break up the net into subnets or to switch off lines and power users. These safety systems are an essential element of power distribution systems, but in spite of all precautions there are occasionally failures in large power systems. The power system is thus a nice example of a complicated distributed system where control is executed on many levels and in many different ways.

(a) F/A-18 "Hornet"

(b) X-45 UCAV

Figure 1.6: Military aerospace systems. (a) The F/A-18 aircraft is one of the first production military fighters to use "fly-by-wire" technology. (b) The X-45 (UCAV) unmanned aerial vehicle is capable of autonomous flight, using inertial measurement sensors and the global positioning system (GPS) to monitor its position relative to a desired trajectory. (Photographs courtesy of NASA Dryden Flight Research Center.)

Aerospace and Transportation

In aerospace, control has been a key technological capability tracing back to the beginning of the 20th century. Indeed, the Wright brothers are correctly famous not for demonstrating simply powered flight but *controlled* powered flight. Their early Wright Flyer incorporated moving control surfaces (vertical fins and canards) and warpable wings that allowed the pilot to regulate the aircraft's flight. In fact, the aircraft itself was not stable, so continuous pilot corrections were mandatory. This early example of controlled flight was followed by a fascinating success story of continuous improvements in flight control technology, culminating in the high-performance, highly reliable automatic flight control systems we see in modern commercial and military aircraft today (Figure 1.6).

Similar success stories for control technology have occurred in many other application areas. Early World War II bombsights and fire control servo systems have evolved into today's highly accurate radar-guided guns and precision-guided weapons. Early failure-prone space missions have evolved into routine launch operations, manned landings on the moon, permanently manned space stations, robotic vehicles roving Mars, orbiting vehicles at the outer planets and a host of commercial and military satellites serving various surveillance, communication, navigation and earth observation needs. Cars have advanced from manually tuned mechanical/pneumatic technology to computer-controlled operation of all major functions, including fuel injection, emission control, cruise control, braking and cabin comfort.

Current research in aerospace and transportation systems is investigating the application of feedback to higher levels of decision making, including logical regulation of operating modes, vehicle configurations, payload configurations and health status. These have historically been performed by human operators, but today that

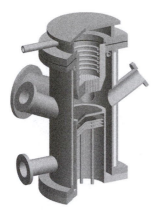
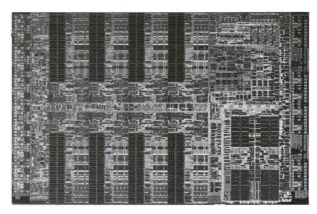

Figure 1.7: Materials processing. Modern materials are processed under carefully controlled conditions, using reactors such as the metal organic chemical vapor deposition (MOCVD) reactor shown on the left, which was for manufacturing superconducting thin films. Using lithography, chemical etching, vapor deposition and other techniques, complex devices can be built, such as the IBM cell processor shown on the right. (MOCVD image courtesy of Bob Kee. IBM cell processor photograph courtesy Tom Way, IBM Corporation; unauthorized use not permitted.)

boundary is moving and control systems are increasingly taking on these functions. Another dramatic trend on the horizon is the use of large collections of distributed entities with local computation, global communication connections, little regularity imposed by the laws of physics and no possibility of imposing centralized control actions. Examples of this trend include the national airspace management problem, automated highway and traffic management and command and control for future battlefields.

Materials and Processing

The chemical industry is responsible for the remarkable progress in developing new materials that are key to our modern society. In addition to the continuing need to improve product quality, several other factors in the process control industry are drivers for the use of control. Environmental statutes continue to place stricter limitations on the production of pollutants, forcing the use of sophisticated pollution control devices. Environmental safety considerations have led to the design of smaller storage capacities to diminish the risk of major chemical leakage, requiring tighter control on upstream processes and, in some cases, supply chains. And large increases in energy costs have encouraged engineers to design plants that are highly integrated, coupling many processes that used to operate independently. All of these trends increase the complexity of these processes and the performance requirements for the control systems, making control system design increasingly challenging. Some examples of materials-processing technology are shown in Figure 1.7.

As in many other application areas, new sensor technology is creating new opportunities for control. Online sensors—including laser backscattering, video microscopy and ultraviolet, infrared and Raman spectroscopy—are becoming more

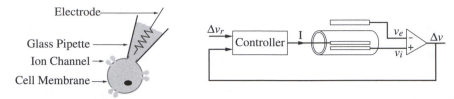

Figure 1.8: The voltage clamp method for measuring ion currents in cells using feedback. A pipet is used to place an electrode in a cell (left and middle) and maintain the potential of the cell at a fixed level. The internal voltage in the cell is v_i, and the voltage of the external fluid is v_e. The feedback system (right) controls the current I into the cell so that the voltage drop across the cell membrane $\Delta v = v_i - v_e$ is equal to its reference value Δv_r. The current I is then equal to the ion current.

robust and less expensive and are appearing in more manufacturing processes. Many of these sensors are already being used by current process control systems, but more sophisticated signal-processing and control techniques are needed to use more effectively the real-time information provided by these sensors. Control engineers also contribute to the design of even better sensors, which are still needed, for example, in the microelectronics industry. As elsewhere, the challenge is making use of the large amounts of data provided by these new sensors in an effective manner. In addition, a control-oriented approach to modeling the essential physics of the underlying processes is required to understand the fundamental limits on observability of the internal state through sensor data.

Instrumentation

The measurement of physical variables is of prime interest in science and engineering. Consider, for example, an accelerometer, where early instruments consisted of a mass suspended on a spring with a deflection sensor. The precision of such an instrument depends critically on accurate calibration of the spring and the sensor. There is also a design compromise because a weak spring gives high sensitivity but low bandwidth.

A different way of measuring acceleration is to use *force feedback*. The spring is replaced by a voice coil that is controlled so that the mass remains at a constant position. The acceleration is proportional to the current through the voice coil. In such an instrument, the precision depends entirely on the calibration of the voice coil and does not depend on the sensor, which is used only as the feedback signal. The sensitivity/bandwidth compromise is also avoided. This way of using feedback has been applied to many different engineering fields and has resulted in instruments with dramatically improved performance. Force feedback is also used in haptic devices for manual control.

Another important application of feedback is in instrumentation for biological systems. Feedback is widely used to measure ion currents in cells using a device called a *voltage clamp*, which is illustrated in Figure 1.8. Hodgkin and Huxley used the voltage clamp to investigate propagation of action potentials in the axon of the giant squid. In 1963 they shared the Nobel Prize in Medicine with Eccles for "their discoveries concerning the ionic mechanisms involved in excitation and inhibition in the peripheral and central portions of the nerve cell membrane." A refinement of the voltage clamp called a *patch clamp* made it possible to measure exactly when a single ion channel is opened or closed. This was developed by Neher and Sakmann, who received the 1991 Nobel Prize in Medicine "for their discoveries concerning the function of single ion channels in cells."

There are many other interesting and useful applications of feedback in scientific instruments. The development of the mass spectrometer is an early example. In a 1935 paper, Nier observed that the deflection of ions depends on both the magnetic and the electric fields [158]. Instead of keeping both fields constant, Nier let the magnetic field fluctuate and the electric field was controlled to keep the ratio between the fields constant. Feedback was implemented using vacuum tube amplifiers. This scheme was crucial for the development of mass spectroscopy.

The Dutch engineer van der Meer invented a clever way to use feedback to maintain a good-quality high-density beam in a particle accelerator [153]. The idea is to sense particle displacement at one point in the accelerator and apply a correcting signal at another point. This scheme, called *stochastic cooling*, was awarded the Nobel Prize in Physics in 1984. The method was essential for the successful experiments at CERN where the existence of the particles W and Z associated with the weak force was first demonstrated.

The 1986 Nobel Prize in Physics—awarded to Binnig and Rohrer for their design of the scanning tunneling microscope—is another example of an innovative use of feedback. The key idea is to move a narrow tip on a cantilever beam across a surface and to register the forces on the tip [34]. The deflection of the tip is measured using tunneling. The tunneling current is used by a feedback system to control the position of the cantilever base so that the tunneling current is constant, an example of force feedback. The accuracy is so high that individual atoms can be registered. A map of the atoms is obtained by moving the base of the cantilever horizontally. The performance of the control system is directly reflected in the image quality and scanning speed. This example is described in additional detail in Chapter 3.

Robotics and Intelligent Machines

The goal of cybernetic engineering, already articulated in the 1940s and even before, has been to implement systems capable of exhibiting highly flexible or "intelligent" responses to changing circumstances. In 1948 the MIT mathematician Norbert Wiener gave a widely read account of cybernetics [200]. A more mathematical treatment of the elements of engineering cybernetics was presented by H. S. Tsien in 1954, driven by problems related to the control of missiles [195]. Together, these works and others of that time form much of the intellectual basis for modern work in robotics and control.

Two accomplishments that demonstrate the successes of the field are the Mars Exploratory Rovers and entertainment robots such as the Sony AIBO, shown in Figure 1.9. The two Mars Exploratory Rovers, launched by the Jet Propulsion

Figure 1.9: Robotic systems. (a) Spirit, one of the two Mars Exploratory Rovers that landed on Mars in January 2004. (b) The Sony AIBO Entertainment Robot, one of the first entertainment robots to be mass-marketed. Both robots make use of feedback between sensors, actuators and computation to function in unknown environments. (Photographs courtesy of Jet Propulsion Laboratory and Sony Electronics, Inc.)

Laboratory (JPL), maneuvered on the surface of Mars for more than 4 years starting in January 2004 and sent back pictures and measurements of their environment. The Sony AIBO robot debuted in June 1999 and was the first "entertainment" robot to be mass-marketed by a major international corporation. It was particularly noteworthy because of its use of artificial intelligence (AI) technologies that allowed it to act in response to external stimulation and its own judgment. This higher level of feedback is a key element in robotics, where issues such as obstacle avoidance, goal seeking, learning and autonomy are prevalent.

Despite the enormous progress in robotics over the last half-century, in many ways the field is still in its infancy. Today's robots still exhibit simple behaviors compared with humans, and their ability to locomote, interpret complex sensory inputs, perform higher-level reasoning and cooperate together in teams is limited. Indeed, much of Wiener's vision for robotics and intelligent machines remains unrealized. While advances are needed in many fields to achieve this vision—including advances in sensing, actuation and energy storage—the opportunity to combine the advances of the AI community in planning, adaptation and learning with the techniques in the control community for modeling, analysis and design of feedback systems presents a renewed path for progress.

Networks and Computing Systems

Control of networks is a large research area spanning many topics, including congestion control, routing, data caching and power management. Several features of these control problems make them very challenging. The dominant feature is the extremely large scale of the system; the Internet is probably the largest feedback control system humans have ever built. Another is the decentralized nature of the control problem: decisions must be made quickly and based only on local informa-

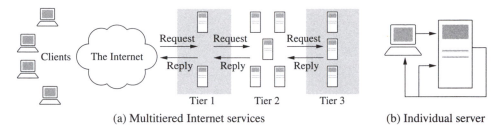

(a) Multitiered Internet services (b) Individual server

Figure 1.10: A multitier system for services on the Internet. In the complete system shown schematically in (a), users request information from a set of computers (tier 1), which in turn collect information from other computers (tiers 2 and 3). The individual server shown in (b) has a set of reference parameters set by a (human) system operator, with feedback used to maintain the operation of the system in the presence of uncertainty. (Based on Hellerstein et al. [97].)

tion. Stability is complicated by the presence of varying time lags, as information about the network state can be observed or relayed to controllers only after a delay, and the effect of a local control action can be felt throughout the network only after substantial delay. Uncertainty and variation in the network, through network topology, transmission channel characteristics, traffic demand and available resources, may change constantly and unpredictably. Other complicating issues are the diverse traffic characteristics — in terms of arrival statistics at both the packet and flow time scales — and the different requirements for quality of service that the network must support.

Related to the control of networks is control of the servers that sit on these networks. Computers are key components of the systems of routers, web servers and database servers used for communication, electronic commerce, advertising and information storage. While hardware costs for computing have decreased dramatically, the cost of operating these systems has increased because of the difficulty in managing and maintaining these complex interconnected systems. The situation is similar to the early phases of process control when feedback was first introduced to control industrial processes. As in process control, there are interesting possibilities for increasing performance and decreasing costs by applying feedback. Several promising uses of feedback in the operation of computer systems are described in the book by Hellerstein et al. [97].

A typical example of a multilayer system for e-commerce is shown in Figure 1.10a. The system has several tiers of servers. The edge server accepts incoming requests and routes them to the HTTP server tier where they are parsed and distributed to the application servers. The processing for different requests can vary widely, and the application servers may also access external servers managed by other organizations.

Control of an individual server in a layer is illustrated in Figure 1.10b. A quantity representing the quality of service or cost of operation — such as response time, throughput, service rate or memory usage — is measured in the computer. The control variables might represent incoming messages accepted, priorities in the operating system or memory allocation. The feedback loop then attempts to maintain quality-of-service variables within a target range of values.

Economics

The economy is a large, dynamical system with many actors: governments, organizations, companies and individuals. Governments control the economy through laws and taxes, the central banks by setting interest rates and companies by setting prices and making investments. Individuals control the economy through purchases, savings and investments. Many efforts have been made to model the system both at the macro level and at the micro level, but this modeling is difficult because the system is strongly influenced by the behaviors of the different actors in the system.

Keynes [122] developed a simple model to understand relations among gross national product, investment, consumption and government spending. One of Keynes' observations was that under certain conditions, e.g., during the 1930s depression, an increase in the investment of government spending could lead to a larger increase in the gross national product. This idea was used by several governments to try to alleviate the depression. Keynes' ideas can be captured by a simple model that is discussed in Exercise 2.4.

A perspective on the modeling and control of economic systems can be obtained from the work of some economists who have received the Sveriges Riksbank Prize in Economics in Memory of Alfred Nobel, popularly called the Nobel Prize in Economics. Paul A. Samuelson received the prize in 1970 for "the scientific work through which he has developed static and dynamic economic theory and actively contributed to raising the level of analysis in economic science." Lawrence Klein received the prize in 1980 for the development of large dynamical models with many parameters that were fitted to historical data [126], e.g., a model of the U.S. economy in the period 1929–1952. Other researchers have modeled other countries and other periods. In 1997 Myron Scholes shared the prize with Robert Merton for a new method to determine the value of derivatives. A key ingredient was a dynamic model of the variation of stock prices that is widely used by banks and investment companies. In 2004 Finn E. Kydland and Edward C. Prescott shared the economics prize "for their contributions to dynamic macroeconomics: the time consistency of economic policy and the driving forces behind business cycles," a topic that is clearly related to dynamics and control.

One of the reasons why it is difficult to model economic systems is that there are no conservation laws. A typical example is that the value of a company as expressed by its stock can change rapidly and erratically. There are, however, some areas with conservation laws that permit accurate modeling. One example is the flow of products from a manufacturer to a retailer as illustrated in Figure 1.11. The products are physical quantities that obey a conservation law, and the system can be modeled by accounting for the number of products in the different inventories. There are considerable economic benefits in controlling supply chains so that products are available to customers while minimizing products that are in storage. The real problems are more complicated than indicated in the figure because there may be

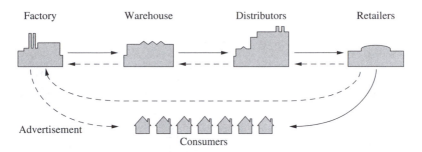

Figure 1.11: Supply chain dynamics (after Forrester [75]). Products flow from the producer to the customer through distributors and retailers as indicated by the solid lines. There are typically many factories and warehouses and even more distributors and retailers. Multiple feedback loops are present as each agent tries to maintain the proper inventory level.

many different products, there may be different factories that are geographically distributed and the factories may require raw material or subassemblies.

Control of supply chains was proposed by Forrester in 1961 [75] and is now growing in importance. Considerable economic benefits can be obtained by using models to minimize inventories. Their use accelerated dramatically when information technology was applied to predict sales, keep track of products and enable just-in-time manufacturing. Supply chain management has contributed significantly to the growing success of global distributors.

Advertising on the Internet is an emerging application of control. With network-based advertising it is easy to measure the effect of different marketing strategies quickly. The response of customers can then be modeled, and feedback strategies can be developed.

Feedback in Nature

Many problems in the natural sciences involve understanding aggregate behavior in complex large-scale systems. This behavior emerges from the interaction of a multitude of simpler systems with intricate patterns of information flow. Representative examples can be found in fields ranging from embryology to seismology. Researchers who specialize in the study of specific complex systems often develop an intuitive emphasis on analyzing the role of feedback (or interconnection) in facilitating and stabilizing aggregate behavior.

While sophisticated theories have been developed by domain experts for the analysis of various complex systems, the development of a rigorous methodology that can discover and exploit common features and essential mathematical structure is just beginning to emerge. Advances in science and technology are creating a new understanding of the underlying dynamics and the importance of feedback in a wide variety of natural and technological systems. We briefly highlight three application areas here.

Biological Systems. A major theme currently of interest to the biology commu-

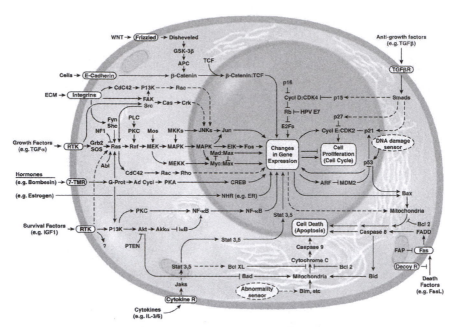

Figure 1.12: The wiring diagram of the growth-signaling circuitry of the mammalian cell [95]. The major pathways that are thought to play a role in cancer are indicated in the diagram. Lines represent interactions between genes and proteins in the cell. Lines ending in arrowheads indicate activation of the given gene or pathway; lines ending in a T-shaped head indicate repression. (Used with permission of Elsevier Ltd. and the authors.)

nity is the science of reverse (and eventually forward) engineering of biological control networks such as the one shown in Figure 1.12. There are a wide variety of biological phenomena that provide a rich source of examples of control, including gene regulation and signal transduction; hormonal, immunological and cardiovascular feedback mechanisms; muscular control and locomotion; active sensing, vision and proprioception; attention and consciousness; and population dynamics and epidemics. Each of these (and many more) provide opportunities to figure out what works, how it works, and what we can do to affect it.

One interesting feature of biological systems is the frequent use of positive feedback to shape the dynamics of the system. Positive feedback can be used to create switchlike behavior through autoregulation of a gene, and to create oscillations such as those present in the cell cycle, central pattern generators or circadian rhythm.

Ecosystems. In contrast to individual cells and organisms, emergent properties of aggregations and ecosystems inherently reflect selection mechanisms that act on multiple levels, and primarily on scales well below that of the system as a whole. Because ecosystems are complex, multiscale dynamical systems, they provide a broad range of new challenges for the modeling and analysis of feedback systems. Recent experience in applying tools from control and dynamical systems to bacterial networks suggests that much of the complexity of these networks is due to the presence of multiple layers of feedback loops that provide robust functionality to the individual cell. Yet in other instances, events at the cell level benefit the colony at the expense of the individual. Systems level analysis can be applied to ecosystems with the goal of understanding the robustness of such systems and the extent to which decisions and events affecting individual species contribute to the robustness and/or fragility of the ecosystem as a whole.

Environmental Science. It is now indisputable that human activities have altered the environment on a global scale. Problems of enormous complexity challenge researchers in this area, and first among these is to understand the feedback systems that operate on the global scale. One of the challenges in developing such an understanding is the multiscale nature of the problem, with detailed understanding of the dynamics of microscale phenomena such as microbiological organisms being a necessary component of understanding global phenomena, such as the carbon cycle.

1.4 Feedback Properties

Feedback is a powerful idea which, as we have seen, is used extensively in natural and technological systems. The principle of feedback is simple: base correcting actions on the difference between desired and actual performance. In engineering, feedback has been rediscovered and patented many times in many different contexts. The use of feedback has often resulted in vast improvements in system capability, and these improvements have sometimes been revolutionary, as discussed above. The reason for this is that feedback has some truly remarkable properties. In this section we will discuss some of the properties of feedback that can be understood intuitively. This intuition will be formalized in subsequent chapters.

Robustness to Uncertainty

One of the key uses of feedback is to provide robustness to uncertainty. By measuring the difference between the sensed value of a regulated signal and its desired value, we can supply a corrective action. If the system undergoes some change that affects the regulated signal, then we sense this change and try to force the system back to the desired operating point. This is precisely the effect that Watt exploited in his use of the centrifugal governor on steam engines.

As an example of this principle, consider the simple feedback system shown in Figure 1.13. In this system, the speed of a vehicle is controlled by adjusting the amount of gas flowing to the engine. Simple *proportional-integral* (PI) feedback is used to make the amount of gas depend on both the error between the current and the desired speed and the integral of that error. The plot on the right shows the results of this feedback for a step change in the desired speed and a variety of different masses for the car, which might result from having a different number of passengers or towing a trailer. Notice that independent of the mass (which varies by a factor of 3!), the steady-state speed of the vehicle always approaches the desired speed and achieves that speed within approximately 5 s. Thus the performance of

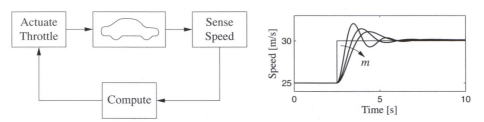

Figure 1.13: A feedback system for controlling the speed of a vehicle. In the block diagram on the left, the speed of the vehicle is measured and compared to the desired speed within the "Compute" block. Based on the difference in the actual and desired speeds, the throttle (or brake) is used to modify the force applied to the vehicle by the engine, drivetrain and wheels. The figure on the right shows the response of the control system to a commanded change in speed from 25 m/s to 30 m/s. The three different curves correspond to differing masses of the vehicle, between 1000 and 3000 kg, demonstrating the robustness of the closed loop system to a very large change in the vehicle characteristics.

the system is robust with respect to this uncertainty.

Another early example of the use of feedback to provide robustness is the negative feedback amplifier. When telephone communications were developed, amplifiers were used to compensate for signal attenuation in long lines. A vacuum tube was a component that could be used to build amplifiers. Distortion caused by the nonlinear characteristics of the tube amplifier together with amplifier drift were obstacles that prevented the development of line amplifiers for a long time. A major breakthrough was the invention of the feedback amplifier in 1927 by Harold S. Black, an electrical engineer at Bell Telephone Laboratories. Black used *negative feedback*, which reduces the gain but makes the amplifier insensitive to variations in tube characteristics. This invention made it possible to build stable amplifiers with linear characteristics despite the nonlinearities of the vacuum tube amplifier.

Design of Dynamics

Another use of feedback is to change the dynamics of a system. Through feedback, we can alter the behavior of a system to meet the needs of an application: systems that are unstable can be stabilized, systems that are sluggish can be made responsive and systems that have drifting operating points can be held constant. Control theory provides a rich collection of techniques to analyze the stability and dynamic response of complex systems and to place bounds on the behavior of such systems by analyzing the gains of linear and nonlinear operators that describe their components.

An example of the use of control in the design of dynamics comes from the area of flight control. The following quote, from a lecture presented by Wilbur Wright to the Western Society of Engineers in 1901 [149], illustrates the role of control in the development of the airplane:

> Men already know how to construct wings or airplanes, which when
> driven through the air at sufficient speed, will not only sustain the

Figure 1.14: Aircraft autopilot system. The Sperry autopilot (left) contained a set of four gyros coupled to a set of air valves that controlled the wing surfaces. The 1912 Curtiss used an autopilot to stabilize the roll, pitch and yaw of the aircraft and was able to maintain level flight as a mechanic walked on the wing (right) [105].

weight of the wings themselves, but also that of the engine, and of the engineer as well. Men also know how to build engines and screws of sufficient lightness and power to drive these planes at sustaining speed ... Inability to balance and steer still confronts students of the flying problem ... When this one feature has been worked out, the age of flying will have arrived, for all other difficulties are of minor importance.

The Wright brothers thus realized that control was a key issue to enable flight. They resolved the compromise between stability and maneuverability by building an airplane, the Wright Flyer, that was unstable but maneuverable. The Flyer had a rudder in the front of the airplane, which made the plane very maneuverable. A disadvantage was the necessity for the pilot to keep adjusting the rudder to fly the plane: if the pilot let go of the stick, the plane would crash. Other early aviators tried to build stable airplanes. These would have been easier to fly, but because of their poor maneuverability they could not be brought up into the air. By using their insight and skillful experiments the Wright brothers made the first successful flight at Kitty Hawk in 1903.

Since it was quite tiresome to fly an unstable aircraft, there was strong motivation to find a mechanism that would stabilize an aircraft. Such a device, invented by Sperry, was based on the concept of feedback. Sperry used a gyro-stabilized pendulum to provide an indication of the vertical. He then arranged a feedback mechanism that would pull the stick to make the plane go up if it was pointing down, and vice versa. The Sperry autopilot was the first use of feedback in aeronautical engineering, and Sperry won a prize in a competition for the safest airplane in Paris in 1914. Figure 1.14 shows the Curtiss seaplane and the Sperry autopilot. The autopilot is a good example of how feedback can be used to stabilize an unstable system and hence "design the dynamics" of the aircraft.

One of the other advantages of designing the dynamics of a device is that it allows for increased modularity in the overall system design. By using feedback to create a system whose response matches a desired profile, we can hide the complexity and variability that may be present inside a subsystem. This allows us to create more complex systems by not having to simultaneously tune the responses of a large number of interacting components. This was one of the advantages of Black's use of negative feedback in vacuum tube amplifiers: the resulting device had a well-defined linear input/output response that did not depend on the individual characteristics of the vacuum tubes being used.

Higher Levels of Automation

A major trend in the use of feedback is its application to higher levels of situational awareness and decision making. This includes not only traditional logical branching based on system conditions but also optimization, adaptation, learning and even higher levels of abstract reasoning. These problems are in the domain of the artificial intelligence community, with an increasing role of dynamics, robustness and interconnection in many applications.

One of the interesting areas of research in higher levels of decision is autonomous control of cars. Early experiments with autonomous driving were performed by Ernst Dickmanns, who in the 1980s equipped cars with cameras and other sensors [60]. In 1994 his group demonstrated autonomous driving with human supervision on a highway near Paris and in 1995 one of his cars drove autonomously (with human supervision) from Munich to Copenhagen at speeds of up to 175 km/hour. The car was able to overtake other vehicles and change lanes automatically.

This application area has been recently explored through the DARPA Grand Challenge, a series of competitions sponsored by the U.S. government to build vehicles that can autonomously drive themselves in desert and urban environments. Caltech competed in the 2005 and 2007 Grand Challenges using a modified Ford E-350 offroad van nicknamed "Alice." It was fully automated, including electronically controlled steering, throttle, brakes, transmission and ignition. Its sensing systems included multiple video cameras scanning at 10–30 Hz, several laser ranging units scanning at 10 Hz and an inertial navigation package capable of providing position and orientation estimates at 5 ms temporal resolution. Computational resources included 12 high-speed servers connected together through a 1-Gb/s Ethernet switch. The vehicle is shown in Figure 1.15, along with a block diagram of its control architecture.

The software and hardware infrastructure that was developed enabled the vehicle to traverse long distances at substantial speeds. In testing, Alice drove itself more than 500 km in the Mojave Desert of California, with the ability to follow dirt roads and trails (if present) and avoid obstacles along the path. Speeds of more than 50 km/h were obtained in the fully autonomous mode. Substantial tuning of the algorithms was done during desert testing, in part because of the lack of systems-level design tools for systems of this level of complexity. Other competitors in the race (including Stanford, which won the 2005 competition) used algorithms for adaptive

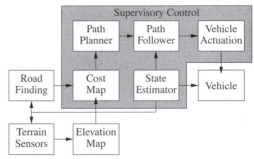

Figure 1.15: DARPA Grand Challenge. "Alice," Team Caltech's entry in the 2005 and 2007 competitions and its networked control architecture [54]. The feedback system fuses data from terrain sensors (cameras and laser range finders) to determine a digital elevation map. This map is used to compute the vehicle's potential speed over the terrain, and an optimization-based path planner then commands a trajectory for the vehicle to follow. A supervisory control module performs higher-level tasks such as handling sensor and actuator failures.

control and learning, increasing the capabilities of their systems in unknown environments. Together, the competitors in the Grand Challenge demonstrated some of the capabilities of the next generation of control systems and highlighted many research directions in control at higher levels of decision making.

Drawbacks of Feedback

While feedback has many advantages, it also has some drawbacks. Chief among these is the possibility of instability if the system is not designed properly. We are all familiar with the effects of *positive feedback* when the amplification on a microphone is turned up too high in a room. This is an example of feedback instability, something that we obviously want to avoid. This is tricky because we must design the system not only to be stable under nominal conditions but also to remain stable under all possible perturbations of the dynamics.

In addition to the potential for instability, feedback inherently couples different parts of a system. One common problem is that feedback often injects measurement noise into the system. Measurements must be carefully filtered so that the actuation and process dynamics do not respond to them, while at the same time ensuring that the measurement signal from the sensor is properly coupled into the closed loop dynamics (so that the proper levels of performance are achieved).

Another potential drawback of control is the complexity of embedding a control system in a product. While the cost of sensing, computation and actuation has decreased dramatically in the past few decades, the fact remains that control systems are often complicated, and hence one must carefully balance the costs and benefits. An early engineering example of this is the use of microprocessor-based feedback systems in automobiles. The use of microprocessors in automotive applications began in the early 1970s and was driven by increasingly strict emissions standards, which could be met only through electronic controls. Early systems were expensive and failed more often than desired, leading to frequent customer dissatisfaction. It was only through aggressive improvements in technology that the performance, reliability and cost of these systems allowed them to be used in a transparent fashion. Even today, the complexity of these systems is such that it is difficult for an individual car owner to fix problems.

Feedforward

Feedback is reactive: there must be an error before corrective actions are taken. However, in some circumstances it is possible to measure a disturbance before it enters the system, and this information can then be used to take corrective action before the disturbance has influenced the system. The effect of the disturbance is thus reduced by measuring it and generating a control signal that counteracts it. This way of controlling a system is called *feedforward*. Feedforward is particularly useful in shaping the response to command signals because command signals are always available. Since feedforward attempts to match two signals, it requires good process models; otherwise the corrections may have the wrong size or may be badly timed.

The ideas of feedback and feedforward are very general and appear in many different fields. In economics, feedback and feedforward are analogous to a market-based economy versus a planned economy. In business, a feedforward strategy corresponds to running a company based on extensive strategic planning, while a feedback strategy corresponds to a reactive approach. In biology, feedforward has been suggested as an essential element for motion control in humans that is tuned during training. Experience indicates that it is often advantageous to combine feedback and feedforward, and the correct balance requires insight and understanding of their respective properties.

Positive Feedback

In most of this text, we will consider the role of *negative feedback*, in which we attempt to regulate the system by reacting to disturbances in a way that decreases the effect of those disturbances. In some systems, particularly biological systems, *positive feedback* can play an important role. In a system with positive feedback, the increase in some variable or signal leads to a situation in which that quantity is further increased through its dynamics. This has a destabilizing effect and is usually accompanied by a saturation that limits the growth of the quantity. Although often considered undesirable, this behavior is used in biological (and engineering) systems to obtain a very fast response to a condition or signal.

One example of the use of positive feedback is to create switching behavior, in which a system maintains a given state until some input crosses a threshold. Hysteresis is often present so that noisy inputs near the threshold do not cause the system to jitter. This type of behavior is called *bistability* and is often associated with memory devices.

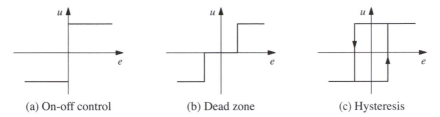

(a) On-off control   (b) Dead zone   (c) Hysteresis

Figure 1.16: Input/output characteristics of on-off controllers. Each plot shows the input on the horizontal axis and the corresponding output on the vertical axis. Ideal on-off control is shown in (a), with modifications for a dead zone (b) or hysteresis (c). Note that for on-off control with hysteresis, the output depends on the value of past inputs.

1.5 Simple Forms of Feedback

The idea of feedback to make corrective actions based on the difference between the desired and the actual values of a quantity can be implemented in many different ways. The benefits of feedback can be obtained by very simple feedback laws such as on-off control, proportional control and proportional-integral-derivative control. In this section we provide a brief preview of some of the topics that will be studied more formally in the remainder of the text.

On-Off Control

A simple feedback mechanism can be described as follows:

$$u = \begin{cases} u_{\max} & \text{if } e > 0 \\ u_{\min} & \text{if } e < 0, \end{cases} \tag{1.1}$$

where the *control error* $e = r - y$ is the difference between the reference signal (or command signal) r and the output of the system y and u is the actuation command. Figure 1.16a shows the relation between error and control. This control law implies that maximum corrective action is always used.

The feedback in equation (1.1) is called *on-off control*. One of its chief advantages is that it is simple and there are no parameters to choose. On-off control often succeeds in keeping the process variable close to the reference, such as the use of a simple thermostat to maintain the temperature of a room. It typically results in a system where the controlled variables oscillate, which is often acceptable if the oscillation is sufficiently small.

Notice that in equation (1.1) the control variable is not defined when the error is zero. It is common to make modifications by introducing either a dead zone or hysteresis (see Figure 1.16b and 1.16c).

PID Control

The reason why on-off control often gives rise to oscillations is that the system overreacts since a small change in the error makes the actuated variable change over the full range. This effect is avoided in *proportional control*, where the characteristic of the controller is proportional to the control error for small errors. This can be achieved with the control law

$$
u = \begin{cases} u_{\max} & \text{if } e \geq e_{\max} \\ k_p e & \text{if } e_{\min} < e < e_{\max} \\ u_{\min} & \text{if } e \leq e_{\min}, \end{cases} \tag{1.2}
$$

where k_p is the controller gain, $e_{\min} = u_{\min}/k_p$ and $e_{\max} = u_{\max}/k_p$. The interval $(e_{\min}, e_{\max})$ is called the *proportional band* because the behavior of the controller is linear when the error is in this interval:

$$
u = k_p(r - y) = k_p e \qquad \text{if } e_{\min} \leq e \leq e_{\max}. \tag{1.3}
$$

While a vast improvement over on-off control, proportional control has the drawback that the process variable often deviates from its reference value. In particular, if some level of control signal is required for the system to maintain a desired value, then we must have $e \neq 0$ in order to generate the requisite input.

This can be avoided by making the control action proportional to the integral of the error:

$$
u(t) = k_i \int_0^t e(\tau) d\tau. \tag{1.4}
$$

This control form is called *integral control*, and k_i is the integral gain. It can be shown through simple arguments that a controller with integral action has zero steady-state error (Exercise 1.5). The catch is that there may not always be a steady state because the system may be oscillating.

An additional refinement is to provide the controller with an anticipative ability by using a prediction of the error. A simple prediction is given by the linear extrapolation

$$
e(t + T_d) \approx e(t) + T_d \frac{de(t)}{dt},
$$

which predicts the error T_d time units ahead. Combining proportional, integral and derivative control, we obtain a controller that can be expressed mathematically as

$$
u(t) = k_p e(t) + k_i \int_0^t e(\tau) \, d\tau + k_d \frac{de(t)}{dt}. \tag{1.5}
$$

The control action is thus a sum of three terms: the past as represented by the integral of the error, the present as represented by the proportional term and the future as represented by a linear extrapolation of the error (the derivative term). This form of feedback is called a *proportional-integral-derivative (PID) controller* and its action is illustrated in Figure 1.17.

A PID controller is very useful and is capable of solving a wide range of control problems. More than 95% of all industrial control problems are solved by PID control, although many of these controllers are actually *proportional-integral* (PI) *controllers* because derivative action is often not included [58]. There are also more advanced controllers, which differ from PID controllers by using more sophisticated methods for prediction.

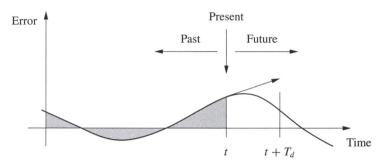

Figure 1.17: Action of a PID controller. At time t, the proportional term depends on the instantaneous value of the error. The integral portion of the feedback is based on the integral of the error up to time t (shaded portion). The derivative term provides an estimate of the growth or decay of the error over time by looking at the rate of change of the error. T_d represents the approximate amount of time in which the error is projected forward (see text).

1.6 Further Reading

The material in this section draws heavily from the report of the Panel on Future Directions on Control, Dynamics and Systems [155]. Several additional papers and reports have highlighted the successes of control [159] and new vistas in control [45, 130, 204]. The early development of control is described by Mayr [148] and in the books by Bennett [28, 29], which cover the period 1800–1955. A fascinating examination of some of the early history of control in the United States has been written by Mindell [152]. A popular book that describes many control concepts across a wide range of disciplines is *Out of Control* by Kelly [121]. There are many textbooks available that describe control systems in the context of specific disciplines. For engineers, the textbooks by Franklin, Powell and Emami-Naeini [79], Dorf and Bishop [61], Kuo and Golnaraghi [133] and Seborg, Edgar and Mellichamp [178] are widely used. More mathematically oriented treatments of control theory include Sontag [182] and Lewis [136]. The book by Hellerstein et al. [97] provides a description of the use of feedback control in computing systems. A number of books look at the role of dynamics and feedback in biological systems, including Milhorn [151] (now out of print), J. D. Murray [154] and Ellner and Guckenheimer [70]. The book by Fradkov [77] and the tutorial article by Bechhoefer [25] cover many specific topics of interest to the physics community.

Exercises

1.1 (Eye motion) Perform the following experiment and explain your results: Holding your head still, move one of your hands left and right in front of your face, following it with your eyes. Record how quickly you can move your hand before you begin to lose track of it. Now hold your hand still and shake your head left to right, once again recording how quickly you can move before losing track of your hand.

1.2 Identify five feedback systems that you encounter in your everyday environment. For each system, identify the sensing mechanism, actuation mechanism and control law. Describe the uncertainty with respect to which the feedback system provides robustness and/or the dynamics that are changed through the use of feedback.

1.3 (Balance systems) Balance yourself on one foot with your eyes closed for 15 s. Using Figure 1.3 as a guide, describe the control system responsible for keeping you from falling down. Note that the "controller" will differ from that in the diagram (unless you are an android reading this in the far future).

1.4 (Cruise control) Download the MATLAB code used to produce simulations for the cruise control system in Figure 1.13 from the companion web site. Using trial and error, change the parameters of the control law so that the overshoot in speed is not more than 1 m/s for a vehicle with mass $m = 1000$ kg.

1.5 (Integral action) We say that a system with a constant input reaches steady state if the output of the system approaches a constant value as time increases. Show that a controller with integral action, such as those given in equations (1.4) and (1.5), gives zero error if the closed loop system reaches steady state.

1.6 Search the web and pick an article in the popular press about a feedback and control system. Describe the feedback system using the terminology given in the article. In particular, identify the control system and describe (a) the underlying process or system being controlled, along with the (b) sensor, (c) actuator and (d) computational element. If the some of the information is not available in the article, indicate this and take a guess at what might have been used.

Chapter Two
System Modeling

... I asked Fermi whether he was not impressed by the agreement between our calculated numbers and his measured numbers. He replied, "How many arbitrary parameters did you use for your calculations?" I thought for a moment about our cut-off procedures and said, "Four." He said, "I remember my friend Johnny von Neumann used to say, with four parameters I can fit an elephant, and with five I can make him wiggle his trunk."

Freeman Dyson on describing the predictions of his model for meson-proton scattering to Enrico Fermi in 1953 [67].

A model is a precise representation of a system's dynamics used to answer questions via analysis and simulation. The model we choose depends on the questions we wish to answer, and so there may be multiple models for a single dynamical system, with different levels of fidelity depending on the phenomena of interest. In this chapter we provide an introduction to the concept of modeling and present some basic material on two specific methods commonly used in feedback and control systems: differential equations and difference equations.

2.1 Modeling Concepts

A *model* is a mathematical representation of a physical, biological or information system. Models allow us to reason about a system and make predictions about how a system will behave. In this text, we will mainly be interested in models of dynamical systems describing the input/output behavior of systems, and we will often work in "state space" form.

Roughly speaking, a dynamical system is one in which the effects of actions do not occur immediately. For example, the velocity of a car does not change immediately when the gas pedal is pushed nor does the temperature in a room rise instantaneously when a heater is switched on. Similarly, a headache does not vanish right after an aspirin is taken, requiring time for it to take effect. In business systems, increased funding for a development project does not increase revenues in the short term, although it may do so in the long term (if it was a good investment). All of these are examples of dynamical systems, in which the behavior of the system evolves with time.

In the remainder of this section we provide an overview of some of the key concepts in modeling. The mathematical details introduced here are explored more fully in the remainder of the chapter.

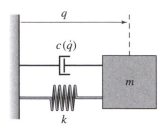

Figure 2.1: Spring–mass system with nonlinear damping. The position of the mass is denoted by q, with $q = 0$ corresponding to the rest position of the spring. The forces on the mass are generated by a linear spring with spring constant k and a damper with force dependent on the velocity $\dot{q}$.

The Heritage of Mechanics

The study of dynamics originated in attempts to describe planetary motion. The basis was detailed observations of the planets by Tycho Brahe and the results of Kepler, who found empirically that the orbits of the planets could be well described by ellipses. Newton embarked on an ambitious program to try to explain why the planets move in ellipses, and he found that the motion could be explained by his law of gravitation and the formula stating that force equals mass times acceleration. In the process he also invented calculus and differential equations.

One of the triumphs of Newton's mechanics was the observation that the motion of the planets could be predicted based on the current positions and velocities of all planets. It was not necessary to know the past motion. The *state* of a dynamical system is a collection of variables that completely characterizes the motion of a system for the purpose of predicting future motion. For a system of planets the state is simply the positions and the velocities of the planets. We call the set of all possible states the *state space*.

A common class of mathematical models for dynamical systems is ordinary differential equations (ODEs). In mechanics, one of the simplest such differential equations is that of a spring–mass system with damping:

$$m\ddot{q} + c(\dot{q}) + kq = 0. \tag{2.1}$$

This system is illustrated in Figure 2.1. The variable $q \in \mathbb{R}$ represents the position of the mass m with respect to its rest position. We use the notation $\dot{q}$ to denote the derivative of q with respect to time (i.e., the velocity of the mass) and $\ddot{q}$ to represent the second derivative (acceleration). The spring is assumed to satisfy Hooke's law, which says that the force is proportional to the displacement. The friction element (damper) is taken as a nonlinear function $c(\dot{q})$, which can model effects such as stiction and viscous drag. The position q and velocity $\dot{q}$ represent the instantaneous state of the system. We say that this system is a *second-order system* since the dynamics depend on the first two derivatives of q.

The evolution of the position and velocity can be described using either a time plot or a phase portrait, both of which are shown in Figure 2.2. The *time plot*, on

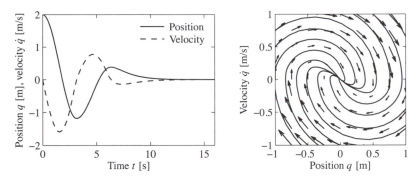

Figure 2.2: Illustration of a state model. A state model gives the rate of change of the state as a function of the state. The plot on the left shows the evolution of the state as a function of time. The plot on the right shows the evolution of the states relative to each other, with the velocity of the state denoted by arrows.

the left, shows the values of the individual states as a function of time. The *phase portrait*, on the right, shows the *vector field* for the system, which gives the state velocity (represented as an arrow) at every point in the state space. In addition, we have superimposed the traces of some of the states from different conditions. The phase portrait gives a strong intuitive representation of the equation as a vector field or a flow. While systems of second order (two states) can be represented in this way, unfortunately it is difficult to visualize equations of higher order using this approach.

The differential equation (2.1) is called an *autonomous* system because there are no external influences. Such a model is natural for use in celestial mechanics because it is difficult to influence the motion of the planets. In many examples, it is useful to model the effects of external disturbances or controlled forces on the system. One way to capture this is to replace equation (2.1) by

$$m\ddot{q} + c(\dot{q}) + kq = u, \tag{2.2}$$

where u represents the effect of external inputs. The model (2.2) is called a *forced* or *controlled differential equation*. It implies that the rate of change of the state can be influenced by the input $u(t)$. Adding the input makes the model richer and allows new questions to be posed. For example, we can examine what influence external disturbances have on the trajectories of a system. Or, in the case where the input variable is something that can be modulated in a controlled way, we can analyze whether it is possible to "steer" the system from one point in the state space to another through proper choice of the input.

The Heritage of Electrical Engineering

A different view of dynamics emerged from electrical engineering, where the design of electronic amplifiers led to a focus on input/output behavior. A system was considered a device that transforms inputs to outputs, as illustrated in Figure 2.3. Conceptually an input/output model can be viewed as a giant table of inputs and

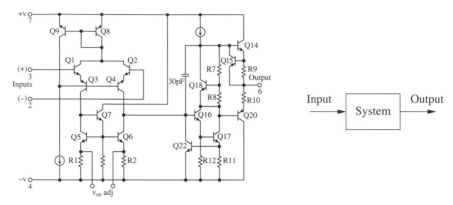

Figure 2.3: Illustration of the input/output view of a dynamical system. The figure on the left shows a detailed circuit diagram for an electronic amplifier; the one on the right is its representation as a block diagram.

outputs. Given an input signal $u(t)$ over some interval of time, the model should produce the resulting output $y(t)$.

The input/output framework is used in many engineering disciplines since it allows us to decompose a system into individual components connected through their inputs and outputs. Thus, we can take a complicated system such as a radio or a television and break it down into manageable pieces such as the receiver, demodulator, amplifier and speakers. Each of these pieces has a set of inputs and outputs and, through proper design, these components can be interconnected to form the entire system.

The input/output view is particularly useful for the special class of *linear time-invariant systems*. This term will be defined more carefully later in this chapter, but roughly speaking a system is linear if the superposition (addition) of two inputs yields an output that is the sum of the outputs that would correspond to individual inputs being applied separately. A system is time-invariant if the output response for a given input does not depend on when that input is applied.

Many electrical engineering systems can be modeled by linear time-invariant systems, and hence a large number of tools have been developed to analyze them. One such tool is the *step response*, which describes the relationship between an input that changes from zero to a constant value abruptly (a step input) and the corresponding output. As we shall see later in the text, the step response is very useful in characterizing the performance of a dynamical system, and it is often used to specify the desired dynamics. A sample step response is shown in Figure 2.4a.

Another way to describe a linear time-invariant system is to represent it by its response to sinusoidal input signals. This is called the *frequency response*, and a rich, powerful theory with many concepts and strong, useful results has emerged. The results are based on the theory of complex variables and Laplace transforms. The basic idea behind frequency response is that we can completely characterize the behavior of a system by its steady-state response to sinusoidal inputs. Roughly

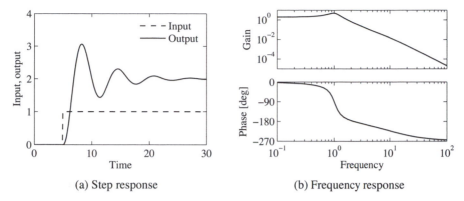

(a) Step response   (b) Frequency response

Figure 2.4: Input/output response of a linear system. The step response (a) shows the output of the system due to an input that changes from 0 to 1 at time $t = 5$ s. The frequency response (b) shows the amplitude gain and phase change due to a sinusoidal input at different frequencies.

speaking, this is done by decomposing any arbitrary signal into a linear combination of sinusoids (e.g., by using the Fourier transform) and then using linearity to compute the output by combining the response to the individual frequencies. A sample frequency response is shown in Figure 2.4b.

The input/output view lends itself naturally to experimental determination of system dynamics, where a system is characterized by recording its response to particular inputs, e.g., a step or a set of sinusoids over a range of frequencies.

The Control View

When control theory emerged as a discipline in the 1940s, the approach to dynamics was strongly influenced by the electrical engineering (input/output) view. A second wave of developments in control, starting in the late 1950s, was inspired by mechanics, where the state space perspective was used. The emergence of space flight is a typical example, where precise control of the orbit of a spacecraft is essential. These two points of view gradually merged into what is today the state space representation of input/output systems.

The development of state space models involved modifying the models from mechanics to include external actuators and sensors and utilizing more general forms of equations. In control, the model given by equation (2.2) was replaced by

$$\frac{dx}{dt} = f(x, u), \qquad y = h(x, u), \tag{2.3}$$

where x is a vector of state variables, u is a vector of control signals and y is a vector of measurements. The term dx/dt represents the derivative of x with respect to time, now considered a vector, and f and h are (possibly nonlinear) mappings of their arguments to vectors of the appropriate dimension. For mechanical systems, the state consists of the position and velocity of the system, so that $x = (q, \dot{q})$ in the case of a damped spring–mass system. Note that in the control formulation we model dynamics as first-order differential equations, but we will see that this can capture the dynamics of higher-order differential equations by appropriate definition of the state and the maps f and h.

Adding inputs and outputs has increased the richness of the classical problems and led to many new concepts. For example, it is natural to ask if possible states x can be reached with the proper choice of u (reachability) and if the measurement y contains enough information to reconstruct the state (observability). These topics will be addressed in greater detail in Chapters 6 and 7.

A final development in building the control point of view was the emergence of disturbances and model uncertainty as critical elements in the theory. The simple way of modeling disturbances as deterministic signals like steps and sinusoids has the drawback that such signals cannot be predicted precisely. A more realistic approach is to model disturbances as random signals. This viewpoint gives a natural connection between prediction and control. The dual views of input/output representations and state space representations are particularly useful when modeling uncertainty since state models are convenient to describe a nominal model but uncertainties are easier to describe using input/output models (often via a frequency response description). Uncertainty will be a constant theme throughout the text and will be studied in particular detail in Chapter 12.

An interesting observation in the design of control systems is that feedback systems can often be analyzed and designed based on comparatively simple models. The reason for this is the inherent robustness of feedback systems. However, other uses of models may require more complexity and more accuracy. One example is feedforward control strategies, where one uses a model to precompute the inputs that cause the system to respond in a certain way. Another area is system validation, where one wishes to verify that the detailed response of the system performs as it was designed. Because of these different uses of models, it is common to use a hierarchy of models having different complexity and fidelity.

 ## Multidomain Modeling

Modeling is an essential element of many disciplines, but traditions and methods from individual disciplines can differ from each other, as illustrated by the previous discussion of mechanical and electrical engineering. A difficulty in systems engineering is that it is frequently necessary to deal with heterogeneous systems from many different domains, including chemical, electrical, mechanical and information systems.

To model such multidomain systems, we start by partitioning a system into smaller subsystems. Each subsystem is represented by balance equations for mass, energy and momentum, or by appropriate descriptions of information processing in the subsystem. The behavior at the interfaces is captured by describing how the variables of the subsystem behave when the subsystems are interconnected. These interfaces act by constraining variables within the individual subsystems to be equal (such as mass, energy or momentum fluxes). The complete model is then obtained by combining the descriptions of the subsystems and the interfaces.

Using this methodology it is possible to build up libraries of subsystems that correspond to physical, chemical and informational components. The procedure mimics the engineering approach where systems are built from subsystems that are themselves built from smaller components. As experience is gained, the components and their interfaces can be standardized and collected in model libraries. In practice, it takes several iterations to obtain a good library that can be reused for many applications.

State models or ordinary differential equations are not suitable for component-based modeling of this form because states may disappear when components are connected. This implies that the internal description of a component may change when it is connected to other components. As an illustration we consider two capacitors in an electrical circuit. Each capacitor has a state corresponding to the voltage across the capacitors, but one of the states will disappear if the capacitors are connected in parallel. A similar situation happens with two rotating inertias, each of which is individually modeled using the angle of rotation and the angular velocity. Two states will disappear when the inertias are joined by a rigid shaft.

This difficulty can be avoided by replacing differential equations by *differential algebraic equations*, which have the form

$$F(z, \dot{z}) = 0,$$

where $z \in \mathbb{R}^n$. A simple special case is

$$\dot{x} = f(x, y), \qquad g(x, y) = 0, \tag{2.4}$$

where $z = (x, y)$ and $F = (\dot{x} - f(x, y), g(x, y))$. The key property is that the derivative $\dot{z}$ is not given explicitly and there may be pure algebraic relations between the components of the vector z.

The model (2.4) captures the examples of the parallel capacitors and the linked rotating inertias. For example, when two capacitors are connected, we simply add the algebraic equation expressing that the voltages across the capacitors are the same.

Modelica is a language that has been developed to support component-based modeling. Differential algebraic equations are used as the basic description, and object-oriented programming is used to structure the models. Modelica is used to model the dynamics of technical systems in domains such as mechanical, electrical, thermal, hydraulic, thermofluid and control subsystems. Modelica is intended to serve as a standard format so that models arising in different domains can be exchanged between tools and users. A large set of free and commercial Modelica component libraries are available and are used by a growing number of people in industry, research and academia. For further information about Modelica, see http://www.modelica.org or Tiller [192].

2.2 State Space Models

In this section we introduce the two primary forms of models that we use in this text: differential equations and difference equations. Both make use of the notions of state, inputs, outputs and dynamics to describe the behavior of a system.

Ordinary Differential Equations

The state of a system is a collection of variables that summarize the past of a system for the purpose of predicting the future. For a physical system the state is composed of the variables required to account for storage of mass, momentum and energy. A key issue in modeling is to decide how accurately this storage has to be represented. The state variables are gathered in a vector $x \in \mathbb{R}^n$ called the *state vector*. The control variables are represented by another vector $u \in \mathbb{R}^p$, and the measured signal by the vector $y \in \mathbb{R}^q$. A system can then be represented by the differential equation

$$\frac{dx}{dt} = f(x, u), \qquad y = h(x, u), \tag{2.5}$$

where $f : \mathbb{R}^n \times \mathbb{R}^p \to \mathbb{R}^n$ and $h : \mathbb{R}^n \times \mathbb{R}^p \to \mathbb{R}^q$ are smooth mappings. We call a model of this form a *state space model*.

The dimension of the state vector is called the *order* of the system. The system (2.5) is called *time-invariant* because the functions f and h do not depend explicitly on time t; there are more general time-varying systems where the functions do depend on time. The model consists of two functions: the function f gives the rate of change of the state vector as a function of state x and control u, and the function h gives the measured values as functions of state x and control u.

A system is called a *linear* state space system if the functions f and h are linear in x and u. A linear state space system can thus be represented by

$$\frac{dx}{dt} = Ax + Bu, \qquad y = Cx + Du, \tag{2.6}$$

where A, B, C and D are constant matrices. Such a system is said to be *linear and time-invariant*, or LTI for short. The matrix A is called the *dynamics matrix*, the matrix B is called the *control matrix*, the matrix C is called the *sensor matrix* and the matrix D is called the *direct term*. Frequently systems will not have a direct term, indicating that the control signal does not influence the output directly.

A different form of linear differential equations, generalizing the second-order dynamics from mechanics, is an equation of the form

$$\frac{d^n y}{dt^n} + a_1 \frac{d^{n-1} y}{dt^{n-1}} + \cdots + a_n y = u, \tag{2.7}$$

where t is the independent (time) variable, $y(t)$ is the dependent (output) variable and $u(t)$ is the input. The notation $d^k y / dt^k$ is used to denote the kth derivative of y with respect to t, sometimes also written as $y^{(k)}$. The controlled differential equation (2.7) is said to be an nth-order system. This system can be converted into state space form by defining

$$
x = \begin{bmatrix} x_1 \\ x_2 \\ \vdots \\ x_{n-1} \\ x_n \end{bmatrix} = \begin{bmatrix} d^{n-1}y/dt^{n-1} \\ d^{n-2}y/dt^{n-2} \\ \vdots \\ dy/dt \\ y \end{bmatrix},
$$

and the state space equations become

$$
\frac{d}{dt}\begin{bmatrix} x_1 \\ x_2 \\ \vdots \\ x_{n-1} \\ x_n \end{bmatrix} = \begin{bmatrix} -a_1 x_1 - \cdots - a_n x_n \\ x_1 \\ \vdots \\ x_{n-2} \\ x_{n-1} \end{bmatrix} + \begin{bmatrix} u \\ 0 \\ \vdots \\ 0 \\ 0 \end{bmatrix}, \qquad y = x_n.
$$

With the appropriate definitions of A, B, C and D, this equation is in linear state space form.

An even more general system is obtained by letting the output be a linear combination of the states of the system, i.e.,

$$
y = b_1 x_1 + b_2 x_2 + \cdots + b_n x_n + du.
$$

This system can be modeled in state space as

$$
\frac{d}{dt}\begin{bmatrix} x_1 \\ x_2 \\ x_3 \\ \vdots \\ x_n \end{bmatrix} = \begin{bmatrix} -a_1 & -a_2 & \cdots & -a_{n-1} & -a_n \\ 1 & 0 & \cdots & 0 & 0 \\ 0 & 1 & & 0 & 0 \\ \vdots & & \ddots & & \vdots \\ 0 & 0 & & 1 & 0 \end{bmatrix} x + \begin{bmatrix} 1 \\ 0 \\ 0 \\ \vdots \\ 0 \end{bmatrix} u,
$$

$$
y = \begin{bmatrix} b_1 & b_2 & \cdots & b_n \end{bmatrix} x + du. \tag{2.8}
$$

This particular form of a linear state space system is called *reachable canonical form* and will be studied in more detail in later chapters.

Example 2.1 Balance systems

An example of a type of system that can be modeled using ordinary differential equations is the class of *balance systems*. A balance system is a mechanical system in which the center of mass is balanced above a pivot point. Some common examples of balance systems are shown in Figure 2.5. The Segway® Personal Transporter (Figure 2.5a) uses a motorized platform to stabilize a person standing on top of it. When the rider leans forward, the transportation device propels itself along the ground but maintains its upright position. Another example is a rocket (Figure 2.5b), in which a gimbaled nozzle at the bottom of the rocket is used to stabilize the body of the rocket above it. Other examples of balance systems include humans or other animals standing upright or a person balancing a stick on their hand.

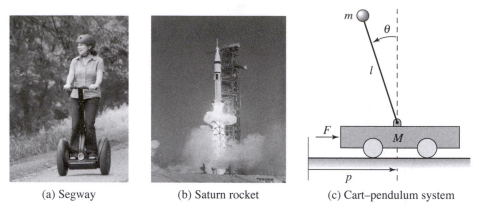

(a) Segway   (b) Saturn rocket   (c) Cart–pendulum system

Figure 2.5: Balance systems. (a) Segway Personal Transporter, (b) Saturn rocket and (c) inverted pendulum on a cart. Each of these examples uses forces at the bottom of the system to keep it upright.

Balance systems are a generalization of the spring–mass system we saw earlier. We can write the dynamics for a mechanical system in the general form

$$M(q)\ddot{q} + C(q, \dot{q}) + K(q) = B(q)u,$$

where $M(q)$ is the inertia matrix for the system, $C(q, \dot{q})$ represents the Coriolis forces as well as the damping, $K(q)$ gives the forces due to potential energy and $B(q)$ describes how the external applied forces couple into the dynamics. The specific form of the equations can be derived using Newtonian mechanics. Note that each of the terms depends on the configuration of the system q and that these terms are often nonlinear in the configuration variables.

Figure 2.5c shows a simplified diagram for a balance system consisting of an inverted pendulum on a cart. To model this system, we choose state variables that represent the position and velocity of the base of the system, p and $\dot{p}$, and the angle and angular rate of the structure above the base, θ and $\dot{\theta}$. We let F represent the force applied at the base of the system, assumed to be in the horizontal direction (aligned with p), and choose the position and angle of the system as outputs. With this set of definitions, the dynamics of the system can be computed using Newtonian mechanics and have the form

$$\begin{bmatrix} (M+m) & -ml\cos\theta \\ -ml\cos\theta & (J+ml^2) \end{bmatrix} \begin{bmatrix} \ddot{p} \\ \ddot{\theta} \end{bmatrix} + \begin{bmatrix} c\dot{p} + ml\sin\theta\,\dot{\theta}^2 \\ \gamma\dot{\theta} - mgl\sin\theta \end{bmatrix} = \begin{bmatrix} F \\ 0 \end{bmatrix}, \qquad (2.9)$$

where M is the mass of the base, m and J are the mass and moment of inertia of the system to be balanced, l is the distance from the base to the center of mass of the balanced body, c and γ are coefficients of viscous friction and g is the acceleration due to gravity.

We can rewrite the dynamics of the system in state space form by defining the state as $x = (p, \theta, \dot{p}, \dot{\theta})$, the input as $u = F$ and the output as $y = (p, \theta)$. If we define the total mass and total inertia as

$$M_t = M + m, \qquad J_t = J + ml^2,$$

the equations of motion then become

$$\frac{d}{dt}\begin{bmatrix} p \\ \theta \\ \dot{p} \\ \dot{\theta} \end{bmatrix} = \begin{bmatrix} \dot{p} \\ \dot{\theta} \\ \dfrac{-mls_\theta\dot{\theta}^2 + mg(ml^2/J_t)s_\theta c_\theta - c\dot{p} - \gamma lmc_\theta\dot{\theta} + u}{M_t - m(ml^2/J_t)c_\theta^2} \\ \dfrac{-ml^2 s_\theta c_\theta\dot{\theta}^2 + M_t gls_\theta - clc_\theta\dot{p} - \gamma(M_t/m)\dot{\theta} + lc_\theta u}{J_t(M_t/m) - m(lc_\theta)^2} \end{bmatrix},$$

$$y = \begin{bmatrix} p \\ \theta \end{bmatrix},$$

where we have used the shorthand $c_\theta = \cos\theta$ and $s_\theta = \sin\theta$.

In many cases, the angle θ will be very close to 0, and hence we can use the approximations $\sin\theta \approx \theta$ and $\cos\theta \approx 1$. Furthermore, if $\dot{\theta}$ is small, we can ignore quadratic and higher terms in $\dot{\theta}$. Substituting these approximations into our equations, we see that we are left with a *linear* state space equation

$$\frac{d}{dt}\begin{bmatrix} p \\ \theta \\ \dot{p} \\ \dot{\theta} \end{bmatrix} = \begin{bmatrix} 0 & 0 & 1 & 0 \\ 0 & 0 & 0 & 1 \\ 0 & m^2l^2g/\mu & -cJ_t/\mu & -\gamma J_t lm/\mu \\ 0 & M_t mgl/\mu & -clm/\mu & -\gamma M_t/\mu \end{bmatrix}\begin{bmatrix} p \\ \theta \\ \dot{p} \\ \dot{\theta} \end{bmatrix} + \begin{bmatrix} 0 \\ 0 \\ J_t/\mu \\ lm/\mu \end{bmatrix} u,$$

$$y = \begin{bmatrix} 1 & 0 & 0 & 0 \\ 0 & 1 & 0 & 0 \end{bmatrix} x,$$

where $\mu = M_t J_t - m^2 l^2$. $\qquad\qquad\qquad \nabla$

Example 2.2 Inverted pendulum

A variation of the previous example is one in which the location of the base p does not need to be controlled. This happens, for example, if we are interested only in stabilizing a rocket's upright orientation without worrying about the location of base of the rocket. The dynamics of this simplified system are given by

$$\frac{d}{dt}\begin{bmatrix} \theta \\ \dot{\theta} \end{bmatrix} = \begin{bmatrix} \dot{\theta} \\ \dfrac{mgl}{J_t}\sin\theta - \dfrac{\gamma}{J_t}\dot{\theta} + \dfrac{l}{J_t}\cos\theta\, u \end{bmatrix}, \qquad y = \theta, \qquad (2.10)$$

where γ is the coefficient of rotational friction, $J_t = J + ml^2$ and u is the force applied at the base. This system is referred to as an *inverted pendulum*. $\qquad \nabla$

Difference Equations

In some circumstances, it is more natural to describe the evolution of a system at discrete instants of time rather than continuously in time. If we refer to each of these times by an integer $k = 0, 1, 2, \ldots$, then we can ask how the state of the system changes for each k. Just as in the case of differential equations, we define the state to be those sets of variables that summarize the past of the system for the purpose of predicting its future. Systems described in this manner are referred to as *discrete-time systems*.

The evolution of a discrete-time system can be written in the form

$$x[k + 1] = f(x[k], u[k]), \qquad y[k] = h(x[k], u[k]), \qquad (2.11)$$

where $x[k] \in \mathbb{R}^n$ is the state of the system at time k (an integer), $u[k] \in \mathbb{R}^p$ is the input and $y[k] \in \mathbb{R}^q$ is the output. As before, f and h are smooth mappings of the appropriate dimension. We call equation (2.11) a *difference equation* since it tells us how $x[k + 1]$ differs from $x[k]$. The state $x[k]$ can be either a scalar- or a vector-valued quantity; in the case of the latter we write $x_j[k]$ for the value of the jth state at time k.

Just as in the case of differential equations, it is often the case that the equations are linear in the state and input, in which case we can describe the system by

$$x[k + 1] = Ax[k] + Bu[k], \qquad y[k] = Cx[k] + Du[k].$$

As before, we refer to the matrices A, B, C and D as the dynamics matrix, the control matrix, the sensor matrix and the direct term. The solution of a linear difference equation with initial condition $x[0]$ and input $u[0], \ldots, u[T]$ is given by

$$x[k] = A^k x_0 + \sum_{j=0}^{k-1} A^{k-j-1} Bu[j],$$

$$k > 0. \qquad (2.12)$$

$$y[k] = C A^k x_0 + \sum_{j=0}^{k-1} C A^{k-j-1} Bu[j] + Du[k],$$

Difference equations are also useful as an approximation of differential equations, as we will show later.

Example 2.3 Predator–prey

As an example of a discrete-time system, consider a simple model for a predator–prey system. The predator–prey problem refers to an ecological system in which we have two species, one of which feeds on the other. This type of system has been studied for decades and is known to exhibit interesting dynamics. Figure 2.6 shows a historical record taken over 90 years for a population of lynxes versus a population of hares [142]. As can been seen from the graph, the annual records of the populations of each species are oscillatory in nature.

A simple model for this situation can be constructed using a discrete-time model by keeping track of the rate of births and deaths of each species. Letting H represent the population of hares and L represent the population of lynxes, we can describe the state in terms of the populations at discrete periods of time. Letting k be the

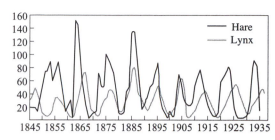

Figure 2.6: Predator versus prey. The photograph on the left shows a Canadian lynx and a snowshoe hare, the lynx's primary prey. The graph on the right shows the populations of hares and lynxes between 1845 and 1935 in a section of the Canadian Rockies [142]. The data were collected on an annual basis over a period of 90 years. (Photograph copyright Tom and Pat Leeson.)

discrete-time index (e.g., the month number), we can write

$$H[k + 1] = H[k] + b_r(u)H[k] - aL[k]H[k],$$
$$L[k + 1] = L[k] + cL[k]H[k] - d_fL[k], \tag{2.13}$$

where $b_r(u)$ is the hare birth rate per unit period and as a function of the food supply u, d_f is the lynx mortality rate and a and c are the interaction coefficients. The interaction term $aL[k]H[k]$ models the rate of predation, which is assumed to be proportional to the rate at which predators and prey meet and is hence given by the product of the population sizes. The interaction term $cL[k]H[k]$ in the lynx dynamics has a similar form and represents the rate of growth of the lynx population. This model makes many simplifying assumptions—such as the fact that hares decrease in number only through predation by lynxes—but it often is sufficient to answer basic questions about the system.

To illustrate the use of this system, we can compute the number of lynxes and hares at each time point from some initial population. This is done by starting with $x[0] = (H_0, L_0)$ and then using equation (2.13) to compute the populations in the following period. By iterating this procedure, we can generate the population over time. The output of this process for a specific choice of parameters and initial conditions is shown in Figure 2.7. While the details of the simulation are different from the experimental data (to be expected given the simplicity of our assumptions), we see qualitatively similar trends and hence we can use the model to help explore the dynamics of the system. ∇

Example 2.4 E-mail server

The IBM Lotus server is an collaborative software system that administers users' e-mail, documents and notes. Client machines interact with end users to provide access to data and applications. The server also handles other administrative tasks. In the early development of the system it was observed that the performance was poor when the central processing unit (CPU) was overloaded because of too many service requests, and mechanisms to control the load were therefore introduced.

The interaction between the client and the server is in the form of remote pro-

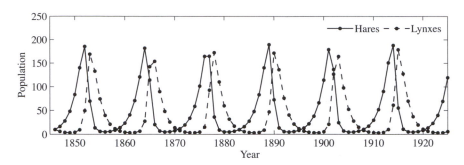

Figure 2.7: Discrete-time simulation of the predator–prey model (2.13). Using the parameters $a = c = 0.014, b_r(u) = 0.6$ and $d = 0.7$ in equation (2.13), the period and magnitude of the lynx and hare population cycles approximately match the data in Figure 2.6.

cedure calls (RPCs). The server maintains a log of statistics of completed requests. The total number of requests being served, called RIS (RPCs in server), is also measured. The load on the server is controlled by a parameter called MaxUsers, which sets the total number of client connections to the server. This parameter is controlled by the system administrator. The server can be regarded as a dynamical system with MaxUsers as the input and RIS as the output. The relationship between input and output was first investigated by exploring the steady-state performance and was found to be linear.

In [97] a dynamic model in the form of a first-order difference equation is used to capture the dynamic behavior of this system. Using system identification techniques, they construct a model of the form

$$y[k + 1] = ay[k] + bu[k],$$

where $u = \text{MaxUsers} - \overline{\text{MaxUsers}}$ and $y = \text{RIS} - \overline{\text{RIS}}$. The parameters $a = 0.43$ and $b = 0.47$ are parameters that describe the dynamics of the system around the operating point, and $\overline{\text{MaxUsers}} = 165$ and $\overline{\text{RIS}} = 135$ represent the nominal operating point of the system. The number of requests was averaged over a sampling period of 60 s.   ∇

Simulation and Analysis

State space models can be used to answer many questions. One of the most common, as we have seen in the previous examples, involves predicting the evolution of the system state from a given initial condition. While for simple models this can be done in closed form, more often it is accomplished through computer simulation. One can also use state space models to analyze the overall behavior of the system without making direct use of simulation.

Consider again the damped spring–mass system from Section 2.1, but this time with an external force applied, as shown in Figure 2.8. We wish to predict the motion of the system for a periodic forcing function, with a given initial condition, and determine the amplitude, frequency and decay rate of the resulting motion.

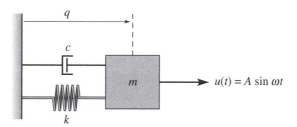

Figure 2.8: A driven spring–mass system with damping. Here we use a linear damping element with coefficient of viscous friction c. The mass is driven with a sinusoidal force of amplitude A.

We choose to model the system with a linear ordinary differential equation. Using Hooke's law to model the spring and assuming that the damper exerts a force that is proportional to the velocity of the system, we have

$$m\ddot{q} + c\dot{q} + kq = u, \tag{2.14}$$

where m is the mass, q is the displacement of the mass, c is the coefficient of viscous friction, k is the spring constant and u is the applied force. In state space form, using $x = (q, \dot{q})$ as the state and choosing $y = q$ as the output, we have

$$\frac{dx}{dt} = \begin{bmatrix} x_2 \\ -\frac{c}{m}x_2 - \frac{k}{m}x_1 + \frac{u}{m} \end{bmatrix}, \qquad y = x_1.$$

We see that this is a linear second-order differential equation with one input u and one output y.

We now wish to compute the response of the system to an input of the form $u = A \sin \omega t$. Although it is possible to solve for the response analytically, we instead make use of a computational approach that does not rely on the specific form of this system. Consider the general state space system

$$\frac{dx}{dt} = f(x, u).$$

Given the state x at time t, we can approximate the value of the state at a short time $h > 0$ later by assuming that the rate of change of $f(x, u)$ is constant over the interval t to $t + h$. This gives

$$x(t + h) = x(t) + hf(x(t), u(t)). \tag{2.15}$$

Iterating this equation, we can thus solve for x as a function of time. This approximation is known as Euler integration and is in fact a difference equation if we let h represent the time increment and write $x[k] = x(kh)$. Although modern simulation tools such as MATLAB and Mathematica use more accurate methods than Euler integration, they still have some of the same basic trade-offs.

Returning to our specific example, Figure 2.9 shows the results of computing $x(t)$ using equation (2.15), along with the analytical computation. We see that as

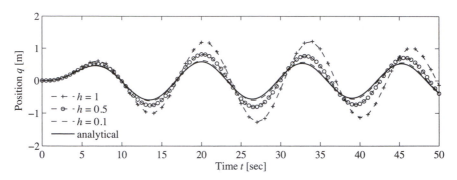

Figure 2.9: Simulation of the forced spring–mass system with different simulation time constants. The dashed line represents the analytical solution. The solid lines represent the approximate solution via the method of Euler integration, using decreasing step sizes.

h gets smaller, the computed solution converges to the exact solution. The form of the solution is also worth noticing: after an initial transient, the system settles into a periodic motion. The portion of the response after the transient is called the *steady-state response* to the input.

In addition to generating simulations, models can also be used to answer other types of questions. Two that are central to the methods described in this text concern the stability of an equilibrium point and the input/output frequency response. We illustrate these two computations through the examples below and return to the general computations in later chapters.

Returning to the damped spring–mass system, the equations of motion with no input forcing are given by

$$\frac{dx}{dt} = \begin{bmatrix} x_2 \\ -\frac{c}{m}x_2 - \frac{k}{m}x_1 \end{bmatrix}, \qquad (2.16)$$

where x_1 is the position of the mass (relative to the rest position) and x_2 is its velocity. We wish to show that if the initial state of the system is away from the rest position, the system will return to the rest position eventually (we will later define this situation to mean that the rest position is *asymptotically stable*). While we could heuristically show this by simulating many, many initial conditions, we seek instead to prove that this is true for *any* initial condition.

To do so, we construct a function $V : \mathbb{R}^n \to \mathbb{R}$ that maps the system state to a positive real number. For mechanical systems, a convenient choice is the energy of the system,

$$V(x) = \frac{1}{2}kx_1^2 + \frac{1}{2}mx_2^2. \qquad (2.17)$$

If we look at the time derivative of the energy function, we see that

$$\frac{dV}{dt} = kx_1\dot{x}_1 + mx_2\dot{x}_2 = kx_1x_2 + mx_2(-\frac{c}{m}x_2 - \frac{k}{m}x_1) = -cx_2^2,$$

which is always either negative or zero. Hence $V(x(t))$ is never increasing and, using a bit of analysis that we will see formally later, the individual states must remain bounded.

If we wish to show that the states eventually return to the origin, we must use a slightly more detailed analysis. Intuitively, we can reason as follows: suppose that for some period of time, $V(x(t))$ stops decreasing. Then it must be true that $\dot{V}(x(t)) = 0$, which in turn implies that $x_2(t) = 0$ for that same period. In that case, $\dot{x}_2(t) = 0$, and we can substitute into the second line of equation (2.16) to obtain

$$0 = \dot{x}_2 = -\frac{c}{m}x_2 - \frac{k}{m}x_1 = \frac{k}{m}x_1.$$

Thus we must have that x_1 also equals zero, and so the only time that $V(x(t))$ can stop decreasing is if the state is at the origin (and hence this system is at its rest position). Since we know that $V(x(t))$ is never increasing (because $\dot{V} \leq 0$), we therefore conclude that the origin is stable (for *any* initial condition).

This type of analysis, called Lyapunov stability analysis, is considered in detail in Chapter 4. It shows some of the power of using models for the analysis of system properties.

Another type of analysis that we can perform with models is to compute the output of a system to a sinusoidal input. We again consider the spring–mass system, but this time keeping the input and leaving the system in its original form:

$$m\ddot{q} + c\dot{q} + kq = u. \tag{2.18}$$

We wish to understand how the system responds to a sinusoidal input of the form

$$u(t) = A \sin \omega t.$$

We will see how to do this analytically in Chapter 6, but for now we make use of simulations to compute the answer.

We first begin with the observation that if $q(t)$ is the solution to equation (2.18) with input $u(t)$, then applying an input $2u(t)$ will give a solution $2q(t)$ (this is easily verified by substitution). Hence it suffices to look at an input with unit magnitude, $A = 1$. A second observation, which we will prove in Chapter 5, is that the long-term response of the system to a sinusoidal input is itself a sinusoid at the same frequency, and so the output has the form

$$q(t) = g(\omega) \sin(\omega t + \varphi(\omega)),$$

where $g(\omega)$ is called the *gain* of the system and $\varphi(\omega)$ is called the *phase* (or phase offset).

To compute the frequency response numerically, we can simulate the system at a set of frequencies $\omega_1, \ldots, \omega_N$ and plot the gain and phase at each of these frequencies. An example of this type of computation is shown in Figure 2.10.

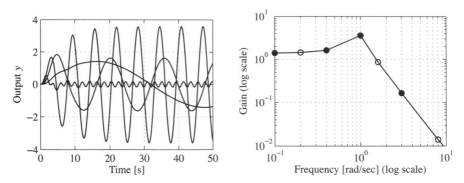

Figure 2.10: A frequency response (gain only) computed by measuring the response of individual sinusoids. The figure on the left shows the response of the system as a function of time to a number of different unit magnitude inputs (at different frequencies). The figure on the right shows this same data in a different way, with the magnitude of the response plotted as a function of the input frequency. The filled circles correspond to the particular frequencies shown in the time responses.

2.3 Modeling Methodology

To deal with large, complex systems, it is useful to have different representations of the system that capture the essential features and hide irrelevant details. In all branches of science and engineering it is common practice to use some graphical description of systems, called *schematic diagrams*. They can range from stylistic pictures to drastically simplified standard symbols. These pictures make it possible to get an overall view of the system and to identify the individual components. Examples of such diagrams are shown in Figure 2.11. Schematic diagrams are useful because they give an overall picture of a system, showing different subprocesses and their interconnection and indicating variables that can be manipulated and signals that can be measured.

Block Diagrams

A special graphical representation called a *block diagram* has been developed in control engineering. The purpose of a block diagram is to emphasize the information flow and to hide details of the system. In a block diagram, different process elements are shown as boxes, and each box has inputs denoted by lines with arrows pointing toward the box and outputs denoted by lines with arrows going out of the box. The inputs denote the variables that influence a process, and the outputs denote the signals that we are interested in or signals that influence other subsystems. Block diagrams can also be organized in hierarchies, where individual blocks may themselves contain more detailed block diagrams.

Figure 2.12 shows some of the notation that we use for block diagrams. Signals are represented as lines, with arrows to indicate inputs and outputs. The first diagram is the representation for a summation of two signals. An input/output response is represented as a rectangle with the system name (or mathematical description) in

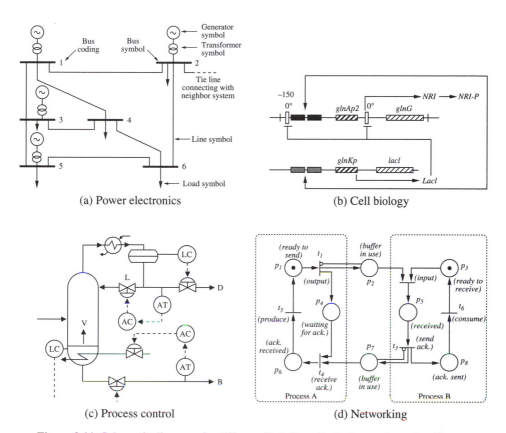

Figure 2.11: Schematic diagrams for different disciplines. Each diagram is used to illustrate the dynamics of a feedback system: (a) electrical schematics for a power system [132], (b) a biological circuit diagram for a synthetic clock circuit [21], (c) a process diagram for a distillation column [178] and (d) a Petri net description of a communication protocol.

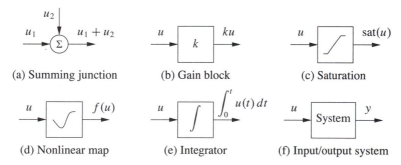

Figure 2.12: Standard block diagram elements. The arrows indicate the the inputs and outputs of each element, with the mathematical operation corresponding to the blocked labeled at the output. The system block (f) represents the full input/output response of a dynamical system.

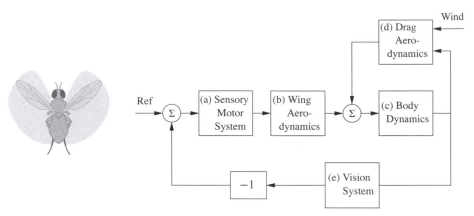

Figure 2.13: A block diagram representation of the flight control system for an insect flying against the wind. The mechanical portion of the model consists of the rigid-body dynamics of the fly, the drag due to flying through the air and the forces generated by the wings. The motion of the body causes the visual environment of the fly to change, and this information is then used to control the motion of the wings (through the sensory motor system), closing the loop.

the block. Two special cases are a proportional gain, which scales the input by a multiplicative factor, and an integrator, which outputs the integral of the input signal.

Figure 2.13 illustrates the use of a block diagram, in this case for modeling the flight response of a fly. The flight dynamics of an insect are incredibly intricate, involving careful coordination of the muscles within the fly to maintain stable flight in response to external stimuli. One known characteristic of flies is their ability to fly upwind by making use of the optical flow in their compound eyes as a feedback mechanism. Roughly speaking, the fly controls its orientation so that the point of contraction of the visual field is centered in its visual field.

To understand this complex behavior, we can decompose the overall dynamics of the system into a series of interconnected subsystems (or *blocks*). Referring to Figure 2.13, we can model the insect navigation system through an interconnection of five blocks. The sensory motor system (a) takes the information from the visual system (e) and generates muscle commands that attempt to steer the fly so that the point of contraction is centered. These muscle commands are converted into forces through the flapping of the wings (b) and the resulting aerodynamic forces that are produced. The forces from the wings are combined with the drag on the fly (d) to produce a net force on the body of the fly. The wind velocity enters through the drag aerodynamics. Finally, the body dynamics (c) describe how the fly translates and rotates as a function of the net forces that are applied to it. The insect position, speed and orientation are fed back to the drag aerodynamics and vision system blocks as inputs.

Each of the blocks in the diagram can itself be a complicated subsystem. For example, the visual system of a fruit fly consists of two complicated compound eyes (with about 700 elements per eye), and the sensory motor system has about 200,000 neurons that are used to process information. A more detailed block diagram of the insect flight control system would show the interconnections between these elements, but here we have used one block to represent how the motion of the fly affects the output of the visual system, and a second block to represent how the visual field is processed by the fly's brain to generate muscle commands. The choice of the level of detail of the blocks and what elements to separate into different blocks often depends on experience and the questions that one wants to answer using the model. One of the powerful features of block diagrams is their ability to hide information about the details of a system that may not be needed to gain an understanding of the essential dynamics of the system.

Modeling from Experiments

Since control systems are provided with sensors and actuators, it is also possible to obtain models of system dynamics from experiments on the process. The models are restricted to input/output models since only these signals are accessible to experiments, but modeling from experiments can also be combined with modeling from physics through the use of feedback and interconnection.

A simple way to determine a system's dynamics is to observe the response to a step change in the control signal. Such an experiment begins by setting the control signal to a constant value; then when steady state is established, the control signal is changed quickly to a new level and the output is observed. The experiment gives the step response of the system, and the shape of the response gives useful information about the dynamics. It immediately gives an indication of the response time, and it tells if the system is oscillatory or if the response is monotone.

Example 2.5 Spring–mass system
Consider the spring–mass system from Section 2.1, whose dynamics are given by

$$m\ddot{q} + c\dot{q} + kq = u. \tag{2.19}$$

We wish to determine the constants m, c and k by measuring the response of the system to a step input of magnitude F_0.

We will show in Chapter 6 that when $c^2 < 4km$, the step response for this system from the rest configuration is given by

$$q(t) = \frac{F_0}{k}\left(1 - \exp\left(-\frac{ct}{2m}\right)\sin(\omega_d t + \varphi)\right), \qquad \omega_d = \frac{\sqrt{4km - c^2}}{2m},$$
$$\varphi = \tan^{-1}\left(\sqrt{4km - c^2}\right).$$

From the form of the solution, we see that the form of the response is determined by the parameters of the system. Hence, by measuring certain features of the step response we can determine the parameter values.

Figure 2.14 shows the response of the system to a step of magnitude $F_0 = 20$ N, along with some measurements. We start by noting that the steady-state position

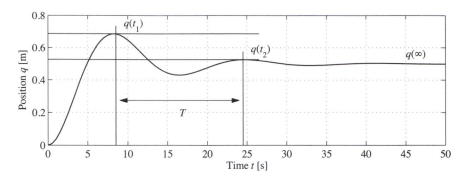

Figure 2.14: Step response for a spring–mass system. The magnitude of the step input is $F_0 = 20$ N. The period of oscillation T is determined by looking at the time between two subsequent local maxima in the response. The period combined with the steady-state value $q(\infty)$ and the relative decrease between local maxima can be used to estimate the parameters in a model of the system.

of the mass (after the oscillations die down) is a function of the spring constant k:

$$q(\infty) = \frac{F_0}{k}, \tag{2.20}$$

where F_0 is the magnitude of the applied force ($F_0 = 1$ for a unit step input). The parameter $1/k$ is called the *gain* of the system. The period of the oscillation can be measured between two peaks and must satisfy

$$\frac{2\pi}{T} = \frac{\sqrt{4km - c^2}}{2m}. \tag{2.21}$$

Finally, the rate of decay of the oscillations is given by the exponential factor in the solution. Measuring the amount of decay between two peaks, we have

$$\log\left(q(t_1) - \frac{F_0}{k}\right) - \log\left(q(t_2) - \frac{F_0}{k}\right) = \frac{c}{2m}(t_2 - t_1). \tag{2.22}$$

Using this set of three equations, we can solve for the parameters and determine that for the step response in Figure 2.14 we have $m \approx 250$ kg, $c \approx 60$ N s/m and $k \approx 40$ N/m. ∇

Modeling from experiments can also be done using many other signals. Sinusoidal signals are commonly used (particularly for systems with fast dynamics) and precise measurements can be obtained by exploiting correlation techniques. An indication of nonlinearities can be obtained by repeating experiments with input signals having different amplitudes.

Normalization and Scaling

Having obtained a model, it is often useful to scale the variables by introducing dimension-free variables. Such a procedure can often simplify the equations for a system by reducing the number of parameters and reveal interesting properties of the model. Scaling can also improve the numerical conditioning of the model to allow faster and more accurate simulations.

The procedure of scaling is straightforward: choose units for each independent variable and introduce new variables by dividing the variables by the chosen normalization unit. We illustrate the procedure with two examples.

Example 2.6 Spring–mass system
Consider again the spring–mass system introduced earlier. Neglecting the damping, the system is described by

$$m\ddot{q} + kq = u.$$

The model has two parameters m and k. To normalize the model we introduce dimension-free variables $x = q/l$ and $\tau = \omega_0 t$, where $\omega_0 = \sqrt{k/m}$ and l is the chosen length scale. We scale force by $ml\omega_0^2$ and introduce $v = u/(ml\omega_0^2)$. The scaled equation then becomes

$$\frac{d^2x}{d\tau^2} = \frac{d^2q/l}{d(\omega_0 t)^2} = \frac{1}{ml\omega_0^2}(-kq + u) = -x + v,$$

which is the normalized undamped spring–mass system. Notice that the normalized model has no parameters, while the original model had two parameters m and k. Introducing the scaled, dimension-free state variables $z_1 = x = q/l$ and $z_2 = dx/d\tau = \dot{q}/(l\omega_0)$, the model can be written as

$$\frac{d}{dt}\begin{bmatrix} z_1 \\ z_2 \end{bmatrix} = \begin{bmatrix} 0 & 1 \\ -1 & 0 \end{bmatrix}\begin{bmatrix} z_1 \\ z_2 \end{bmatrix} + \begin{bmatrix} 0 \\ v \end{bmatrix}.$$

This simple linear equation describes the dynamics of any spring–mass system, independent of the particular parameters, and hence gives us insight into the fundamental dynamics of this oscillatory system. To recover the physical frequency of oscillation or its magnitude, we must invert the scaling we have applied. ∇

Example 2.7 Balance system
Consider the balance system described in Section 2.1. Neglecting damping by putting $c = 0$ and $\gamma = 0$ in equation (2.9), the model can be written as

$$(M + m)\frac{d^2q}{dt^2} - ml\cos\theta\frac{d^2\theta}{dt^2} + ml\sin\theta\left(\frac{dq}{dt}\right)^2 = F,$$

$$-ml\cos\theta\frac{d^2q}{dt^2} + (J + ml^2)\frac{d^2\theta}{dt^2} - mgl\sin\theta = 0.$$

Let $\omega_0 = \sqrt{mgl/(J + ml^2)}$, choose the length scale as l, let the time scale be $1/\omega_0$, choose the force scale as $(M + m)l\omega_0^2$ and introduce the scaled variables $\tau = \omega_0 t$, $x = q/l$ and $u = F/((M + m)l\omega_0^2)$. The equations then become

$$\frac{d^2x}{d\tau^2} - \alpha\cos\theta\frac{d^2\theta}{d\tau^2} + \alpha\sin\theta\left(\frac{d\theta}{d\tau}\right)^2 = u, \quad -\beta\cos\theta\frac{d^2x}{d\tau^2} + \frac{d^2\theta}{d\tau^2} - \sin\theta = 0,$$

where $\alpha = m/(M+m)$ and $\beta = ml^2/(J+ml^2)$. Notice that the original model has five parameters m, M, J, l and g but the normalized model has only two parameters

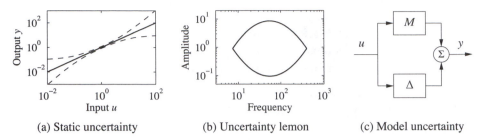

(a) Static uncertainty   (b) Uncertainty lemon   (c) Model uncertainty

Figure 2.15: Characterization of model uncertainty. Uncertainty of a static system is illustrated in (a), where the solid line indicates the nominal input/output relationship and the dashed lines indicate the range of possible uncertainty. The uncertainty lemon [83] in (b) is one way to capture uncertainty in dynamical systems emphasizing that a model is valid only in some amplitude and frequency ranges. In (c) a model is represented by a nominal model M and another model Δ representing the uncertainty analogous to the representation of parameter uncertainty.

α and β. If $M \gg m$ and $ml^2 \gg J$, we get $\alpha \approx 0$ and $\beta \approx 1$ and the model can be approximated by

$$\frac{d^2 x}{d\tau^2} = u, \qquad \frac{d^2 \theta}{d\tau^2} - \sin\theta = u\cos\theta.$$

The model can be interpreted as a mass combined with an inverted pendulum driven by the same input. ∇

Model Uncertainty

Reducing uncertainty is one of the main reasons for using feedback, and it is therefore important to characterize uncertainty. When making measurements, there is a good tradition to assign both a nominal value and a measure of uncertainty. It is useful to apply the same principle to modeling, but unfortunately it is often difficult to express the uncertainty of a model quantitatively.

For a static system whose input/output relation can be characterized by a function, uncertainty can be expressed by an uncertainty band as illustrated in Figure 2.15a. At low signal levels there are uncertainties due to sensor resolution, friction and quantization. Some models for queuing systems or cells are based on averages that exhibit significant variations for small populations. At large signal levels there are saturations or even system failures. The signal ranges where a model is reasonably accurate vary dramatically between applications, but it is rare to find models that are accurate for signal ranges larger than 10^4.

Characterization of the uncertainty of a dynamic model is much more difficult. We can try to capture uncertainties by assigning uncertainties to parameters of the model, but this is often not sufficient. There may be errors due to phenomena that have been neglected, e.g., small time delays. In control the ultimate test is how well a control system based on the model performs, and time delays can be important. There is also a frequency aspect. There are slow phenomena, such as aging, that can cause changes or drift in the systems. There are also high-frequency effects: a resistor will no longer be a pure resistance at very high frequencies, and a beam has stiffness and will exhibit additional dynamics when subject to high-frequency excitation. The *uncertainty lemon* [83] shown in Figure 2.15b is one way to conceptualize the uncertainty of a system. It illustrates that a model is valid only in certain amplitude and frequency ranges.

We will introduce some formal tools for representing uncertainty in Chapter 12 using figures such as Figure 2.15c. These tools make use of the concept of a transfer function, which describes the frequency response of an input/output system. For now, we simply note that one should always be careful to recognize the limits of a model and not to make use of models outside their range of applicability. For example, one can describe the uncertainty lemon and then check to make sure that signals remain in this region. In early analog computing, a system was simulated using operational amplifiers, and it was customary to give alarms when certain signal levels were exceeded. Similar features can be included in digital simulation.

2.4 Modeling Examples

In this section we introduce additional examples that illustrate some of the different types of systems for which one can develop differential equation and difference equation models. These examples are specifically chosen from a range of different fields to highlight the broad variety of systems to which feedback and control concepts can be applied. A more detailed set of applications that serve as running examples throughout the text are given in the next chapter.

Motion Control Systems

Motion control systems involve the use of computation and feedback to control the movement of a mechanical system. Motion control systems range from nanopositioning systems (atomic force microscopes, adaptive optics), to control systems for the read/write heads in a disk drive of a CD player, to manufacturing systems (transfer machines and industrial robots), to automotive control systems (antilock brakes, suspension control, traction control), to air and space flight control systems (airplanes, satellites, rockets and planetary rovers).

Example 2.8 Vehicle steering—the bicycle model
A common problem in motion control is to control the trajectory of a vehicle through an actuator that causes a change in the orientation. A steering wheel on an automobile and the front wheel of a bicycle are two examples, but similar dynamics occur in the steering of ships or control of the pitch dynamics of an aircraft. In many cases, we can understand the basic behavior of these systems through the use of a simple model that captures the basic kinematics of the system.

Consider a vehicle with two wheels as shown in Figure 2.16. For the purpose of steering we are interested in a model that describes how the velocity of the vehicle

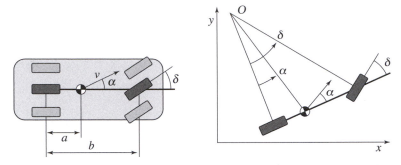

Figure 2.16: Vehicle steering dynamics. The left figure shows an overhead view of a vehicle with four wheels. The wheel base is b and the center of mass at a distance a forward of the rear wheels. By approximating the motion of the front and rear pairs of wheels by a single front wheel and a single rear wheel, we obtain an abstraction called the *bicycle model*, shown on the right. The steering angle is δ and the velocity at the center of mass has the angle α relative the length axis of the vehicle. The position of the vehicle is given by (x, y) and the orientation (heading) by θ.

depends on the steering angle δ. To be specific, consider the velocity v at the center of mass, a distance a from the rear wheel, and let b be the wheel base, as shown in Figure 2.16. Let x and y be the coordinates of the center of mass, θ the heading angle and α the angle between the velocity vector v and the centerline of the vehicle. Since $b = r_a \tan \delta$ and $a = r_a \tan \alpha$, it follows that $\tan \alpha = (a/b) \tan \delta$ and we get the following relation between α and the steering angle δ:

$$\alpha(\delta) = \arctan\left(\frac{a \tan \delta}{b}\right). \tag{2.23}$$

Assume that the wheels are rolling without slip and that the velocity of the rear wheel is v_0. The vehicle speed at its center of mass is $v = v_0 / \cos \alpha$, and we find that the motion of this point is given by

$$\begin{aligned}
\frac{dx}{dt} &= v \cos(\alpha + \theta) = v_0 \frac{\cos(\alpha + \theta)}{\cos \alpha}, \\
\frac{dy}{dt} &= v \sin(\alpha + \theta) = v_0 \frac{\sin(\alpha + \theta)}{\cos \alpha}.
\end{aligned} \tag{2.24}$$

To see how the angle θ is influenced by the steering angle, we observe from Figure 2.16 that the vehicle rotates with the angular velocity v_0/r_a around the point O. Hence

$$\frac{d\theta}{dt} = \frac{v_0}{r_a} = \frac{v_0}{b} \tan \delta. \tag{2.25}$$

Equations (2.23)–(2.25) can be used to model an automobile under the assumptions that there is no slip between the wheels and the road and that the two front wheels can be approximated by a single wheel at the center of the car. The assumption of no slip can be relaxed by adding an extra state variable, giving a more realistic model. Such a model also describes the steering dynamics of ships as well

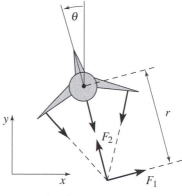

(a) Harrier "jump jet"   (b) Simplified model

Figure 2.17: Vectored thrust aircraft. The Harrier AV-8B military aircraft (a) redirects its engine thrust downward so that it can "hover" above the ground. Some air from the engine is diverted to the wing tips to be used for maneuvering. As shown in (b), the net thrust on the aircraft can be decomposed into a horizontal force F_1 and a vertical force F_2 acting at a distance r from the center of mass.

as the pitch dynamics of aircraft and missiles. It is also possible to choose coordinates so that the reference point is at the rear wheels (corresponding to setting $\alpha = 0$), a model often referred to as the *Dubins car* [66].

Figure 2.16 represents the situation when the vehicle moves forward and has front-wheel steering. The case when the vehicle reverses is obtained by changing the sign of the velocity, which is equivalent to a vehicle with rear-wheel steering.

$$\nabla$$

Example 2.9 Vectored thrust aircraft
Consider the motion of vectored thrust aircraft, such as the Harrier "jump jet" shown Figure 2.17a. The Harrier is capable of vertical takeoff by redirecting its thrust downward and through the use of smaller maneuvering thrusters located on its wings. A simplified model of the Harrier is shown in Figure 2.17b, where we focus on the motion of the vehicle in a vertical plane through the wings of the aircraft. We resolve the forces generated by the main downward thruster and the maneuvering thrusters as a pair of forces F_1 and F_2 acting at a distance r below the aircraft (determined by the geometry of the thrusters).

Let (x, y, θ) denote the position and orientation of the center of mass of the aircraft. Let m be the mass of the vehicle, J the moment of inertia, g the gravitational constant and c the damping coefficient. Then the equations of motion for the vehicle are given by

$$m\ddot{x} = F_1 \cos\theta - F_2 \sin\theta - c\dot{x},$$
$$m\ddot{y} = F_1 \sin\theta + F_2 \cos\theta - mg - c\dot{y}, \qquad (2.26)$$
$$J\ddot{\theta} = rF_1.$$

It is convenient to redefine the inputs so that the origin is an equilibrium point of the

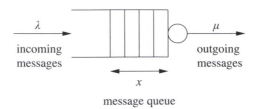

message queue

Figure 2.18: Schematic diagram of a queuing system. Messages arrive at rate λ and are stored in a queue. Messages are processed and removed from the queue at rate μ. The average size of the queue is given by $x \in \mathbb{R}$.

system with zero input. Letting $u_1 = F_1$ and $u_2 = F_2 - mg$, the equations become

$$m\ddot{x} = -mg\sin\theta - c\dot{x} + u_1\cos\theta - u_2\sin\theta,$$
$$m\ddot{y} = mg(\cos\theta - 1) - c\dot{y} + u_1\sin\theta + u_2\cos\theta, \qquad (2.27)$$
$$J\ddot{\theta} = ru_1.$$

These equations describe the motion of the vehicle as a set of three coupled second-order differential equations. ∇

Information Systems

Information systems range from communication systems like the Internet to software systems that manipulate data or manage enterprisewide resources. Feedback is present in all these systems, and designing strategies for routing, flow control and buffer management is a typical problem. Many results in queuing theory emerged from design of telecommunication systems and later from development of the Internet and computer communication systems [32, 127, 177]. Management of queues to avoid congestion is a central problem and we will therefore start by discussing the modeling of queuing systems.

Example 2.10 Queuing systems
A schematic picture of a simple queue is shown in Figure 2.18. Requests arrive and are then queued and processed. There can be large variations in arrival rates and service rates, and the queue length builds up when the arrival rate is larger than the service rate. When the queue becomes too large, service is denied using an admission control policy.

The system can be modeled in many different ways. One way is to model each incoming request, which leads to an event-based model where the state is an integer that represents the queue length. The queue changes when a request arrives or a request is serviced. The statistics of arrival and servicing are typically modeled as random processes. In many cases it is possible to determine statistics of quantities like queue length and service time, but the computations can be quite complicated.

A significant simplification can be obtained by using a *flow model*. Instead of keeping track of each request we instead view service and requests as flows, similar to what is done when replacing molecules by a continuum when analyzing

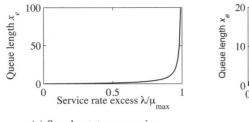
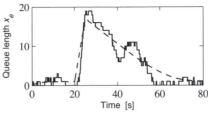

(a) Steady-state queue size   (b) Overload condition

Figure 2.19: Queuing dynamics. (a) The steady-state queue length as a function of λ/μ_{max}.
(b) The behavior of the queue length when there is a temporary overload in the system. The
solid line shows a realization of an event-based simulation, and the dashed line shows the
behavior of the flow model (2.29).

fluids. Assuming that the average queue length x is a continuous variable and that
arrivals and services are flows with rates λ and μ, the system can be modeled by
the first-order differential equation

$$\frac{dx}{dt} = \lambda - \mu = \lambda - \mu_{max} f(x), \quad x \geq 0, \tag{2.28}$$

where μ_{max} is the maximum service rate and $f(x)$ is a number between 0 and 1
that describes the effective service rate as a function of the queue length.

It is natural to assume that the effective service rate depends on the queue length
because larger queues require more resources. In steady state we have $f(x) = \lambda/\mu_{max}$, and we assume that the queue length goes to zero when λ/μ_{max} goes to zero
and that it goes to infinity when λ/μ_{max} goes to 1. This implies that $f(0) = 0$ and
that $f(\infty) = 1$. In addition, if we assume that the effective service rate deteriorates
monotonically with queue length, then the function $f(x)$ is monotone and concave.
A simple function that satisfies the basic requirements is $f(x) = x/(1+x)$, which
gives the model

$$\frac{dx}{dt} = \lambda - \mu_{max} \frac{x}{x+1}. \tag{2.29}$$

This model was proposed by Agnew [5]. It can be shown that if arrival and service processes are Poisson processes, the average queue length is given by equation (2.29) and that equation (2.29) is a good approximation even for short queue
lengths; see Tipper [193].

To explore the properties of the model (2.29) we will first investigate the equilibrium value of the queue length when the arrival rate λ is constant. Setting the
derivative dx/dt to zero in equation (2.29) and solving for x, we find that the queue
length x approaches the steady-state value

$$x_e = \frac{\lambda}{\mu_{max} - \lambda}. \tag{2.30}$$

Figure 2.19a shows the steady-state queue length as a function of λ/μ_{max}, the
effective service rate excess. Notice that the queue length increases rapidly as λ

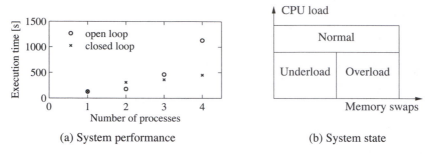

(a) System performance   (b) System state

Figure 2.20: Illustration of feedback in the virtual memory system of the IBM/370. (a) The effect of feedback on execution times in a simulation, following [43]. Results with no feedback are shown with o, and results with feedback with x. Notice the dramatic decrease in execution time for the system with feedback. (b) How the three states are obtained based on process measurements.

approaches μ_{max}. To have a queue length less than 20 requires $\lambda/\mu_{max} < 0.95$. The average time to service a request is $T_s = (x+1)/\mu_{max}$, and it increases dramatically as λ approaches μ_{max}.

Figure 2.19b illustrates the behavior of the server in a typical overload situation. The maximum service rate is $\mu_{max} = 1$, and the arrival rate starts at $\lambda = 0.5$. The arrival rate is increased to $\lambda = 4$ at time 20, and it returns to $\lambda = 0.5$ at time 25. The figure shows that the queue builds up quickly and clears very slowly. Since the response time is proportional to queue length, it means that the quality of service is poor for a long period after an overload. This behavior is called the *rush-hour effect* and has been observed in web servers and many other queuing systems such as automobile traffic.

The dashed line in Figure 2.19b shows the behavior of the flow model, which describes the average queue length. The simple model captures behavior qualitatively, but there are variations from sample to sample when the queue length is short.   ∇

Many complex systems use discrete control actions. Such systems can be modeled by characterizing the situations that correspond to each control action, as illustrated in the following example.

Example 2.11 Virtual memory paging control
An early example of the use of feedback in computer systems was applied in the operating system OS/VS for the IBM 370 [43, 55]. The system used virtual memory, which allows programs to address more memory than is physically available as fast memory. Data in current fast memory (random access memory, RAM) is accessed directly, but data that resides in slower memory (disk) is automatically loaded into fast memory. The system is implemented in such a way that it appears to the programmer as a single large section of memory. The system performed very well in many situations, but very long execution times were encountered in overload situations, as shown by the open circles in Figure 2.20a. The difficulty was resolved with a simple discrete feedback system. The load of the central processing unit

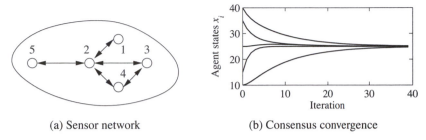

(a) Sensor network   (b) Consensus convergence

Figure 2.21: Consensus protocols for sensor networks. (a) A simple sensor network with five nodes. In this network, node 1 communicates with node 2 and node 2 communicates with nodes 1, 3, 4, 5, etc. (b) A simulation demonstrating the convergence of the consensus protocol (2.31) to the average value of the initial conditions.

(CPU) was measured together with the number of page swaps between fast memory and slow memory. The operating region was classified as being in one of three states: normal, underload or overload. The normal state is characterized by high CPU activity, the underload state is characterized by low CPU activity and few page replacements, the overload state has moderate to low CPU load but many page replacements; see Figure 2.20b. The boundaries between the regions and the time for measuring the load were determined from simulations using typical loads. The control strategy was to do nothing in the normal load condition, to exclude a process from memory in the overload condition and to allow a new process or a previously excluded process in the underload condition. The crosses in Figure 2.20a show the effectiveness of the simple feedback system in simulated loads. Similar principles are used in many other situations, e.g., in fast, on-chip cache memory.

∇

Example 2.12 Consensus protocols in sensor networks

Sensor networks are used in a variety of applications where we want to collect and aggregate information over a region of space using multiple sensors that are connected together via a communications network. Examples include monitoring environmental conditions in a geographical area (or inside a building), monitoring the movement of animals or vehicles and monitoring the resource loading across a group of computers. In many sensor networks the computational resources are distributed along with the sensors, and it can be important for the set of distributed agents to reach a consensus about a certain property, such as the average temperature in a region or the average computational load among a set of computers.

We model the connectivity of the sensor network using a graph, with nodes corresponding to the sensors and edges corresponding to the existence of a direct communications link between two nodes. We use the notation $\mathcal{N}_i$ to represent the set of neighbors of a node i. For example, in the network shown in Figure 2.21a $\mathcal{N}_2 = \{1, 3, 4, 5\}$ and $\mathcal{N}_3 = \{2, 4\}$.

To solve the consensus problem, let x_i be the state of the ith sensor, corresponding to that sensor's estimate of the average value that we are trying to compute. We initialize the state to the value of the quantity measured by the individual sensor.

The consensus protocol (algorithm) can now be realized as a local update law

$$x_i[k+1] = x_i[k] + \gamma \sum_{j \in \mathcal{N}_i} (x_j[k] - x_i[k]). \tag{2.31}$$

This protocol attempts to compute the average by updating the local state of each agent based on the value of its neighbors. The combined dynamics of all agents can be written in the form

$$x[k+1] = x[k] - \gamma (D - A)x[k], \tag{2.32}$$

where A is the adjacency matrix and D is a diagonal matrix with entries corresponding to the number of neighbors of each node. The constant γ describes the rate at which the estimate of the average is updated based on information from neighboring nodes. The matrix $L := D - A$ is called the *Laplacian* of the graph.

The equilibrium points of equation (2.32) are the set of states such that $x_e[k+1] = x_e[k]$. It can be shown that $x_e = (\alpha, \alpha, \ldots, \alpha)$ is an equilibrium state for the system, corresponding to each sensor having an identical estimate α for the average. Furthermore, we can show that α is indeed the average value of the initial states. Since there can be cycles in the graph, it is possible that the state of the system could enter into an infinite loop and never converge to the desired consensus state. A formal analysis requires tools that will be introduced later in the text, but it can be shown that for any connected graph we can always find a γ such that the states of the individual agents converge to the average. A simulation demonstrating this property is shown in Figure 2.21b.   ∇

Biological Systems

Biological systems provide perhaps the richest source of feedback and control examples. The basic problem of homeostasis, in which a quantity such as temperature or blood sugar level is regulated to a fixed value, is but one of the many types of complex feedback interactions that can occur in molecular machines, cells, organisms and ecosystems.

Example 2.13 Transcriptional regulation
Transcription is the process by which messenger RNA (mRNA) is generated from a segment of DNA. The promoter region of a gene allows transcription to be controlled by the presence of other proteins, which bind to the promoter region and either repress or activate RNA polymerase, the enzyme that produces an mRNA transcript from DNA. The mRNA is then translated into a protein according to its nucleotide sequence. This process is illustrated in Figure 2.22.

A simple model of the transcriptional regulation process is through the use of a Hill function [56, 154]. Consider the regulation of a protein A with a concentration given by p_a and a corresponding mRNA concentration m_a. Let B be a second protein with concentration p_b that represses the production of protein A through

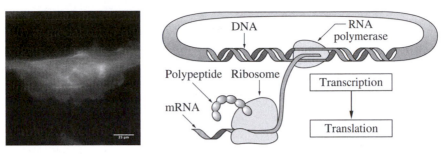

Figure 2.22: Biological circuitry. The cell on the left is a bovine pulmonary cell, stained so that the nucleus, actin and chromatin are visible. The figure on the right gives an overview of the process by which proteins in the cell are made. RNA is transcribed from DNA by an RNA polymerase enzyme. The RNA is then translated into a protein by an organelle called a ribosome.

transcriptional regulation. The resulting dynamics of p_a and m_a can be written as

$$\frac{dm_a}{dt} = \frac{\alpha_{ab}}{1 + k_{ab}p_b^{n_{ab}}} + \alpha_{a0} - \gamma_a m_a, \qquad \frac{dp_a}{dt} = \beta_a m_a - \delta_a p_a, \qquad (2.33)$$

where $\alpha_{ab} + \alpha_{a0}$ is the unregulated transcription rate, γ_a represents the rate of degradation of mRNA, α_{ab}, k_{ab} and n_{ab} are parameters that describe how B represses A, β_a represents the rate of production of the protein from its corresponding mRNA and δ_a represents the rate of degradation of the protein A. The parameter α_{a0} describes the "leakiness" of the promoter, and n_{ab} is called the Hill coefficient and relates to the cooperativity of the promoter.

A similar model can be used when a protein activates the production of another protein rather than repressing it. In this case, the equations have the form

$$\frac{dm_a}{dt} = \frac{\alpha_{ab}k_{ab}p_b^{n_{ab}}}{1 + k_{ab}p_b^{n_{ab}}} + \alpha_{a0} - \gamma_a m_a, \qquad \frac{dp_a}{dt} = \beta_a m_a - \delta_a p_a, \qquad (2.34)$$

where the variables are the same as described previously. Note that in the case of the activator, if p_b is zero, then the production rate is α_{a0} (versus $\alpha_{ab} + \alpha_{a0}$ for the repressor). As p_b gets large, the first term in the expression for $\dot{m}_a$ approaches 1 and the transcription rate becomes $\alpha_{ab} + \alpha_{a0}$ (versus α_{a0} for the repressor). Thus we see that the activator and repressor act in opposite fashion from each other.

As an example of how these models can be used, we consider the model of a "repressilator," originally due to Elowitz and Leibler [71]. The repressilator is a synthetic circuit in which three proteins each repress another in a cycle. This is shown schematically in Figure 2.23a, where the three proteins are TetR, λ cI and LacI. The basic idea of the repressilator is that if TetR is present, then it represses the production of λ cI. If λ cI is absent, then LacI is produced (at the unregulated transcription rate), which in turn represses TetR. Once TetR is repressed, then λ cI is no longer repressed, and so on. If the dynamics of the circuit are designed properly, the resulting protein concentrations will oscillate.

We can model this system using three copies of equation (2.33), with A and

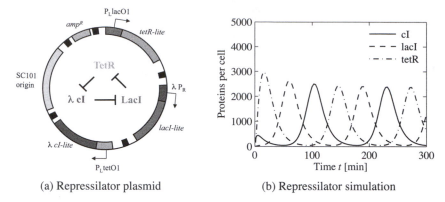

(a) Repressilator plasmid   (b) Repressilator simulation

Figure 2.23: The repressilator genetic regulatory network. (a) A schematic diagram of the repressilator, showing the layout of the genes in the plasmid that holds the circuit as well as the circuit diagram (center). (b) A simulation of a simple model for the repressilator, showing the oscillation of the individual protein concentrations. (Figure courtesy M. Elowitz.)

B replaced by the appropriate combination of TetR, cI and LacI. The state of the system is then given by $x = (m_{\text{TetR}}, p_{\text{TetR}}, m_{\text{cI}}, p_{\text{cI}}, m_{\text{LacI}}, p_{\text{LacI}})$. Figure 2.23b shows the traces of the three protein concentrations for parameters $n = 2, \alpha = 0.5$, $k = 6.25 \times 10^{-4}, \alpha_0 = 5 \times 10^{-4}, \gamma = 5.8 \times 10^{-3}, \beta = 0.12$ and $\delta = 1.2 \times 10^{-3}$ with initial conditions $x(0) = (1, 0, 0, 200, 0, 0)$ (following [71]). ∇

Example 2.14 Wave propagation in neuronal networks
The dynamics of the membrane potential in a cell are a fundamental mechanism in understanding signaling in cells, particularly in neurons and muscle cells. The Hodgkin–Huxley equations give a simple model for studying propagation waves in networks of neurons. The model for a single neuron has the form

$$C \frac{dV}{dt} = -I_{\text{Na}} - I_{\text{K}} - I_{\text{leak}} + I_{\text{input}},$$

where V is the membrane potential, C is the capacitance, I_{Na} and I_{K} are the current caused by the transport of sodium and potassium across the cell membrane, I_{leak} is a leakage current and I_{input} is the external stimulation of the cell. Each current obeys Ohm's law, i.e.,

$$I = g(V - E),$$

where g is the conductance and E is the equilibrium voltage. The equilibrium voltage is given by Nernst's law,

$$E = \frac{RT}{nF} \log \frac{c_e}{c_i},$$

where R is Boltzmann's constant, T is the absolute temperature, F is Faraday's constant, n is the charge (or valence) of the ion and c_i and c_e are the ion concentrations inside the cell and in the external fluid. At 20 °C we have $RT/F = 20$ mV.

The Hodgkin–Huxley model was originally developed as a means to predict the quantitative behavior of the squid giant axon [100]. Hodgkin and Huxley shared the 1963 Nobel Prize in Physiology (along with J. C. Eccles) for analysis of the electrical and chemical events in nerve cell discharges. The voltage clamp described in Section 1.3 was a key element in Hodgkin and Huxley's experiments.   ∇

2.5 Further Reading

Modeling is ubiquitous in engineering and science and has a long history in applied mathematics. For example, the Fourier series was introduced by Fourier when he modeled heat conduction in solids [76]. Models of dynamics have been developed in many different fields, including mechanics [12, 86], heat conduction [50], fluids [37], vehicles [1, 38, 69], robotics [156, 183], circuits [92], power systems [132], acoustics [30] and micromechanical systems [179]. Control theory requires modeling from many different domains, and most control theory texts contain several chapters on modeling using ordinary differential equations and difference equations (see, for example, [79]). A classic book on the modeling of physical systems, especially mechanical, electrical and thermofluid systems, is Cannon [49]. The book by Aris [11] is highly original and has a detailed discussion of the use of dimension-free variables. Two of the authors' favorite books on modeling of biological systems are J. D. Murray [154] and Wilson [203].

Exercises

2.1 (Chain of integrators form) Consider the linear ordinary differential equation (2.7). Show that by choosing a state space representation with $x_1 = y$, the dynamics can be written as

$$
A = \begin{bmatrix} 0 & 1 & & 0 \\ 0 & \ddots & \ddots & 0 \\ 0 & \cdots & 0 & 1 \\ -a_n & -a_{n-1} & & -a_1 \end{bmatrix}, \quad B = \begin{bmatrix} 0 \\ 0 \\ \vdots \\ 1 \end{bmatrix}, \quad C = \begin{bmatrix} 1 & \cdots & 0 & 0 \end{bmatrix}.
$$

This canonical form is called the *chain of integrators* form.

2.2 (Inverted pendulum) Use the equations of motion for a balance system to derive a dynamic model for the inverted pendulum described in Example 2.2 and verify that for small θ the dynamics are approximated by equation (2.10).

2.3 (Discrete-time dynamics) Consider the following discrete-time system

$$
x[k+1] = Ax[k] + Bu[k], \qquad y[k] = Cx[k],
$$

where

$$
x = \begin{bmatrix} x_1 \\ x_2 \end{bmatrix}, \quad A = \begin{bmatrix} a_{11} & a_{12} \\ 0 & a_{22} \end{bmatrix}, \quad B = \begin{bmatrix} 0 \\ 1 \end{bmatrix}, \quad C = \begin{bmatrix} 1 & 0 \end{bmatrix}.
$$

In this problem, we will explore some of the properties of this discrete-time system as a function of the parameters, the initial conditions and the inputs.

(a) For the case when $a_{12} = 0$ and $u = 0$, give a closed form expression for the output of the system.

(b) A discrete system is in *equilibrium* when $x[k + 1] = x[k]$ for all k. Let $u = r$ be a constant input and compute the resulting equilibrium point for the system. Show that if $|a_{ii}| < 1$ for all i, all initial conditions give solutions that converge to the equilibrium point.

(c) Write a computer program to plot the output of the system in response to a unit step input, $u[k] = 1, k \geq 0$. Plot the response of your system with $x[0] = 0$ and A given by $a_{11} = 0.5, a_{12} = 1$ and $a_{22} = 0.25$.

2.4 (Keynesian economics) Keynes' simple model for an economy is given by

$$Y[k] = C[k] + I[k] + G[k],$$

where Y, C, I and G are gross national product (GNP), consumption, investment and government expenditure for year k. Consumption and investment are modeled by difference equations of the form

$$C[k + 1] = aY[k], \qquad I[k + 1] = b(C[k + 1] - C[k]),$$

where a and b are parameters. The first equation implies that consumption increases with GNP but that the effect is delayed. The second equation implies that investment is proportional to the rate of change of consumption.

   Show that the equilibrium value of the GNP is given by

$$Y_e = \frac{1}{1 - a}(I_e + G_e),$$

where the parameter $1/(1 - a)$ is the Keynes multiplier (the gain from I or G to Y). With $a = 0.25$ an increase of government expenditure will result in a fourfold increase of GNP. Also show that the model can be written as the following discrete-time state model:

$$\begin{bmatrix} C[k + 1] \\ I[k + 1] \end{bmatrix} = \begin{bmatrix} a & a \\ ab - b & ab \end{bmatrix} \begin{bmatrix} C[k] \\ I[k] \end{bmatrix} + \begin{bmatrix} a \\ ab \end{bmatrix} G[k],$$

$$Y[k] = C[k] + I[k] + G[k].$$

 2.5 (Least squares system identification) Consider a nonlinear differential equation that can be written in the form

$$\frac{dx}{dt} = \sum_{i=1}^{M} \alpha_i f_i(x),$$

where $f_i(x)$ are known nonlinear functions and α_i are unknown, but constant, parameters. Suppose that we have measurements (or estimates) of the full state x

at time instants $t_1, t_2, \ldots, t_N$, with $N > M$. Show that the parameters α_i can be determined by finding the least squares solution to a linear equation of the form

$$H\alpha = b,$$

where $\alpha \in \mathbb{R}^M$ is the vector of all parameters and $H \in \mathbb{R}^{N \times M}$ and $b \in \mathbb{R}^N$ are appropriately defined.

2.6 (Normalized oscillator dynamics) Consider a damped spring–mass system with dynamics

$$m\ddot{q} + c\dot{q} + kq = F.$$

Let $\omega_0 = \sqrt{k/m}$ be the natural frequency and $\zeta = c/(2\sqrt{km})$ be the damping ratio.

(a) Show that by rescaling the equations, we can write the dynamics in the form

$$\ddot{q} + 2\zeta\omega_0\dot{q} + \omega_0^2 q = \omega_0^2 u, \tag{2.35}$$

where $u = F/k$. This form of the dynamics is that of a linear oscillator with natural frequency ω_0 and damping ratio ζ.

(b) Show that the system can be further normalized and written in the form

$$\frac{dz_1}{d\tau} = z_2, \qquad \frac{dz_2}{d\tau} = -z_1 - 2\zeta z_2 + v. \tag{2.36}$$

The essential dynamics of the system are governed by a single damping parameter ζ. The *Q-value* defined as $Q = 1/2\zeta$ is sometimes used instead of ζ.

2.7 (Electric generator) An electric generator connected to a strong power grid can be modeled by a momentum balance for the rotor of the generator:

$$J\frac{d^2\varphi}{dt^2} = P_m - P_e = P_m - \frac{EV}{X}\sin\varphi,$$

where J is the effective moment of inertia of the generator, φ the angle of rotation, P_m the mechanical power that drives the generator, P_e is the active electrical power, E the generator voltage, V the grid voltage and X the reactance of the line. Assuming that the line dynamics are much faster than the rotor dynamics, $P_e = VI = (EV/X)\sin\varphi$, where I is the current component in phase with the voltage E and φ is the phase angle between voltages E and V. Show that the dynamics of the electric generator has a normalized form that is similar to the dynamics of a pendulum with forcing at the pivot.

2.8 (Admission control for a queue) Consider the queuing system described in Example 2.10. The long delays created by temporary overloads can be reduced by rejecting requests when the queue gets large. This allows requests that are accepted to be serviced quickly and requests that cannot be accommodated to receive a rejection quickly so that they can try another server. Consider an admission control system described by

$$\frac{dx}{dt} = \lambda u - \mu_{\max}\frac{x}{x+1}, \qquad u = \mathrm{sat}_{(0,1)}(k(r-x)), \tag{2.37}$$

where the controller is a simple proportional control with saturation ($sat_{(a,b)}$ is defined by equation (3.9)) and r is the desired (reference) queue length. Use a simulation to show that this controller reduces the rush-hour effect and explain how the choice of r affects the system dynamics.

2.9 (Biological switch) A genetic switch can be formed by connecting two repressors together in a cycle as shown below.

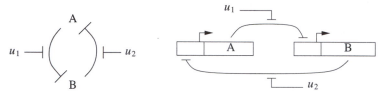

Using the models from Example 2.13—assuming that the parameters are the same for both genes and that the mRNA concentrations reach steady state quickly—show that the dynamics can be written in normalized coordinates as

$$\frac{dz_1}{d\tau} = \frac{\mu}{1 + z_2^n} - z_1 - v_1, \qquad \frac{dz_2}{d\tau} = \frac{\mu}{1 + z_1^n} - z_2 - v_2, \qquad (2.38)$$

where z_1 and z_2 are scaled versions of the protein concentrations and the time scale has also been changed. Show that $\mu \approx 200$ using the parameters in Example 2.13, and use simulations to demonstrate the switch-like behavior of the system.

2.10 (Motor drive) Consider a system consisting of a motor driving two masses that are connected by a torsional spring, as shown in the diagram below.

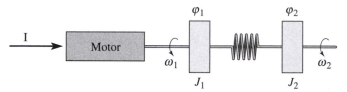

This system can represent a motor with a flexible shaft that drives a load. Assuming that the motor delivers a torque that is proportional to the current, the dynamics of the system can be described by the equations

$$J_1 \frac{d^2\varphi_1}{dt^2} + c\left(\frac{d\varphi_1}{dt} - \frac{d\varphi_2}{dt}\right) + k(\varphi_1 - \varphi_2) = k_I I,$$
$$J_2 \frac{d^2\varphi_2}{dt^2} + c\left(\frac{d\varphi_2}{dt} - \frac{d\varphi_1}{dt}\right) + k(\varphi_2 - \varphi_1) = T_d. \qquad (2.39)$$

Similar equations are obtained for a robot with flexible arms and for the arms of DVD and optical disk drives.

Derive a state space model for the system by introducing the (normalized) state variables $x_1 = \varphi_1$, $x_2 = \varphi_2$, $x_3 = \omega_1/\omega_0$, and $x_4 = \omega_2/\omega_0$, where $\omega_0 = \sqrt{k(J_1 + J_2)/(J_1 J_2)}$ is the undamped natural frequency of the system when the control signal is zero.

Chapter Three
Examples

... Don't apply any model until you understand the simplifying assumptions on which it is based, and you can test their validity. Catch phrase: use only as directed. Don't limit yourself to a single model: More than one model may be useful for understanding different aspects of the same phenomenon. Catch phrase: legalize polygamy."

Saul Golomb, "Mathematical Models—Uses and Limitations," 1970 [87].

In this chapter we present a collection of examples spanning many different fields of science and engineering. These examples will be used throughout the text and in exercises to illustrate different concepts. First-time readers may wish to focus on only a few examples with which they have had the most prior experience or insight to understand the concepts of state, input, output and dynamics in a familiar setting.

3.1 Cruise Control

The cruise control system of a car is a common feedback system encountered in everyday life. The system attempts to maintain a constant velocity in the presence of disturbances primarily caused by changes in the slope of a road. The controller compensates for these unknowns by measuring the speed of the car and adjusting the throttle appropriately.

To model the system we start with the block diagram in Figure 3.1. Let v be the speed of the car and v_r the desired (reference) speed. The controller, which typically is of the proportional-integral (PI) type described briefly in Chapter 1, receives the signals v and v_r and generates a control signal u that is sent to an actuator that controls the throttle position. The throttle in turn controls the torque T delivered by the engine, which is transmitted through the gears and the wheels, generating a force F that moves the car. There are disturbance forces F_d due to variations in the slope of the road, the rolling resistance and aerodynamic forces. The cruise controller also has a human–machine interface that allows the driver to set and modify the desired speed. There are also functions that disconnect the cruise control when the brake is touched.

The system has many individual components—actuator, engine, transmission, wheels and car body—and a detailed model can be very complicated. In spite of this, the model required to design the cruise controller can be quite simple.

To develop a mathematical model we start with a force balance for the car body. Let v be the speed of the car, m the total mass (including passengers), F the force generated by the contact of the wheels with the road, and F_d the disturbance force

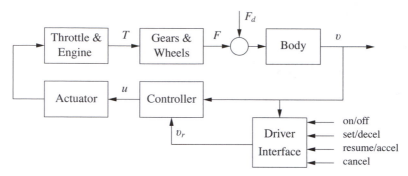

Figure 3.1: Block diagram of a cruise control system for an automobile. The throttle-controlled engine generates a torque T that is transmitted to the ground through the gearbox and wheels. Combined with the external forces from the environment, such as aerodynamic drag and gravitational forces on hills, the net force causes the car to move. The velocity of the car v is measured by a control system that adjusts the throttle through an actuation mechanism. A driver interface allows the system to be turned on and off and the reference speed v_r to be established.

due to gravity, friction and aerodynamic drag. The equation of motion of the car is simply

$$m\frac{dv}{dt} = F - F_d. \tag{3.1}$$

The force F is generated by the engine, whose torque is proportional to the rate of fuel injection, which is itself proportional to a control signal $0 \le u \le 1$ that controls the throttle position. The torque also depends on engine speed ω. A simple representation of the torque at full throttle is given by the torque curve

$$T(\omega) = T_m\left(1 - \beta\left(\frac{\omega}{\omega_m} - 1\right)^2\right), \tag{3.2}$$

where the maximum torque T_m is obtained at engine speed ω_m. Typical parameters are $T_m = 190$ Nm, $\omega_m = 420$ rad/s (about 4000 RPM) and $\beta = 0.4$. Let n be the gear ratio and r the wheel radius. The engine speed is related to the velocity through the expression

$$\omega = \frac{n}{r}v =: \alpha_n v,$$

and the driving force can be written as

$$F = \frac{nu}{r}T(\omega) = \alpha_n u T(\alpha_n v).$$

Typical values of α_n for gears 1 through 5 are $\alpha_1 = 40, \alpha_2 = 25, \alpha_3 = 16, \alpha_4 = 12$ and $\alpha_5 = 10$. The inverse of α_n has a physical interpretation as the *effective wheel radius*. Figure 3.2 shows the torque as a function of engine speed and vehicle speed. The figure shows that the effect of the gear is to "flatten" the torque curve so that an almost full torque can be obtained almost over the whole speed range.

The disturbance force F_d has three major components: F_g, the forces due to

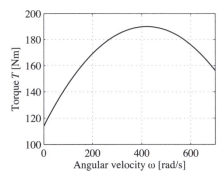
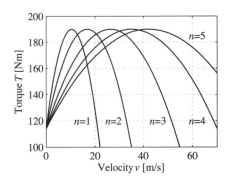

Figure 3.2: Torque curves for typical car engine. The graph on the left shows the torque generated by the engine as a function of the angular velocity of the engine, while the curve on the right shows torque as a function of car speed for different gears.

gravity; F_r, the forces due to rolling friction; and F_a, the aerodynamic drag. Letting the slope of the road be θ, gravity gives the force $F_g = mg \sin \theta$, as illustrated in Figure 3.3a, where $g = 9.8$ m/s^2 is the gravitational constant. A simple model of rolling friction is

$$F_r = mgC_r \, \text{sgn}(v),$$

where C_r is the coefficient of rolling friction and $\text{sgn}(v)$ is the sign of v (± 1) or zero if $v = 0$. A typical value for the coefficient of rolling friction is $C_r = 0.01$. Finally, the aerodynamic drag is proportional to the square of the speed:

$$F_a = \frac{1}{2}\rho C_d A v^2,$$

where ρ is the density of air, C_d is the shape-dependent aerodynamic drag coefficient and A is the frontal area of the car. Typical parameters are $\rho = 1.3$ kg/m^3, $C_d = 0.32$ and $A = 2.4$ m^2.

Summarizing, we find that the car can be modeled by

$$m\frac{dv}{dt} = \alpha_n u T(\alpha_n v) - mgC_r \, \text{sgn}(v) - \frac{1}{2}\rho C_d A v^2 - mg \sin \theta, \qquad (3.3)$$

where the function T is given by equation (3.2). The model (3.3) is a dynamical system of first order. The state is the car velocity v, which is also the output. The input is the signal u that controls the throttle position, and the disturbance is the force F_d, which depends on the slope of the road. The system is nonlinear because of the torque curve, the gravity term and the nonlinear character of rolling friction and aerodynamic drag. There can also be variations in the parameters; e.g., the mass of the car depends on the number of passengers and the load being carried in the car.

We add to this model a feedback controller that attempts to regulate the speed of the car in the presence of disturbances. We shall use a proportional-integral

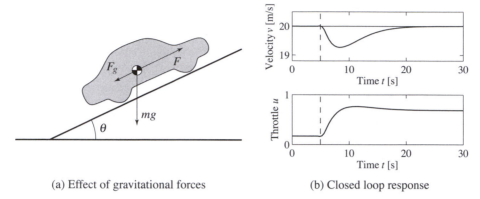

(a) Effect of gravitational forces   (b) Closed loop response

Figure 3.3: Car with cruise control encountering a sloping road. A schematic diagram is shown in (a), and (b) shows the response in speed and throttle when a slope of 4° is encountered. The hill is modeled as a net change of 4° in hill angle θ, with a linear change in the angle between $t = 5$ and $t = 6$. The PI controller has proportional gain is $k_p = 0.5$, and the integral gain is $k_i = 0.1$.

controller, which has the form

$$u(t) = k_p e(t) + k_i \int_0^t e(\tau)\, d\tau.$$

This controller can itself be realized as an input/output dynamical system by defining a controller state z and implementing the differential equation

$$\frac{dz}{dt} = v_r - v, \qquad u = k_p(v_r - v) + k_i z, \tag{3.4}$$

where v_r is the desired (reference) speed. As discussed briefly in Section 1.5, the integrator (represented by the state z) ensures that in steady state the error will be driven to zero, even when there are disturbances or modeling errors. (The design of PI controllers is the subject of Chapter 10.) Figure 3.3b shows the response of the closed loop system, consisting of equations (3.3) and (3.4), when it encounters a hill. The figure shows that even if the hill is so steep that the throttle changes from 0.17 to almost full throttle, the largest speed error is less than 1 m/s, and the desired velocity is recovered after 20 s.

Many approximations were made when deriving the model (3.3). It may seem surprising that such a seemingly complicated system can be described by the simple model (3.3). It is important to make sure that we restrict our use of the model to the uncertainty lemon conceptualized in Figure 2.15b. The model is not valid for very rapid changes of the throttle because we have ignored the details of the engine dynamics, neither is it valid for very slow changes because the properties of the engine will change over the years. Nevertheless the model is very useful for the design of a cruise control system. As we shall see in later chapters, the reason for this is the inherent robustness of feedback systems: even if the model is not perfectly accurate, we can use it to design a controller and make use of the feedback in the

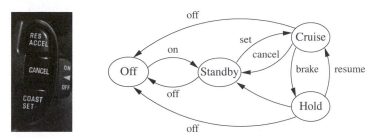

Figure 3.4: Finite state machine for cruise control system. The figure on the left shows some typical buttons used to control the system. The controller can be in one of four modes, corresponding to the nodes in the diagram on the right. Transition between the modes is controlled by pressing one of the five buttons on the cruise control interface: on, off, set, resume or cancel.

controller to manage the uncertainty in the system.

The cruise control system also has a human–machine interface that allows the driver to communicate with the system. There are many different ways to implement this system; one version is illustrated in Figure 3.4. The system has four buttons: on-off, set/decelerate, resume/accelerate and cancel. The operation of the system is governed by a finite state machine that controls the modes of the PI controller and the reference generator. Implementation of controllers and reference generators will be discussed more fully in Chapter 10.

The use of control in automotive systems goes well beyond the simple cruise control system described here. Applications include emissions control, traction control, power control (especially in hybrid vehicles) and adaptive cruise control. Many automotive applications are discussed in detail in the book by Kiencke and Nielsen [124] and in the survey papers by Powers et al. [22, 166].

3.2 Bicycle Dynamics

The bicycle is an interesting dynamical system with the feature that one of its key properties is due to a feedback mechanism that is created by the design of the front fork. A detailed model of a bicycle is complex because the system has many degrees of freedom and the geometry is complicated. However, a great deal of insight can be obtained from simple models.

To derive the equations of motion we assume that the bicycle rolls on the horizontal xy plane. Introduce a coordinate system that is fixed to the bicycle with the ξ-axis through the contact points of the wheels with the ground, the η-axis horizontal and the ζ-axis vertical, as shown in Figure 3.5. Let v_0 be the velocity of the bicycle at the rear wheel, b the wheel base, φ the tilt angle and δ the steering angle. The coordinate system rotates around the point O with the angular velocity $\omega = v_0\delta/b$, and an observer fixed to the bicycle experiences forces due to the motion of the coordinate system.

The tilting motion of the bicycle is similar to an inverted pendulum, as shown in

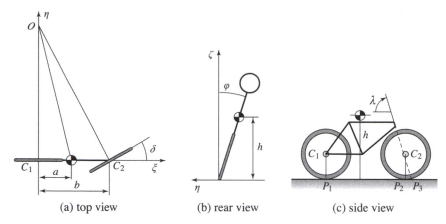

(a) top view   (b) rear view   (c) side view

Figure 3.5: Schematic views of a bicycle. The steering angle is δ, and the roll angle is φ. The center of mass has height h and distance a from a vertical through the contact point P_1 of the rear wheel. The wheel base is b, and the trail is c.

the rear view in Figure 3.5b. To model the tilt, consider the rigid body obtained when the wheels, the rider and the front fork assembly are fixed to the bicycle frame. Let m be the total mass of the system, J the moment of inertia of this body with respect to the ξ-axis and D the product of inertia with respect to the $\xi\zeta$ axes. Furthermore, let the ξ and ζ coordinates of the center of mass with respect to the rear wheel contact point, P_1, be a and h, respectively. We have $J \approx mh^2$ and $D = mah$. The torques acting on the system are due to gravity and centripetal action. Assuming that the steering angle δ is small, the equation of motion becomes

$$J\frac{d^2\varphi}{dt^2} - \frac{Dv_0}{b}\frac{d\delta}{dt} = mgh\sin\varphi + \frac{mv_0^2 h}{b}\delta. \qquad (3.5)$$

The term $mgh\sin\varphi$ is the torque generated by gravity. The terms containing δ and its derivative are the torques generated by steering, with the term $(Dv_0/b)\,d\delta/dt$ due to inertial forces and the term $(mv_0^2 h/b)\,\delta$ due to centripetal forces.

The steering angle is influenced by the torque the rider applies to the handle bar. Because of the tilt of the steering axis and the shape of the front fork, the contact point of the front wheel with the road P_2 is behind the axis of rotation of the front wheel assembly, as shown in Figure 3.5c. The distance c between the contact point of the front wheel P_2 and the projection of the axis of rotation of the front fork assembly P_3 is called the *trail*. The steering properties of a bicycle depend critically on the trail. A large trail increases stability but makes the steering less agile.

A consequence of the design of the front fork is that the steering angle δ is influenced both by steering torque T and by the tilt of the frame φ. This means that a bicycle with a front fork is a *feedback system* as illustrated by the block diagram in Figure 3.6. The steering angle δ influences the tilt angle φ, and the tilt angle influences the steering angle, giving rise to the circular causality that is characteristic of reasoning about feedback. For a front fork with a positive trail, the

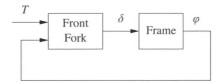

Figure 3.6: Block diagram of a bicycle with a front fork. The steering torque applied to the handlebars is T, the roll angle is φ and the steering angle is δ. Notice that the front fork creates a feedback from the roll angle φ to the steering angle δ that under certain conditions can stabilize the system.

bicycle will steer into the lean, creating a centrifugal force that attempts to diminish the lean. Under certain conditions, the feedback can actually stabilize the bicycle. A crude empirical model is obtained by assuming that the block B can be modeled as the static system

$$\delta = k_1 T - k_2 \varphi. \tag{3.6}$$

This model neglects the dynamics of the front fork, the tire–road interaction and the fact that the parameters depend on the velocity. A more accurate model, called the *Whipple model*, is obtained using the rigid-body dynamics of the front fork and the frame. Assuming small angles, this model becomes

$$M \begin{bmatrix} \ddot{\varphi} \\ \ddot{\delta} \end{bmatrix} + C v_0 \begin{bmatrix} \dot{\varphi} \\ \dot{\delta} \end{bmatrix} + (K_0 + K_2 v_0^2) \begin{bmatrix} \varphi \\ \delta \end{bmatrix} = \begin{bmatrix} 0 \\ T \end{bmatrix}, \tag{3.7}$$

where the elements of the 2×2 matrices M, C, K_0 and K_2 depend on the geometry and the mass distribution of the bicycle. Note that this has a form somewhat similar to that of the spring–mass system introduced in Chapter 2 and the balance system in Example 2.1. Even this more complex model is inaccurate because the interaction between the tire and the road is neglected; taking this into account requires two additional state variables. Again, the uncertainty lemon in Figure 2.15b provides a framework for understanding the validity of the model under these assumptions.

Interesting presentations on the development of the bicycle are given in the books by D. Wilson [202] and Herlihy [98]. The model (3.7) was presented in a paper by Whipple in 1899 [197]. More details on bicycle modeling are given in the paper [17], which has many references.

3.3 Operational Amplifier Circuits

An operational amplifier (op amp) is a modern implementation of Black's feedback amplifier. It is a universal component that is widely used for instrumentation, control and communication. It is also a key element in analog computing. Schematic diagrams of the operational amplifier are shown in Figure 3.7. The amplifier has one inverting input (v_-), one noninverting input (v_+) and one output (v_{out}). There are also connections for the supply voltages, e_- and e_+, and a zero adjustment (offset null). A simple model is obtained by assuming that the input currents i_- and i_+ are

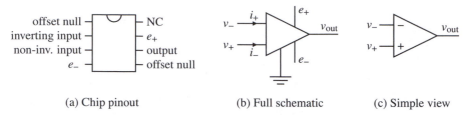

(a) Chip pinout   (b) Full schematic   (c) Simple view

Figure 3.7: An operational amplifier and two schematic diagrams. (a) The amplifier pin connections on an integrated circuit chip. (b) A schematic with all connections. (c) Only the signal connections.

zero and that the output is given by the static relation

$$v_{\text{out}} = \text{sat}_{(v_{\min}, v_{\max})}\big(k(v_+ - v_-)\big),\tag{3.8}$$

where sat denotes the saturation function

$$\text{sat}_{(a,b)}(x) = \begin{cases} a & \text{if } x < a \\ x & \text{if } a \le x \le b \\ b & \text{if } x > b. \end{cases}\tag{3.9}$$

We assume that the gain k is large, in the range of 10^6–10^8, and the voltages $v_{\min}$ and $v_{\max}$ satisfy

$$e_- \le v_{\min} < v_{\max} \le e_+$$

and hence are in the range of the supply voltages. More accurate models are obtained by replacing the saturation function with a smooth function as shown in Figure 3.8. For small input signals the amplifier characteristic (3.8) is linear:

$$v_{\text{out}} = k(v_+ - v_-) =: -kv.\tag{3.10}$$

Since the open loop gain k is very large, the range of input signals where the system is linear is very small.

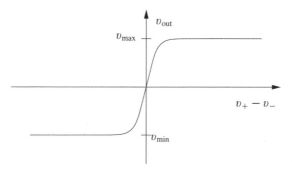

Figure 3.8: Input/output characteristics of an operational amplifier. The differential input is given by $v_+ - v_-$. The output voltage is a linear function of the input in a small range around 0, with saturation at $v_{\min}$ and $v_{\max}$. In the linear regime the op amp has high gain.

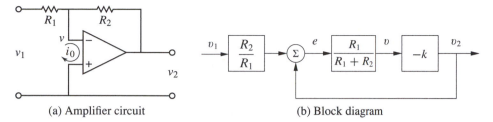

(a) Amplifier circuit   (b) Block diagram

Figure 3.9: Stable amplifier using an op amp. The circuit (a) uses negative feedback around an operational amplifier and has a corresponding block diagram (b). The resistors R_1 and R_2 determine the gain of the amplifier.

A simple amplifier is obtained by arranging feedback around the basic operational amplifier as shown in Figure 3.9a. To model the feedback amplifier in the linear range, we assume that the current $i_0 = i_- + i_+$ is zero and that the gain of the amplifier is so large that the voltage $v = v_- - v_+$ is also zero. It follows from Ohm's law that the currents through resistors R_1 and R_2 are given by

$$\frac{v_1}{R_1} = -\frac{v_2}{R_2},$$

and hence the closed loop gain of the amplifier is

$$\frac{v_2}{v_1} = -k_{cl}, \qquad \text{where} \quad k_{cl} = \frac{R_2}{R_1}. \tag{3.11}$$

A more accurate model is obtained by continuing to neglect the current i_0 but assuming that the voltage v is small but not negligible. The current balance is then

$$\frac{v_1 - v}{R_1} = \frac{v - v_2}{R_2}. \tag{3.12}$$

Assuming that the amplifier operates in the linear range and using equation (3.10), the gain of the closed loop system becomes

$$k_{cl} = -\frac{v_2}{v_1} = \frac{R_2}{R_1} \frac{kR_1}{R_1 + R_2 + kR_1} \tag{3.13}$$

If the open loop gain k of the operational amplifier is large, the closed loop gain k_{cl} is the same as in the simple model given by equation (3.11). Notice that the closed loop gain depends only on the passive components and that variations in k have only a marginal effect on the closed loop gain. For example if $k = 10^6$ and $R_2/R_1 = 100$, a variation of k by 100% gives only a variation of 0.01% in the closed loop gain. The drastic reduction in sensitivity is a nice illustration of how feedback can be used to make precise systems from uncertain components. In this particular case, feedback is used to trade high gain and low robustness for low gain and high robustness. Equation (3.13) was the formula that inspired Black when he invented the feedback amplifier [35] (see the quote at the beginning of Chapter 12).

It is instructive to develop a block diagram for the feedback amplifier in Figure 3.9a. To do this we will represent the pure amplifier with input v and output v_2

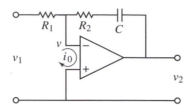

Figure 3.10: Circuit diagram of a PI controller obtained by feedback around an operational amplifier. The capacitor C is used to store charge and represents the integral of the input.

as one block. To complete the block diagram, we must describe how v depends on v_1 and v_2. Solving equation (3.12) for v gives

$$v = \frac{R_2}{R_1 + R_2} v_1 + \frac{R_1}{R_1 + R_2} v_2 = \frac{R_1}{R_1 + R_2} \left(\frac{R_2}{R_1} v_1 + v_2 \right),$$

and we obtain the block diagram shown in Figure 3.9b. The diagram clearly shows that the system has feedback and that the gain from v_2 to v is $R_1/(R_1 + R_2)$, which can also be read from the circuit diagram in Figure 3.9a. If the loop is stable and the gain of the amplifier is large, it follows that the error e is small, and we find that $v_2 = -(R_2/R_1)v_1$. Notice that the resistor R_1 appears in two blocks in the block diagram. This situation is typical in electrical circuits, and it is one reason why block diagrams are not always well suited for some types of physical modeling.

The simple model of the amplifier given by equation (3.10) provides qualitative insight, but it neglects the fact that the amplifier is a dynamical system. A more realistic model is

$$\frac{dv_{\text{out}}}{dt} = -av_{\text{out}} - bv. \tag{3.14}$$

The parameter b that has dimensions of frequency and is called the *gain-bandwidth product* of the amplifier. Whether a more complicated model is used depends on the questions to be answered and the required size of the uncertainty lemon. The model (3.14) is still not valid for very high or very low frequencies since drift causes deviations at low frequencies and there are additional dynamics that appear at frequencies close to b. The model is also not valid for large signals—an upper limit is given by the voltage of the power supply, typically in the range of 5–10 V— neither is it valid for very low signals because of electrical noise. These effects can be added, if needed, but increase the complexity of the analysis.

The operational amplifier is very versatile, and many different systems can be built by combining it with resistors and capacitors. In fact, any linear system can be implemented by combining operational amplifiers with resistors and capacitors. Exercise 3.5 shows how a second-order oscillator is implemented, and Figure 3.10 shows the circuit diagram for an analog proportional-integral controller. To develop a simple model for the circuit we assume that the current i_0 is zero and that the open loop gain k is so large that the input voltage v is negligible. The current i through the capacitor is $i = C dv_c/dt$, where v_c is the voltage across the capacitor. Since the same current goes through the resistor R_1, we get

$$i = \frac{v_1}{R_1} = C\frac{dv_c}{dt},$$

which implies that

$$v_c(t) = \frac{1}{C}\int i(t)\,dt = \frac{1}{R_1 C}\int_0^t v_1(\tau)d\tau.$$

The output voltage is thus given by

$$v_2(t) = -R_2 i - v_c = -\frac{R_2}{R_1}v_1(t) - \frac{1}{R_1 C}\int_0^t v_1(\tau)d\tau,$$

which is the input/output relation for a PI controller.

The development of operational amplifiers was pioneered by Philbrick [139, 165], and their usage is described in many textbooks (e.g., [53]). Good information is also available from suppliers [112, 145].

3.4 Computing Systems and Networks

The application of feedback to computing systems follows the same principles as the control of physical systems, but the types of measurements and control inputs that can be used are somewhat different. Measurements (sensors) are typically related to resource utilization in the computing system or network and can include quantities such as the processor load, memory usage or network bandwidth. Control variables (actuators) typically involve setting limits on the resources available to a process. This might be done by controlling the amount of memory, disk space or time that a process can consume, turning on or off processing, delaying availability of a resource or rejecting incoming requests to a server process. Process modeling for networked computing systems is also challenging, and empirical models based on measurements are often used when a first-principles model is not available.

Web Server Control

Web servers respond to requests from the Internet and provide information in the form of web pages. Modern web servers start multiple processes to respond to requests, with each process assigned to a single source until no further requests are received from that source for a predefined period of time. Processes that are idle become part of a pool that can be used to respond to new requests. To provide a fast response to web requests, it is important that the web server processes do not overload the server's computational capabilities or exhaust its memory. Since other processes may be running on the server, the amount of available processing power and memory is uncertain, and feedback can be used to provide good performance in the presence of this uncertainty.

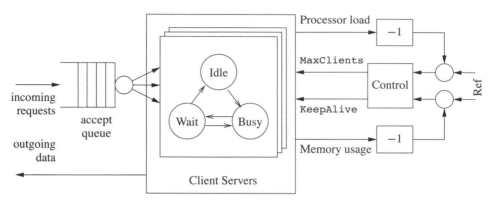

Figure 3.11: Feedback control of a web server. Connection requests arrive on an input queue, where they are sent to a server process. A finite state machine keeps track of the state of the individual server processes and responds to requests. A control algorithm can modify the server's operation by controlling parameters that affect its behavior, such as the maximum number of requests that can be serviced at a single time (`MaxClients`) or the amount of time that a connection can remain idle before it is dropped (`KeepAlive`).

Figure 3.11 illustrates the use of feedback to modulate the operation of an Apache web server. The web server operates by placing incoming connection requests on a queue and then starting a subprocess to handle requests for each accepted connection. This subprocess responds to requests from a given connection as they come in, alternating between a `Busy` state and a `Wait` state. (Keeping the subprocess active between requests is known as the *persistence* of the connection and provides a substantial reduction in latency to requests for multiple pieces of information from a single site.) If no requests are received for a sufficiently long period of time, controlled by the `KeepAlive` parameter, then the connection is dropped and the subprocess enters an `Idle` state, where it can be assigned another connection. A maximum of `MaxClients` simultaneous requests will be served, with the remainder remaining on the incoming request queue.

The parameters that control the server represent a trade-off between performance (how quickly requests receive a response) and resource usage (the amount of processing power and memory used by the server). Increasing the `MaxClients` parameter allows connection requests to be pulled off of the queue more quickly but increases the amount of processing power and memory usage that is required. Increasing the `KeepAlive` timeout means that individual connections can remain idle for a longer period of time, which decreases the processing load on the machine but increases the size of the queue (and hence the amount of time required for a user to initiate a connection). Successful operation of a busy server requires a proper choice of these parameters, often based on trial and error.

To model the dynamics of this system in more detail, we create a discrete-time model with states given by the average processor load x_{cpu} and the percentage memory usage x_{mem}. The inputs to the system are taken as the maximum number of clients u_{mc} and the keep-alive time u_{ka}. If we assume a linear model around the equilibrium point, the dynamics can be written as

$$\begin{bmatrix} x_{\text{cpu}}[k+1] \\ x_{\text{mem}}[k+1] \end{bmatrix} = \begin{bmatrix} A_{11} & A_{12} \\ A_{21} & A_{22} \end{bmatrix} \begin{bmatrix} x_{\text{cpu}}[k] \\ x_{\text{mem}}[k] \end{bmatrix} + \begin{bmatrix} B_{11} & B_{12} \\ B_{21} & B_{22} \end{bmatrix} \begin{bmatrix} u_{\text{ka}}[k] \\ u_{\text{mc}}[k] \end{bmatrix}, \quad (3.15)$$

where the coefficients of the A and B matrices can be determined based on empirical measurements or detailed modeling of the web server's processing and memory usage. Using system identification, Diao et al. [59, 97] identified the linearized dynamics as

$$A = \begin{bmatrix} 0.54 & -0.11 \\ -0.026 & 0.63 \end{bmatrix}, \qquad B = \begin{bmatrix} -85 & 4.4 \\ -2.5 & 2.8 \end{bmatrix} \times 10^{-4},$$

where the system was linearized about the equilibrium point

$$x_{\text{cpu}} = 0.58, \qquad u_{\text{ka}} = 11 \text{ s}, \qquad x_{\text{mem}} = 0.55, \qquad u_{\text{mc}} = 600.$$

This model shows the basic characteristics that were described above. Looking first at the B matrix, we see that increasing the KeepAlive timeout (first column of the B matrix) decreases both the processor usage and the memory usage since there is more persistence in connections and hence the server spends a longer time waiting for a connection to close rather than taking on a new active connection. The MaxClients connection increases both the processing and memory requirements. Note that the largest effect on the processor load is the KeepAlive timeout. The A matrix tells us how the processor and memory usage evolve in a region of the state space near the equilibrium point. The diagonal terms describe how the individual resources return to equilibrium after a transient increase or decrease. The off-diagonal terms show that there is coupling between the two resources, so that a change in one could cause a later change in the other.

Although this model is very simple, we will see in later examples that it can be used to modify the parameters controlling the server in real time and provide robustness with respect to uncertainties in the load on the machine. Similar types of mechanisms have been used for other types of servers. It is important to remember the assumptions on the model and their role in determining when the model is valid. In particular, since we have chosen to use average quantities over a given sample time, the model will not provide an accurate representation for high-frequency phenomena.

Congestion Control

The Internet was created to obtain a large, highly decentralized, efficient and expandable communication system. The system consists of a large number of interconnected gateways. A message is split into several packets which are transmitted over different paths in the network, and the packages are rejoined to recover the message at the receiver. An acknowledgment ("ack") message is sent back to the sender when a packet is received. The operation of the system is governed by a simple but powerful decentralized control structure that has evolved over time.

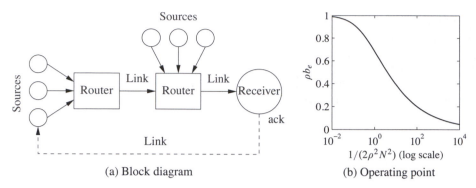

(a) Block diagram   (b) Operating point

Figure 3.12: Internet congestion control. (a) Source computers send information to routers, which forward the information to other routers that eventually connect to the receiving computer. When a packet is received, an acknowledgment packet is sent back through the routers (not shown). The routers buffer information received from the sources and send the data across the outgoing link. (b) The equilibrium buffer size b_e for a set of N identical computers sending packets through a single router with drop probability ρ.

The system has two control mechanisms called *protocols*: the Transmission Control Protocol (TCP) for end-to-end network communication and the Internet Protocol (IP) for routing packets and for host-to-gateway or gateway-to-gateway communication. The current protocols evolved after some spectacular congestion collapses occurred in the mid 1980s, when throughput unexpectedly could drop by a factor of 1000 [108]. The control mechanism in TCP is based on conserving the number of packets in the loop from the sender to the receiver and back to the sender. The sending rate is increased exponentially when there is no congestion, and it is dropped to a low level when there is congestion.

To derive an overall model for congestion control, we model three separate elements of the system: the rate at which packets are sent by individual sources (computers), the dynamics of the queues in the links (routers) and the admission control mechanism for the queues. Figure 3.12a is a block diagram of the system.

The current source control mechanism on the Internet is a protocol known as TCP/Reno [137]. This protocol operates by sending packets to a receiver and waiting to receive an acknowledgment from the receiver that the packet has arrived. If no acknowledgment is sent within a certain timeout period, the packet is retransmitted. To avoid waiting for the acknowledgment before sending the next packet, Reno transmits multiple packets up to a fixed *window* around the latest packet that has been acknowledged. If the window length is chosen properly, packets at the beginning of the window will be acknowledged before the source transmits packets at the end of the window, allowing the computer to continuously stream packets at a high rate.

To determine the size of the window to use, TCP/Reno uses a feedback mechanism in which (roughly speaking) the window size is increased by 1 every time a packet is acknowledged and the window size is cut in half when packets are lost. This mechanism allows a dynamic adjustment of the window size in which each computer acts in a greedy fashion as long as packets are being delivered but backs off quickly when congestion occurs.

A model for the behavior of the source can be developed by describing the dynamics of the window size. Suppose we have N computers and let w_i be the current window size (measured in number of packets) for the ith computer. Let q_i represent the end-to-end probability that a packet will be dropped someplace between the source and the receiver. We can model the dynamics of the window size by the differential equation

$$\frac{dw_i}{dt} = (1 - q_i)\frac{r_i(t - \tau_i)}{w_i} + q_i(-\frac{w_i}{2}r_i(t - \tau_i)), \qquad r_i = \frac{w_i}{\tau_i}, \tag{3.16}$$

where τ_i is the end-to-end transmission time for a packet to reach is destination and the acknowledgment to be sent back and r_i is the resulting rate at which packets are cleared from the list of packets that have been received. The first term in the dynamics represents the increase in window size when a packet is received, and the second term represents the decrease in window size when a packet is lost. Notice that r_i is evaluated at time $t - \tau_i$, representing the time required to receive additional acknowledgments.

The link dynamics are controlled by the dynamics of the router queue and the admission control mechanism for the queue. Assume that we have L links in the network and use l to index the individual links. We model the queue in terms of the current number of packets in the router's buffer b_l and assume that the router can contain a maximum of $b_{l,\max}$ packets and transmits packets at a rate c_l, equal to the capacity of the link. The buffer dynamics can then be written as

$$\frac{db_l}{dt} = s_l - c_l, \qquad s_l = \sum_{\{i:\ l \in L_i\}} r_i(t - \tau_{li}^f), \tag{3.17}$$

where L_i is the set of links that are being used by source i, τ_{li}^f is the time it takes a packet from source i to reach link l and s_l is the total rate at which packets arrive at link l.

The admission control mechanism determines whether a given packet is accepted by a router. Since our model is based on the average quantities in the network and not the individual packets, one simple model is to assume that the probability that a packet is dropped depends on how full the buffer is: $p_l = m_l(b_l, b_{\max})$. For simplicity, we will assume for now that $p_l = \rho_l b_l$ (see Exercise 3.6 for a more detailed model). The probability that a packet is dropped at a given link can be used to determine the end-to-end probability that a packet is lost in transmission:

$$q_i = 1 - \prod_{l \in L_i}(1 - p_l) \approx \sum_{l \in L_i} p_l(t - \tau_{li}^b), \tag{3.18}$$

where τ_{li}^b is the backward delay from link l to source i and the approximation is valid as long as the individual drop probabilities are small. We use the backward delay since this represents the time required for the acknowledgment packet to be received by the source.

Together, equations (3.16), (3.17) and (3.18) represent a model of congestion control dynamics. We can obtain substantial insight by considering a special case in which we have N identical sources and 1 link. In addition, we assume for the moment that the forward and backward time delays can be ignored, in which case the dynamics can be reduced to the form

$$\frac{dw_i}{dt} = \frac{1}{\tau} - \frac{\rho c(2 + w_i^2)}{2}, \qquad \frac{db}{dt} = \sum_{i=1}^{N} \frac{w_i}{\tau} - c, \qquad \tau = \frac{b}{c}, \qquad (3.19)$$

where $w_i \in \mathbb{R}, i = 1, \ldots, N$, are the window sizes for the sources of data, $b \in \mathbb{R}$ is the current buffer size of the router, ρ controls the rate at which packets are dropped and c is the capacity of the link connecting the router to the computers. The variable τ represents the amount of time required for a packet to be processed by a router, based on the size of the buffer and the capacity of the link. Substituting τ into the equations, we write the state space dynamics as

$$\frac{dw_i}{dt} = \frac{c}{b} - \rho c\left(1 + \frac{w_i^2}{2}\right), \qquad \frac{db}{dt} = \sum_{i=1}^{N} \frac{cw_i}{b} - c. \qquad (3.20)$$

More sophisticated models can be found in [101, 137].

The nominal operating point for the system can be found by setting $\dot{w}_i = \dot{b} = 0$:

$$0 = \frac{c}{b} - \rho c\left(1 + \frac{w_i^2}{2}\right), \qquad 0 = \sum_{i=1}^{N} \frac{cw_i}{b} - c.$$

Exploiting the fact that all of the source dynamics are identical, it follows that all of the w_i should be the same, and it can be shown that there is a unique equilibrium satisfying the equations

$$w_{i,e} = \frac{b_e}{N} = \frac{c\tau_e}{N}, \qquad \frac{1}{2\rho^2 N^2}(\rho b_e)^3 + (\rho b_e) - 1 = 0. \qquad (3.21)$$

The solution for the second equation is a bit messy but can easily be determined numerically. A plot of its solution as a function of $1/(2\rho^2 N^2)$ is shown in Figure 3.12b. We also note that at equilibrium we have the following additional equalities:

$$\tau_e = \frac{b_e}{c} = \frac{Nw_e}{c}, \qquad q_e = Np_e = N\rho b_e, \qquad r_e = \frac{w_e}{\tau_e}. \qquad (3.22)$$

Figure 3.13 shows a simulation of 60 sources communicating across a single link, with 20 sources dropping out at $t = 500$ ms and the remaining sources increasing their rates (window sizes) to compensate. Note that the buffer size and window sizes automatically adjust to match the capacity of the link.

A comprehensive treatment of computer networks is given in the textbook by Tannenbaum [189]. A good presentation of the ideas behind the control principles for the Internet is given by one of its designers, Van Jacobson, in [108]. F. Kelly [120] presents an early effort on the analysis of the system. The book by Hellerstein et al. [97] gives many examples of the use of feedback in computer systems.

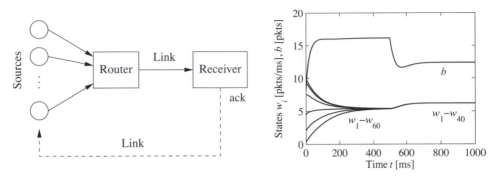

Figure 3.13: Internet congestion control for N identical sources across a single link. As shown on the left, multiple sources attempt to communicate through a router across a single link. An "ack" packet sent by the receiver acknowledges that the message was received; otherwise the message packet is resent and the sending rate is slowed down at the source. The simulation on the right is for 60 sources starting random rates, with 20 sources dropping out at $t = 500$ ms. The buffer size is shown at the top, and the individual source rates for 6 of the sources are shown at the bottom.

3.5 Atomic Force Microscopy

The 1986 Nobel Prize in Physics was shared by Gerd Binnig and Heinrich Rohrer for their design of the *scanning tunneling microscope*. The idea of the instrument is to bring an atomically sharp tip so close to a conducting surface that tunneling occurs. An image is obtained by traversing the tip across the sample and measuring the tunneling current as a function of tip position. This invention has stimulated the development of a family of instruments that permit visualization of surface structure at the nanometer scale, including the *atomic force microscope* (AFM), where a sample is probed by a tip on a cantilever. An AFM can operate in two modes. In *tapping mode* the cantilever is vibrated, and the amplitude of vibration is controlled by feedback. In *contact mode* the cantilever is in contact with the sample, and its bending is controlled by feedback. In both cases control is actuated by a piezo element that controls the vertical position of the cantilever base (or the sample). The control system has a direct influence on picture quality and scanning rate.

A schematic picture of an atomic force microscope is shown in Figure 3.14a. A microcantilever with a tip having a radius of the order of 10 nm is placed close to the sample. The tip can be moved vertically and horizontally using a piezoelectric scanner. It is clamped to the sample surface by attractive van der Waals forces and repulsive Pauli forces. The cantilever tilt depends on the topography of the surface and the position of the cantilever base, which is controlled by the piezo element. The tilt is measured by sensing the deflection of the laser beam using a photodiode. The signal from the photodiode is amplified and sent to a controller that drives the amplifier for the vertical position of the cantilever. By controlling the piezo element so that the deflection of the cantilever is constant, the signal driving the

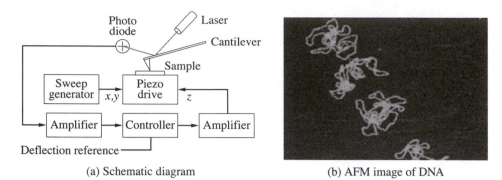

(a) Schematic diagram   (b) AFM image of DNA

Figure 3.14: Atomic force microscope. (a) A schematic diagram of an atomic force micro-scope, consisting of a piezo drive that scans the sample under the AFM tip. A laser reflects off of the cantilever and is used to measure the detection of the tip through a feedback controller. (b) An AFM image of strands of DNA. (Image courtesy Veeco Instruments.)

vertical deflection of the piezo element is a measure of the atomic forces between the cantilever tip and the atoms of the sample. An image of the surface is obtained by scanning the cantilever along the sample. The resolution makes it possible to see the structure of the sample on the atomic scale, as illustrated in Figure 3.14b, which shows an AFM image of DNA.

The horizontal motion of an AFM is typically modeled as a spring–mass system with low damping. The vertical motion is more complicated. To model the system, we start with the block diagram shown in Figure 3.15. Signals that are easily acces-sible are the input voltage u to the power amplifier that drives the piezo element, the voltage v applied to the piezo element and the output voltage y of the signal amplifier for the photodiode. The controller is a PI controller implemented by a computer, which is connected to the system by analog-to-digital (A/D) and digital-to-analog (D/A) converters. The deflection of the cantilever φ is also shown in the figure. The desired reference value for the deflection is an input to the computer.

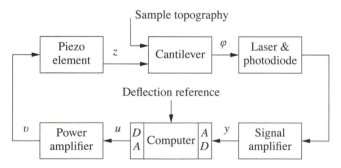

Figure 3.15: Block diagram of the system for vertical positioning of the cantilever for an atomic force microscope in contact mode. The control system attempts to keep the can-tilever deflection equal to its reference value. Cantilever deflection is measured, amplified and converted to a digital signal, then compared with its reference value. A correcting signal is generated by the computer, converted to analog form, amplified and sent to the piezo element.

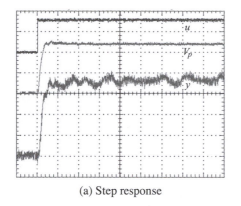

(a) Step response

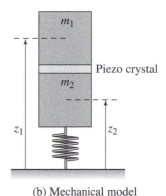

(b) Mechanical model

Figure 3.16: Modeling of an atomic force microscope. (a) A measured step response. The top curve shows the voltage u applied to the drive amplifier (50 mV/div), the middle curve is the output V_p of the power amplifier (500 mV/div) and the bottom curve is the output y of the signal amplifier (500 mV/div). The time scale is 25 μs/div. Data have been supplied by Georg Schitter. (b) A simple mechanical model for the vertical positioner and the piezo crystal.

There are several different configurations that have different dynamics. Here we will discuss a high-performance system from [176] where the cantilever base is positioned vertically using a piezo stack. We begin the modeling with a simple experiment on the system. Figure 3.16a shows a step response of a scanner from the input voltage u to the power amplifier to the output voltage y of the signal amplifier for the photodiode. This experiment captures the dynamics of the chain of blocks from u to y in the block diagram in Figure 3.15. Figure 3.16a shows that the system responds quickly but that there is a poorly damped oscillatory mode with a period of about 35 μs. A primary task of the modeling is to understand the origin of the oscillatory behavior. To do so we will explore the system in more detail.

The natural frequency of the clamped cantilever is typically several hundred kilohertz, which is much higher than the observed oscillation of about 30 kHz. As a first approximation we will model it as a static system. Since the deflections are small, we can assume that the bending φ of the cantilever is proportional to the difference in height between the cantilever tip at the probe and the piezo scanner. A more accurate model can be obtained by modeling the cantilever as a spring–mass system of the type discussed in Chapter 2.

Figure 3.16a also shows that the response of the power amplifier is fast. The photodiode and the signal amplifier also have fast responses and can thus be modeled as static systems. The remaining block is a piezo system with suspension. A schematic mechanical representation of the vertical motion of the scanner is shown in Figure 3.16b. We will model the system as two masses separated by an ideal piezo element. The mass m_1 is half of the piezo system, and the mass m_2 is the other half of the piezo system plus the mass of the support.

A simple model is obtained by assuming that the piezo crystal generates a force F between the masses and that there is a damping c in the spring. Let the positions of the center of the masses be z_1 and z_2. A momentum balance gives the following model for the system:

$$m_1 \frac{d^2 z_1}{dt^2} = F, \qquad m_2 \frac{d^2 z_2}{dt^2} = -c_2 \frac{dz_2}{dt} - k_2 z_2 - F.$$

Let the elongation of the piezo element $l = z_1 - z_2$ be the control variable and the height z_1 of the cantilever base be the output. Eliminating the variable F in equations (3.23) and substituting $z_1 - l$ for z_2 gives the model

$$(m_1 + m_2) \frac{d^2 z_1}{dt^2} + c_2 \frac{dz_1}{dt} + k_2 z_1 = m_2 \frac{d^2 l}{dt^2} + c_2 \frac{dl}{dt} + k_2 l. \qquad (3.23)$$

Summarizing, we find that a simple model of the system is obtained by modeling the piezo by (3.23) and all the other blocks by static models. Introducing the linear equations $l = k_3 u$ and $y = k_4 z_1$, we now have a complete model relating the output y to the control signal u. A more accurate model can be obtained by introducing the dynamics of the cantilever and the power amplifier. As in the previous examples, the concept of the uncertainty lemon in Figure 2.15b provides a framework for describing the uncertainty: the model will be accurate up to the frequencies of the fastest modeled modes and over a range of motion in which linearized stiffness models can be used.

The experimental results in Figure 3.16a can be explained qualitatively as follows. When a voltage is applied to the piezo, it expands by l_0, the mass m_1 moves up and the mass m_2 moves down instantaneously. The system settles after a poorly damped oscillation.

It is highly desirable to design a control system for the vertical motion so that it responds quickly with little oscillation. The instrument designer has several choices: to accept the oscillation and have a slow response time, to design a control system that can damp the oscillations or to redesign the mechanics to give resonances of higher frequency. The last two alternatives give a faster response and faster imaging.

Since the dynamic behavior of the system changes with the properties of the sample, it is necessary to tune the feedback loop. In simple systems this is currently done manually by adjusting parameters of a PI controller. There are interesting possibilities for making AFM systems easier to use by introducing automatic tuning and adaptation.

The book by Sarid [173] gives a broad coverage of atomic force microscopes. The interaction of atoms close to surfaces is fundamental to solid state physics, see Kittel [125]. The model discussed in this section is based on Schitter [175].

3.6 Drug Administration

The phrase "Take two pills three times a day" is a recommendation with which we are all familiar. Behind this recommendation is a solution of an open loop control problem. The key issue is to make sure that the concentration of a medicine in a part of the body is sufficiently high to be effective but not so high that it will

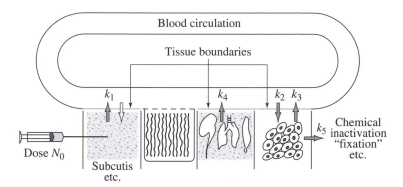

Figure 3.17: Abstraction used to compartmentalize the body for the purpose of describing drug distribution (based on Teorell [190]). The body is abstracted by a number of compartments with perfect mixing, and the complex transport processes are approximated by assuming that the flow is proportional to the concentration differences in the compartments. The constants k_i parameterize the rates of flow between different compartments.

cause undesirable side effects. The control action is quantized, *take two pills*, and sampled, *every 8 hours*. The prescriptions are based on simple models captured in empirical tables, and the dose is based on the age and weight of the patient.

Drug administration is a control problem. To solve it we must understand how a drug spreads in the body after it is administered. This topic, called *pharmacokinetics*, is now a discipline of its own, and the models used are called *compartment models*. They go back to the 1920s when Widmark modeled the propagation of alcohol in the body [199]. Compartment models are now important for the screening of all drugs used by humans. The schematic diagram in Figure 3.17 illustrates the idea of a compartment model. The body is viewed as a number of compartments like blood plasma, kidney, liver and tissues that are separated by membranes. It is assumed that there is perfect mixing so that the drug concentration is constant in each compartment. The complex transport processes are approximated by assuming that the flow rates between the compartments are proportional to the concentration differences in the compartments.

To describe the effect of a drug it is necessary to know both its concentration and how it influences the body. The relation between concentration c and its effect e is typically nonlinear. A simple model is

$$e = \frac{c}{c_0 + c} e_{\max}. \tag{3.24}$$

The effect is linear for low concentrations, and it saturates at high concentrations. The relation can also be dynamic, and it is then called *pharmacodynamics*.

Compartment Models

The simplest dynamic model for drug administration is obtained by assuming that the drug is evenly distributed in a single compartment after it has been administered and that the drug is removed at a rate proportional to the concentration. The compartments behave like stirred tanks with perfect mixing. Let c be the concentration, V the volume and q the outflow rate. Converting the description of the system into differential equations gives the model

$$V\frac{dc}{dt} = -qc, \quad c \geq 0. \tag{3.25}$$

This equation has the solution $c(t) = c_0 e^{-qt/V} = c_0 e^{-kt}$, which shows that the concentration decays exponentially with the time constant $T = V/q$ after an injection. The input is introduced implicitly as an initial condition in the model (3.25). More generally, the way the input enters the model depends on how the drug is administered. For example, the input can be represented as a mass flow into the compartment where the drug is injected. A pill that is dissolved can also be interpreted as an input in terms of a mass flow rate.

The model (3.25) is called a a *one-compartment model* or a *single-pool model*. The parameter q/V is called the elimination rate constant. This simple model is often used to model the concentration in the blood plasma. By measuring the concentration at a few times, the initial concentration can be obtained by extrapolation. If the total amount of injected substance is known, the volume V can then be determined as $V = m/c_0$; this volume is called the *apparent volume of distribution*. This volume is larger than the real volume if the concentration in the plasma is lower than in other parts of the body. The model (3.25) is very simple, and there are large individual variations in the parameters. The parameters V and q are often normalized by dividing by the weight of the person. Typical parameters for aspirin are $V = 0.2$ L/kg and $q = 0.01$ (L/h)/kg. These numbers can be compared with a blood volume of 0.07 L/kg, a plasma volume of 0.05 L/kg, an intracellular fluid volume of 0.4 L/kg and an outflow of 0.0015 L/ min /kg.

The simple one-compartment model captures the gross behavior of drug distribution, but it is based on many simplifications. Improved models can be obtained by considering the body as composed of several compartments. Examples of such systems are shown in Figure 3.18, where the compartments are represented as circles and the flows by arrows.

Modeling will be illustrated using the two-compartment model in Figure 3.18a. We assume that there is perfect mixing in each compartment and that the transport between the compartments is driven by concentration differences. We further assume that a drug with concentration c_0 is injected in compartment 1 at a volume flow rate of u and that the concentration in compartment 2 is the output. Let c_1 and c_2 be the concentrations of the drug in the compartments and let V_1 and V_2 be the volumes of the compartments. The mass balances for the compartments are

$$V_1\frac{dc_1}{dt} = q(c_2 - c_1) - q_0 c_1 + c_0 u, \quad c_1 \geq 0,$$
$$V_2\frac{dc_2}{dt} = q(c_1 - c_2), \quad c_2 \geq 0, \tag{3.26}$$
$$y = c_2.$$

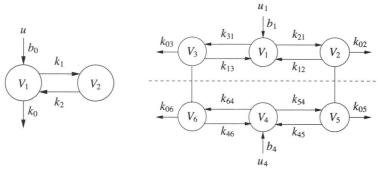

(a) Two compartment model   (b) Thyroid hormone model

Figure 3.18: Schematic diagrams of compartment models. (a) A simple two-compartment model. Each compartment is labeled by its volume, and arrows indicate the flow of chemical into, out of and between compartments. (b) A system with six compartments used to study the metabolism of thyroid hormone [85]. The notation k_{ij} denotes the transport from compartment j to compartment i.

Introducing the variables $k_0 = q_0/V_1, k_1 = q/V_1, k_2 = q/V_2$ and $b_0 = c_0/V_1$ and using matrix notation, the model can be written as

$$\frac{dc}{dt} = \begin{bmatrix} -k_0 - k_1 & k_1 \\ k_2 & -k_2 \end{bmatrix} c + \begin{bmatrix} b_0 \\ 0 \end{bmatrix} u, \qquad y = \begin{bmatrix} 0 & 1 \end{bmatrix} c. \qquad (3.27)$$

Comparing this model with its graphical representation in Figure 3.18a, we find that the mathematical representation (3.27) can be written by inspection.

It should also be emphasized that simple compartment models such as the one in equation (3.27) have a limited range of validity. Low-frequency limits exist because the human body changes with time, and since the compartment model uses average concentrations, they will not accurately represent rapid changes. There are also nonlinear effects that influence transportation between the compartments.

Compartment models are widely used in medicine, engineering and environmental science. An interesting property of these systems is that variables like concentration and mass are always positive. An essential difficulty in compartment modeling is deciding how to divide a complex system into compartments. Compartment models can also be nonlinear, as illustrated in the next section.

Insulin–glucose Dynamics

It is essential that the blood glucose concentration in the body is kept within a narrow range (0.7–1.1 g/L). Glucose concentration is influenced by many factors like food intake, digestion and exercise. A schematic picture of the relevant parts of the body is shown in Figures 3.19a and b.

There is a sophisticated mechanism that regulates glucose concentration. Glucose concentration is maintained by the pancreas, which secretes the hormones insulin and glucagon. Glucagon is released into the bloodstream when the glucose

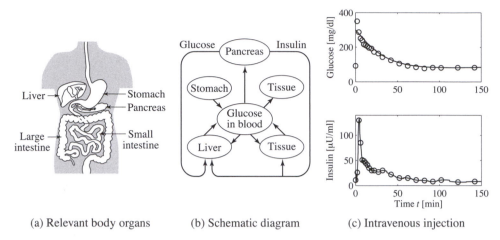

(a) Relevant body organs   (b) Schematic diagram   (c) Intravenous injection

Figure 3.19: Insulin–glucose dynamics. (a) Sketch of body parts involved in the control of glucose. (b) Schematic diagram of the system. (c) Responses of insulin and glucose when glucose in injected intravenously. From [164].

level is low. It acts on cells in the liver that release glucose. Insulin is secreted when the glucose level is high, and the glucose level is lowered by causing the liver and other cells to take up more glucose. In diseases like juvenile diabetes the pancreas is unable to produce insulin and the patient must inject insulin into the body to maintain a proper glucose level.

The mechanisms that regulate glucose and insulin are complicated; dynamics with time scales that range from seconds to hours have been observed. Models of different complexity have been developed. The models are typically tested with data from experiments where glucose is injected intravenously and insulin and glucose concentrations are measured at regular time intervals.

A relatively simple model called the *minimal model* was developed by Bergman and coworkers [31]. This models uses two compartments, one representing the concentration of glucose in the bloodstream and the other representing the concentration of insulin in the interstitial fluid. Insulin in the bloodstream is considered an input. The reaction of glucose to insulin can be modeled by the equations

$$\frac{dx_1}{dt} = -(p_1 + x_2)x_1 + p_1 g_e, \qquad \frac{dx_2}{dt} = -p_2 x_2 + p_3(u - i_e), \qquad (3.28)$$

where g_e and i_e represent the equilibrium values of glucose and insulin, x_1 is the concentration of glucose and x_2 is proportional to the concentration of interstitial insulin. Notice the presence of the term $x_2 x_1$ in the first equation. Also notice that the model does not capture the complete feedback loop because it does not describe how the pancreas reacts to the glucose. Figure 3.19c shows a fit of the model to a test on a normal person where glucose was injected intravenously at time $t = 0$. The glucose concentration rises rapidly, and the pancreas responds with a rapid spikelike injection of insulin. The glucose and insulin levels then gradually approach the equilibrium values.

Models of the type in equation (3.28) and more complicated models having many compartments have been developed and fitted to experimental data. A difficulty in modeling is that there are significant variations in model parameters over time and for different patients. For example, the parameter p_1 in equation (3.28) has been reported to vary with an order of magnitude for healthy individuals. The models have been used for diagnosis and to develop schemes for the treatment of persons with diseases. Attempts to develop a fully automatic artificial pancreas have been hampered by the lack of reliable sensors.

The papers by Widmark and Tandberg [199] and Teorell [190] are classics in pharmacokinetics, which is now an established discipline with many textbooks [62, 109, 84]. Because of its medical importance, pharmacokinetics is now an essential component of drug development. The book by Riggs [168] is a good source for the modeling of physiological systems, and a more mathematical treatment is given in [119]. Compartment models are discussed in [85]. The problem of determining rate coefficients from experimental data is discussed in [26] and [85]. There are many publications on the insulin–glucose model. The minimal model is discussed in [52, 31] and more recent references are [143, 72].

3.7 Population Dynamics

Population growth is a complex dynamic process that involves the interaction of one or more species with their environment and the larger ecosystem. The dynamics of population groups are interesting and important in many different areas of social and environmental policy. There are examples where new species have been introduced into new habitats, sometimes with disastrous results. There have also been attempts to control population growth both through incentives and through legislation. In this section we describe some of the models that can be used to understand how populations evolve with time and as a function of their environments.

Logistic Growth Model

Let x be the population of a species at time t. A simple model is to assume that the birth rates and mortality rates are proportional to the total population. This gives the linear model

$$\frac{dx}{dt} = bx - dx = (b - d)x = rx, \qquad x \geq 0, \tag{3.29}$$

where birth rate b and mortality rate d are parameters. The model gives an exponential increase if $b > d$ or an exponential decrease if $b < d$. A more realistic model is to assume that the birth rate decreases when the population is large. The following modification of the model (3.29) has this property:

$$\frac{dx}{dt} = rx(1 - \frac{x}{k}), \qquad x \geq 0, \tag{3.30}$$

where k is the *carrying capacity* of the environment. The model (3.30) is called the *logistic growth model*.

Predator–Prey Models

A more sophisticated model of population dynamics includes the effects of competing populations, where one species may feed on another. This situation, referred to as the *predator–prey problem*, was introduced in Example 2.3, where we developed a discrete-time model that captured some of the features of historical records of lynx and hare populations.

In this section, we replace the difference equation model used there with a more sophisticated differential equation model. Let $H(t)$ represent the number of hares (prey) and let $L(t)$ represent the number of lynxes (predator). The dynamics of the system are modeled as

$$\frac{dH}{dt} = rH\left(1 - \frac{H}{k}\right) - \frac{aHL}{c + H}, \qquad H \geq 0,$$
$$\frac{dL}{dt} = b\frac{aHL}{c + H} - dL, \qquad L \geq 0. \tag{3.31}$$

In the first equation, r represents the growth rate of the hares, k represents the maximum population of the hares (in the absence of lynxes), a represents the interaction term that describes how the hares are diminished as a function of the lynx population and c controls the prey consumption rate for low hare population. In the second equation, b represents the growth coefficient of the lynxes and d represents the mortality rate of the lynxes. Note that the hare dynamics include a term that resembles the logistic growth model (3.30).

Of particular interest are the values at which the population values remain constant, called *equilibrium points*. The equilibrium points for this system can be determined by setting the right-hand side of the above equations to zero. Letting H_e and L_e represent the equilibrium state, from the second equation we have

$$L_e = 0 \quad \text{or} \quad H_e^* = \frac{cd}{ab - d}. \tag{3.32}$$

Substituting this into the first equation, we have that for $L_e = 0$ either $H_e = 0$ or $H_e = k$. For $L_e \neq 0$, we obtain

$$L_e^* = \frac{rH_e(c + H_e)}{aH_e}\left(1 - \frac{H_e}{k}\right) = \frac{bcr(abk - cd - dk)}{(ab - d)^2 k}. \tag{3.33}$$

Thus, we have three possible equilibrium points $x_e = (L_e, H_e)$:

$$x_e = \begin{bmatrix} 0 \\ 0 \end{bmatrix}, \qquad x_e = \begin{bmatrix} k \\ 0 \end{bmatrix}, \qquad x_e = \begin{bmatrix} H_e^* \\ L_e^* \end{bmatrix},$$

where H_e^* and L_e^* are given in equations (3.32) and (3.33). Note that the equilibrium populations may be negative for some parameter values, corresponding to a nonachievable equilibrium point.

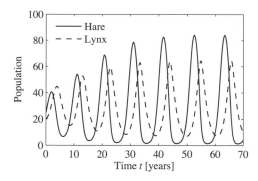
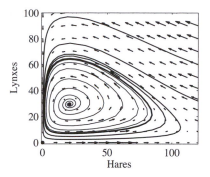

Figure 3.20: Simulation of the predator–prey system. The figure on the left shows a simulation of the two populations as a function of time. The figure on the right shows the populations plotted against each other, starting from different values of the population. The oscillation seen in both figures is an example of a *limit cycle*. The parameter values used for the simulations are $a = 3.2, b = 0.6, c = 50, d = 0.56, k = 125$ and $r = 1.6$.

Figure 3.20 shows a simulation of the dynamics starting from a set of population values near the nonzero equilibrium values. We see that for this choice of parameters, the simulation predicts an oscillatory population count for each species, reminiscent of the data shown in Figure 2.6.

Volume I of the two-volume set by J. D. Murray [154] give a broad coverage of population dynamics.

Exercises

3.1 (Cruise control) Consider the cruise control example described in Section 3.1. Build a simulation that re-creates the response to a hill shown in Figure 3.3b and show the effects of increasing and decreasing the mass of the car by 25%. Redesign the controller (using trial and error is fine) so that it returns to within 1% of the desired speed within 3 s of encountering the beginning of the hill.

3.2 (Bicycle dynamics) Show that the dynamics of a bicycle frame given by equation (3.5) can be approximated in state space form as

$$\frac{d}{dt}\begin{bmatrix} x_1 \\ x_2 \end{bmatrix} = \begin{bmatrix} 0 & 1 \\ mgh/J & 0 \end{bmatrix} \begin{bmatrix} x_1 \\ x_2 \end{bmatrix} + \begin{bmatrix} Dv_0/(bJ) \\ mv_0^2 h/(bJ) \end{bmatrix} u,$$

$$y = \begin{bmatrix} 1 & 0 \end{bmatrix} x,$$

where the input u is the steering angle δ and the output y is the tilt angle φ. What do the states x_1 and x_2 represent?

3.3 (Bicycle steering) Combine the bicycle model given by equation (3.5) and the model for steering kinematics in Example 2.8 to obtain a model that describes the path of the center of mass of the bicycle.

3.4 (Operational amplifier circuit) Consider the op amp circuit shown below.

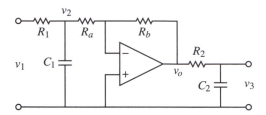

Show that the dynamics can be written in state space form as

$$
\frac{dx}{dt} =
\begin{bmatrix}
-\dfrac{1}{R_1 C_1} - \dfrac{1}{R_a C_1} & 0 \\[2ex]
\dfrac{R_b}{R_a}\dfrac{1}{R_2 C_2} & -\dfrac{1}{R_2 C_2}
\end{bmatrix}
x +
\begin{bmatrix}
\dfrac{1}{R_1 C_1} \\[2ex]
0
\end{bmatrix}
u, \quad y = \begin{bmatrix} 0 & 1 \end{bmatrix} x,
$$

where $u = v_1$ and $y = v_3$. (Hint: Use v_2 and v_3 as your state variables.)

3.5 (Operational amplifier oscillator) The op amp circuit shown below is an implementation of an oscillator.

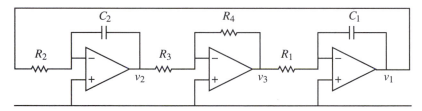

Show that the dynamics can be written in state space form as

$$
\frac{dx}{dt} =
\begin{bmatrix}
0 & \dfrac{R_4}{R_1 R_3 C_1} \\[2ex]
-\dfrac{1}{R_2 C_2} & 0
\end{bmatrix}
x,
$$

where the state variables represent the voltages across the capacitors $x_1 = v_1$ and $x_2 = v_2$.

3.6 (Congestion control using RED [138]) A number of improvements can be made to the model for Internet congestion control presented in Section 3.4. To ensure that the router's buffer size remains positive, we can modify the buffer dynamics to satisfy

$$
\frac{db_l}{dt} =
\begin{cases}
s_l - c_l & b_l > 0 \\
\text{sat}_{(0,\infty)}(s_l - c_l) & b_l = 0.
\end{cases}
$$

In addition, we can model the drop probability of a packet based on how close we are to the buffer limits, a mechanism known as random early detection (RED):

$$p_l = m_l(a_l) = \begin{cases} 0 & a_l(t) \leq b_l^{\text{lower}} \\ \rho_l r_i(t) - \rho_l b_l^{\text{lower}} & b_l^{\text{lower}} < a_l(t) < b_l^{\text{upper}} \\ \eta_l r_i(t) - (1 - 2b_l^{\text{upper}}) & b_l^{\text{upper}} \leq a_l(t) < 2b_l^{\text{upper}} \\ 1 & a_l(t) \geq 2b_l^{\text{upper}}, \end{cases}$$

$$\frac{da_l}{dt} = -\alpha_l c_l(a_l - b_l),$$

where α_l, b_l^{upper}, b_l^{lower} and p_l^{upper} are parameters for the RED protocol.

Using the model above, write a simulation for the system and find a set of parameter values for which there is a stable equilibrium point and a set for which the system exhibits oscillatory solutions. The following sets of parameters should be explored:

$$N = 20, 30, \ldots, 60, \qquad b_l^{\text{lower}} = 40 \text{ pkts}, \qquad \rho_l = 0.1,$$
$$c = 8, 9, \ldots, 15 \text{ pkts/ms}, \qquad b_l^{\text{upper}} = 540 \text{ pkts}, \qquad \alpha_l = 10^{-4},$$
$$\tau = 55, 60, \ldots, 100 \text{ ms}.$$

3.7 (Atomic force microscope with piezo tube) A schematic diagram of an AFM where the vertical scanner is a piezo tube with preloading is shown below.

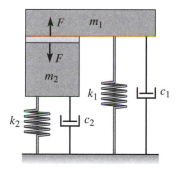

Show that the dynamics can be written as

$$(m_1 + m_2)\frac{d^2 z_1}{dt^2} + (c_1 + c_2)\frac{dz_1}{dt} + (k_1 + k_2)z_1 = m_2\frac{d^2 l}{dt^2} + c_2\frac{dl}{dt} + k_2 l.$$

Are there parameter values that make the dynamics particularly simple?

3.8 (Drug administration) The metabolism of alcohol in the body can be modeled by the nonlinear compartment model

$$V_b\frac{dc_b}{dt} = q(c_l - c_b) + q_{iv}, \qquad V_l\frac{dc_l}{dt} = q(c_b - c_l) - q_{\max}\frac{c_l}{c_0 + c_l} + q_{gi},$$

where $V_b = 48$ L and $V_l = 0.6$ L are the apparent volumes of distribution of body water and liver water, c_b and c_l are the concentrations of alcohol in the compartments, q_{iv} and q_{gi} are the injection rates for intravenous and gastrointestinal intake, $q = 1.5$ L/min is the total hepatic blood flow, $q_{max} = 2.75$ mmol/min and $c_0 = 0.1$ mmol/L. Simulate the system and compute the concentration in the blood for oral and intravenous doses of 12 g and 40 g of alcohol.

3.9 (Population dynamics) Consider the model for logistic growth given by equation (3.30). Show that the maximum growth rate occurs when the size of the population is half of the steady-state value.

3.10 (Fisheries management) The dynamics of a commercial fishery can be described by the following simple model:

$$\frac{dx}{dt} = f(x) - h(x, u), \qquad y = bh(x, u) - cu$$

where x is the total biomass, $f(x) = rx(1 - x/k)$ is the growth rate and $h(x, u) = axu$ is the harvesting rate. The output y is the rate of revenue, and the parameters a, b and c are constants representing the price of fish and the cost of fishing. Show that there is an equilibrium where the steady-state biomass is $x_e = c/(ab)$. Compare with the situation when the biomass is regulated to a constant value and find the maximum sustainable return in that case.

Chapter Four
Dynamic Behavior

It Don't Mean a Thing If It Ain't Got That Swing.

Duke Ellington (1899–1974)

In this chapter we present a broad discussion of the behavior of dynamical systems focused on systems modeled by nonlinear differential equations. This allows us to consider equilibrium points, stability, limit cycles and other key concepts in understanding dynamic behavior. We also introduce some methods for analyzing the global behavior of solutions.

4.1 Solving Differential Equations

In the last two chapters we saw that one of the methods of modeling dynamical systems is through the use of ordinary differential equations (ODEs). A state space, input/output system has the form

$$\frac{dx}{dt} = f(x, u), \qquad y = h(x, u), \tag{4.1}$$

where $x = (x_1, \ldots, x_n) \in \mathbb{R}^n$ is the state, $u \in \mathbb{R}^p$ is the input and $y \in \mathbb{R}^q$ is the output. The smooth maps $f : \mathbb{R}^n \times \mathbb{R}^p \to \mathbb{R}^n$ and $h : \mathbb{R}^n \times \mathbb{R}^p \to \mathbb{R}^q$ represent the dynamics and measurements for the system. In general, they can be nonlinear functions of their arguments. We will sometimes focus on single-input, single-output (SISO) systems, for which $p = q = 1$.

We begin by investigating systems in which the input has been set to a function of the state, $u = \alpha(x)$. This is one of the simplest types of feedback, in which the system regulates its own behavior. The differential equations in this case become

$$\frac{dx}{dt} = f(x, \alpha(x)) =: F(x). \tag{4.2}$$

To understand the dynamic behavior of this system, we need to analyze the features of the solutions of equation (4.2). While in some simple situations we can write down the solutions in analytical form, often we must rely on computational approaches. We begin by describing the class of solutions for this problem.

We say that $x(t)$ is a *solution* of the differential equation (4.2) on the time interval $t_0 \in \mathbb{R}$ to $t_f \in \mathbb{R}$ if

$$\frac{dx(t)}{dt} = F(x(t)) \quad \text{for all } t_0 < t < t_f.$$

A given differential equation may have many solutions. We will most often be interested in the *initial value problem*, where $x(t)$ is prescribed at a given time $t_0 \in \mathbb{R}$ and we wish to find a solution valid for all *future* time $t > t_0$.

We say that $x(t)$ is a solution of the differential equation (4.2) with initial value $x_0 \in \mathbb{R}^n$ at $t_0 \in \mathbb{R}$ if

$$x(t_0) = x_0 \quad \text{and} \quad \frac{dx(t)}{dt} = F(x(t)) \quad \text{for all } t_0 < t < t_f.$$

For most differential equations we will encounter, there is a *unique* solution that is defined for $t_0 < t < t_f$. The solution may be defined for all time $t > t_0$, in which case we take $t_f = \infty$. Because we will primarily be interested in solutions of the initial value problem for ODEs, we will usually refer to this simply as the solution of an ODE.

We will typically assume that t_0 is equal to 0. In the case when F is independent of time (as in equation (4.2)), we can do so without loss of generality by choosing a new independent (time) variable, $\tau = t - t_0$ (Exercise 4.1).

Example 4.1 Damped oscillator
Consider a damped linear oscillator with dynamics of the form

$$\ddot{q} + 2\zeta\omega_0\dot{q} + \omega_0^2 q = 0,$$

where q is the displacement of the oscillator from its rest position. These dynamics are equivalent to those of a spring–mass system, as shown in Exercise 2.6. We assume that $\zeta < 1$, corresponding to a lightly damped system (the reason for this particular choice will become clear later). We can rewrite this in state space form by setting $x_1 = q$ and $x_2 = \dot{q}/\omega_0$, giving

$$\frac{dx_1}{dt} = \omega_0 x_2, \qquad \frac{dx_2}{dt} = -\omega_0 x_1 - 2\zeta\omega_0 x_2.$$

In vector form, the right-hand side can be written as

$$F(x) = \begin{bmatrix} \omega_0 x_2 \\ -\omega_0 x_1 - 2\zeta\omega_0 x_2 \end{bmatrix}.$$

The solution to the initial value problem can be written in a number of different ways and will be explored in more detail in Chapter 5. Here we simply assert that the solution can be written as

$$x_1(t) = e^{-\zeta\omega_0 t}\left(x_{10}\cos\omega_d t + \frac{1}{\omega_d}(\omega_0\zeta x_{10} + x_{20})\sin\omega_d t\right),$$

$$x_2(t) = e^{-\zeta\omega_0 t}\left(x_{20}\cos\omega_d t - \frac{1}{\omega_d}(\omega_0^2 x_{10} + \omega_0\zeta x_{20})\sin\omega_d t\right),$$

where $x_0 = (x_{10}, x_{20})$ is the initial condition and $\omega_d = \omega_0\sqrt{1 - \zeta^2}$. This solution can be verified by substituting it into the differential equation. We see that the solution is explicitly dependent on the initial condition, and it can be shown that this solution is unique. A plot of the initial condition response is shown in Figure 4.1.

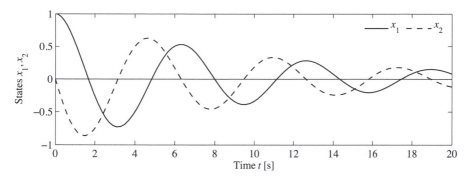

Figure 4.1: Response of the damped oscillator to the initial condition $x_0 = (1, 0)$. The solution is unique for the given initial conditions and consists of an oscillatory solution for each state, with an exponentially decaying magnitude.

We note that this form of the solution holds only for $0 < \zeta < 1$, corresponding to an "underdamped" oscillator. ▽

Without imposing some mathematical conditions on the function F, the differential equation (4.2) may not have a solution for all t, and there is no guarantee that the solution is unique. We illustrate these possibilities with two examples.

Example 4.2 Finite escape time
Let $x \in \mathbb{R}$ and consider the differential equation

$$\frac{dx}{dt} = x^2 \tag{4.3}$$

with the initial condition $x(0) = 1$. By differentiation we can verify that the function

$$x(t) = \frac{1}{1-t}$$

satisfies the differential equation and that it also satisfies the initial condition. A graph of the solution is given in Figure 4.2a; notice that the solution goes to infinity as t goes to 1. We say that this system has *finite escape time*. Thus the solution exists only in the time interval $0 \le t < 1$. ▽

Example 4.3 Nonunique solution
Let $x \in \mathbb{R}$ and consider the differential equation

$$\frac{dx}{dt} = 2\sqrt{x} \tag{4.4}$$

with initial condition $x(0) = 0$. We can show that the function

$$x(t) = \begin{cases} 0 & \text{if } 0 \le t \le a \\ (t-a)^2 & \text{if } t > a \end{cases}$$

satisfies the differential equation for all values of the parameter $a \ge 0$. To see this,

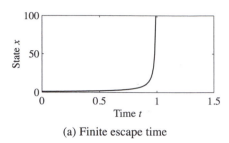
(a) Finite escape time

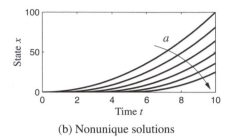
(b) Nonunique solutions

Figure 4.2: Existence and uniqueness of solutions. Equation (4.3) has a solution only for time $t < 1$, at which point the solution goes to ∞, as shown in (a). Equation (4.4) is an example of a system with many solutions, as shown in (b). For each value of a, we get a different solution starting from the same initial condition.

we differentiate $x(t)$ to obtain

$$\frac{dx}{dt} = \begin{cases} 0 & \text{if } 0 \le t \le a \\ 2(t - a) & \text{if } t > a, \end{cases}$$

and hence $\dot{x} = 2\sqrt{x}$ for all $t \ge 0$ with $x(0) = 0$. A graph of some of the possible solutions is given in Figure 4.2b. Notice that in this case there are many solutions to the differential equation. ∇

These simple examples show that there may be difficulties even with simple differential equations. Existence and uniqueness can be guaranteed by requiring that the function F have the property that for some fixed $c \in \mathbb{R}$,

$$\|F(x) - F(y)\| < c\|x - y\| \quad \text{for all } x, y,$$

which is called *Lipschitz continuity*. A sufficient condition for a function to be Lipschitz is that the Jacobian $\partial F / \partial x$ is uniformly bounded for all x. The difficulty in Example 4.2 is that the derivative $\partial F / \partial x$ becomes large for large x, and the difficulty in Example 4.3 is that the derivative $\partial F / \partial x$ is infinite at the origin.

4.2 Qualitative Analysis

The qualitative behavior of nonlinear systems is important in understanding some of the key concepts of stability in nonlinear dynamics. We will focus on an important class of systems known as planar dynamical systems. These systems have two state variables $x \in \mathbb{R}^2$, allowing their solutions to be plotted in the (x_1, x_2) plane. The basic concepts that we describe hold more generally and can be used to understand dynamical behavior in higher dimensions.

Phase Portraits

A convenient way to understand the behavior of dynamical systems with state $x \in \mathbb{R}^2$ is to plot the phase portrait of the system, briefly introduced in Chapter 2.

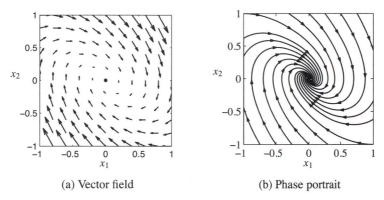

(a) Vector field　　　　　　　(b) Phase portrait

Figure 4.3: Phase portraits. (a) This plot shows the vector field for a planar dynamical system. Each arrow shows the velocity at that point in the state space. (b) This plot includes the solutions (sometimes called streamlines) from different initial conditions, with the vector field superimposed.

We start by introducing the concept of a *vector field*. For a system of ordinary differential equations

$$\frac{dx}{dt} = F(x),$$

the right-hand side of the differential equation defines at every $x \in \mathbb{R}^n$ a velocity $F(x) \in \mathbb{R}^n$. This velocity tells us how x changes and can be represented as a vector $F(x) \in \mathbb{R}^n$.

For planar dynamical systems, each state corresponds to a point in the plane and $F(x)$ is a vector representing the velocity of that state. We can plot these vectors on a grid of points in the plane and obtain a visual image of the dynamics of the system, as shown in Figure 4.3a. The points where the velocities are zero are of particular interest since they define stationary points of the flow: if we start at such a state, we stay at that state.

A *phase portrait* is constructed by plotting the flow of the vector field corresponding to the planar dynamical system. That is, for a set of initial conditions, we plot the solution of the differential equation in the plane $\mathbb{R}^2$. This corresponds to following the arrows at each point in the phase plane and drawing the resulting trajectory. By plotting the solutions for several different initial conditions, we obtain a phase portrait, as show in Figure 4.3b. Phase portraits are also sometimes called *phase plane diagrams*.

Phase portraits give insight into the dynamics of the system by showing the solutions plotted in the (two-dimensional) state space of the system. For example, we can see whether all trajectories tend to a single point as time increases or whether there are more complicated behaviors. In the example in Figure 4.3, corresponding to a damped oscillator, the solutions approach the origin for all initial conditions. This is consistent with our simulation in Figure 4.1, but it allows us to infer the behavior for all initial conditions rather than a single initial condition. However, the phase portrait does not readily tell us the rate of change of the states (although

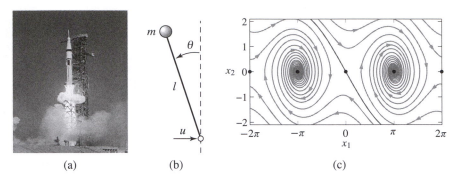

(a)   (b)   (c)

Figure 4.4: Equilibrium points for an inverted pendulum. An inverted pendulum is a model for a class of balance systems in which we wish to keep a system upright, such as a rocket (a). Using a simplified model of an inverted pendulum (b), we can develop a phase portrait that shows the dynamics of the system (c). The system has multiple equilibrium points, marked by the solid dots along the $x_2 = 0$ line.

this can be inferred from the lengths of the arrows in the vector field plot).

Equilibrium Points and Limit Cycles

An *equilibrium point* of a dynamical system represents a stationary condition for the dynamics. We say that a state x_e is an equilibrium point for a dynamical system

$$\frac{dx}{dt} = F(x)$$

if $F(x_e) = 0$. If a dynamical system has an initial condition $x(0) = x_e$, then it will stay at the equilibrium point: $x(t) = x_e$ for all $t \geq 0$, where we have taken $t_0 = 0$.

Equilibrium points are one of the most important features of a dynamical system since they define the states corresponding to constant operating conditions. A dynamical system can have zero, one or more equilibrium points.

Example 4.4 Inverted pendulum

Consider the inverted pendulum in Figure 4.4, which is a part of the balance system we considered in Chapter 2. The inverted pendulum is a simplified version of the problem of stabilizing a rocket: by applying forces at the base of the rocket, we seek to keep the rocket stabilized in the upright position. The state variables are the angle $\theta = x_1$ and the angular velocity $d\theta/dt = x_2$, the control variable is the acceleration u of the pivot and the output is the angle θ.

For simplicity we assume that $mgl/J_t = 1$ and $ml/J_t = 1$, so that the dynamics (equation (2.10)) become

$$\frac{dx}{dt} = \begin{bmatrix} x_2 \\ \sin x_1 - cx_2 + u \cos x_1 \end{bmatrix}. \tag{4.5}$$

This is a nonlinear time-invariant system of second order. This same set of equations can also be obtained by appropriate normalization of the system dynamics as illustrated in Example 2.7.

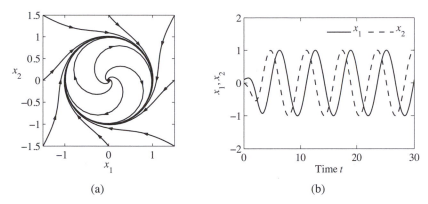

(a)   (b)

Figure 4.5: Phase portrait and time domain simulation for a system with a limit cycle. The phase portrait (a) shows the states of the solution plotted for different initial conditions. The limit cycle corresponds to a closed loop trajectory. The simulation (b) shows a single solution plotted as a function of time, with the limit cycle corresponding to a steady oscillation of fixed amplitude.

We consider the open loop dynamics by setting $u = 0$. The equilibrium points for the system are given by

$$x_e = \begin{bmatrix} \pm n\pi \\ 0 \end{bmatrix},$$

where $n = 0, 1, 2, \ldots$. The equilibrium points for n even correspond to the pendulum pointing up and those for n odd correspond to the pendulum hanging down. A phase portrait for this system (without corrective inputs) is shown in Figure 4.4c. The phase portrait shows $-2\pi \leq x_1 \leq 2\pi$, so five of the equilibrium points are shown. ∇

Nonlinear systems can exhibit rich behavior. Apart from equilibria they can also exhibit stationary periodic solutions. This is of great practical value in generating sinusoidally varying voltages in power systems or in generating periodic signals for animal locomotion. A simple example is given in Exercise 4.12, which shows the circuit diagram for an electronic oscillator. A normalized model of the oscillator is given by the equation

$$\frac{dx_1}{dt} = x_2 + x_1(1 - x_1^2 - x_2^2), \qquad \frac{dx_2}{dt} = -x_1 + x_2(1 - x_1^2 - x_2^2). \qquad (4.6)$$

The phase portrait and time domain solutions are given in Figure 4.5. The figure shows that the solutions in the phase plane converge to a circular trajectory. In the time domain this corresponds to an oscillatory solution. Mathematically the circle is called a *limit cycle*. More formally, we call an isolated solution $x(t)$ a limit cycle of period $T > 0$ if $x(t + T) = x(t)$ for all $t \in \mathbb{R}$.

There are methods for determining limit cycles for second-order systems, but for general higher-order systems we have to resort to computational analysis. Computer algorithms find limit cycles by searching for periodic trajectories in state space that

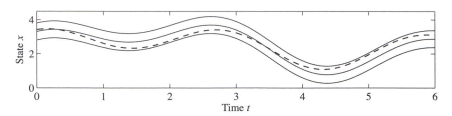

Figure 4.6: Illustration of Lyapunov's concept of a stable solution. The solution represented by the solid line is stable if we can guarantee that all solutions remain within a tube of diameter ϵ by choosing initial conditions sufficiently close the solution.

satisfy the dynamics of the system. In many situations, stable limit cycles can be found by simulating the system with different initial conditions.

4.3 Stability

The stability of a solution determines whether or not solutions nearby the solution remain close, get closer or move further away. We now give a formal definition of stability and describe tests for determining whether a solution is stable.

Definitions

Let $x(t; a)$ be a solution to the differential equation with initial condition a. A solution is *stable* if other solutions that start near a stay close to $x(t; a)$. Formally, we say that the solution $x(t; a)$ is stable if for all $\epsilon > 0$, there exists a $\delta > 0$ such that

$$\|b - a\| < \delta \quad \Longrightarrow \quad \|x(t; b) - x(t; a)\| < \epsilon \quad \text{for all } t > 0.$$

Note that this definition does not imply that $x(t; b)$ approaches $x(t; a)$ as time increases but just that it stays nearby. Furthermore, the value of δ may depend on ϵ, so that if we wish to stay very close to the solution, we may have to start very, very close ($\delta \ll \epsilon$). This type of stability, which is illustrated in Figure 4.6, is also called *stability in the sense of Lyapunov*. If a solution is stable in this sense and the trajectories do not converge, we say that the solution is *neutrally stable*.

An important special case is when the solution $x(t; a) = x_e$ is an equilibrium solution. Instead of saying that the solution is stable, we simply say that the equilibrium point is stable. An example of a neutrally stable equilibrium point is shown in Figure 4.7. From the phase portrait, we see that if we start near the equilibrium point, then we stay near the equilibrium point. Indeed, for this example, given any ϵ that defines the range of possible initial conditions, we can simply choose $\delta = \epsilon$ to satisfy the definition of stability since the trajectories are perfect circles.

A solution $x(t; a)$ is *asymptotically stable* if it is stable in the sense of Lyapunov and also $x(t; b) \to x(t; a)$ as $t \to \infty$ for b sufficiently close to a. This corresponds to the case where all nearby trajectories converge to the stable solution for large time. Figure 4.8 shows an example of an asymptotically stable equilibrium point. Note

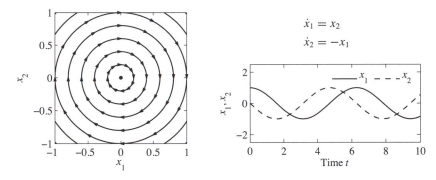

$$\dot{x}_1 = x_2$$
$$\dot{x}_2 = -x_1$$

Figure 4.7: Phase portrait and time domain simulation for a system with a single stable equilibrium point. The equilibrium point x_e at the origin is stable since all trajectories that start near x_e stay near x_e.

from the phase portraits that not only do all trajectories stay near the equilibrium point at the origin, but that they also all approach the origin as t gets large (the directions of the arrows on the phase portrait show the direction in which the trajectories move).

A solution $x(t; a)$ is *unstable* if it is not stable. More specifically, we say that a solution $x(t; a)$ is unstable if given some $\epsilon > 0$, there does *not* exist a $\delta > 0$ such that if $\|b - a\| < \delta$, then $\|x(t; b) - x(t; a)\| < \epsilon$ for all t. An example of an unstable equilibrium point is shown in Figure 4.9.

The definitions above are given without careful description of their domain of applicability. More formally, we define a solution to be *locally stable* (or *locally asymptotically stable*) if it is stable for all initial conditions $x \in B_r(a)$, where

$$B_r(a) = \{x : \|x - a\| < r\}$$

is a ball of radius r around a and $r > 0$. A system is *globally stable* if it is stable for all $r > 0$. Systems whose equilibrium points are only locally stable can have

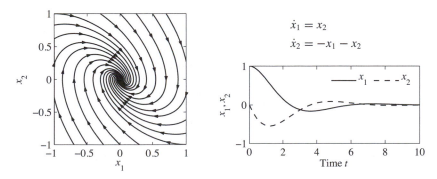

$$\dot{x}_1 = x_2$$
$$\dot{x}_2 = -x_1 - x_2$$

Figure 4.8: Phase portrait and time domain simulation for a system with a single asymptotically stable equilibrium point. The equilibrium point x_e at the origin is asymptotically stable since the trajectories converge to this point as $t \to \infty$.

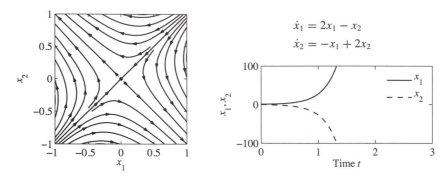

$$\dot{x}_1 = 2x_1 - x_2$$
$$\dot{x}_2 = -x_1 + 2x_2$$

Figure 4.9: Phase portrait and time domain simulation for a system with a single unstable equilibrium point. The equilibrium point x_e at the origin is unstable since not all trajectories that start near x_e stay near x_e. The sample trajectory on the right shows that the trajectories very quickly depart from zero.

interesting behavior away from equilibrium points, as we explore in the next section.

For planar dynamical systems, equilibrium points have been assigned names based on their stability type. An asymptotically stable equilibrium point is called a *sink* or sometimes an *attractor*. An unstable equilibrium point can be either a *source*, if all trajectories lead away from the equilibrium point, or a *saddle*, if some trajectories lead to the equilibrium point and others move away (this is the situation pictured in Figure 4.9). Finally, an equilibrium point that is stable but not asymptotically stable (i.e., neutrally stable, such as the one in Figure 4.7) is called a *center*.

Example 4.5 Congestion control

The model for congestion control in a network consisting of N identical computers connected to a single router, introduced in Section 3.4, is given by

$$\frac{dw}{dt} = \frac{c}{b} - \rho c \left(1 + \frac{w^2}{2} \right), \qquad \frac{db}{dt} = N \frac{wc}{b} - c,$$

where w is the window size and b is the buffer size of the router. Phase portraits are shown in Figure 4.10 for two different sets of parameter values. In each case we see that the system converges to an equilibrium point in which the buffer is below its full capacity of 500 packets. The equilibrium size of the buffer represents a balance between the transmission rates for the sources and the capacity of the link. We see from the phase portraits that the equilibrium points are asymptotically stable since all initial conditions result in trajectories that converge to these points. ∇

Stability of Linear Systems

A linear dynamical system has the form

$$\frac{dx}{dt} = Ax, \quad x(0) = x_0, \tag{4.7}$$

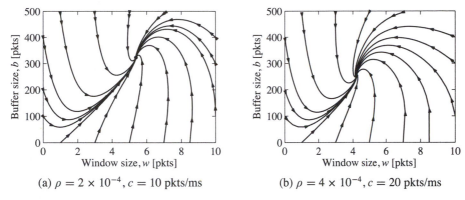

(a) $\rho = 2 \times 10^{-4}, c = 10$ pkts/ms $\qquad$ (b) $\rho = 4 \times 10^{-4}, c = 20$ pkts/ms

Figure 4.10: Phase portraits for a congestion control protocol running with $N = 60$ identical source computers. The equilibrium values correspond to a fixed window at the source, which results in a steady-state buffer size and corresponding transmission rate. A faster link (b) uses a smaller buffer size since it can handle packets at a higher rate.

where $A \in \mathbb{R}^{n \times n}$ is a square matrix, corresponding to the dynamics matrix of a linear control system (2.6). For a linear system, the stability of the equilibrium at the origin can be determined from the eigenvalues of the matrix A:

$$\lambda(A) = \{s \in \mathbb{C} : \det(sI - A) = 0\}.$$

The polynomial $\det(sI - A)$ is the *characteristic polynomial* and the eigenvalues are its roots. We use the notation λ_j for the jth eigenvalue of A, so that $\lambda_j \in \lambda(A)$. In general λ can be complex-valued, although if A is real-valued, then for any eigenvalue λ, its complex conjugate λ^* will also be an eigenvalue. The origin is always an equilibrium for a linear system. Since the stability of a linear system depends only on the matrix A, we find that stability is a property of the system. For a linear system we can therefore talk about the stability of the system rather than the stability of a particular solution or equilibrium point.

The easiest class of linear systems to analyze are those whose system matrices are in diagonal form. In this case, the dynamics have the form

$$\frac{dx}{dt} = \begin{bmatrix} \lambda_1 & & & 0 \\ & \lambda_2 & & \\ & & \ddots & \\ 0 & & & \lambda_n \end{bmatrix} x. \qquad (4.8)$$

It is easy to see that the state trajectories for this system are independent of each other, so that we can write the solution in terms of n individual systems $\dot{x}_j = \lambda_j x_j$. Each of these scalar solutions is of the form

$$x_j(t) = e^{\lambda_j t} x(0).$$

We see that the equilibrium point $x_e = 0$ is stable if $\lambda_j \leq 0$ and asymptotically stable if $\lambda_j < 0$.

Another simple case is when the dynamics are in the block diagonal form

$$
\frac{dx}{dt} = \begin{bmatrix}
\sigma_1 & \omega_1 & & 0 & 0 \\
-\omega_1 & \sigma_1 & & 0 & 0 \\
0 & 0 & \ddots & \vdots & \vdots \\
0 & 0 & & \sigma_m & \omega_m \\
0 & 0 & & -\omega_m & \sigma_m
\end{bmatrix} x.
$$

In this case, the eigenvalues can be shown to be $\lambda_j = \sigma_j \pm i\omega_j$. We once again can separate the state trajectories into independent solutions for each pair of states, and the solutions are of the form

$$
x_{2j-1}(t) = e^{\sigma_j t}\left(x_{2j-1}(0)\cos\omega_j t + x_{2j}(0)\sin\omega_j t\right),
$$
$$
x_{2j}(t) = e^{\sigma_j t}\left(-x_{2j-1}(0)\sin\omega_j t + x_{2j}(0)\cos\omega_j t\right),
$$

where $j = 1, 2, \ldots, m$. We see that this system is asymptotically stable if and only if $\sigma_j = \operatorname{Re}\lambda_j < 0$. It is also possible to combine real and complex eigenvalues in (block) diagonal form, resulting in a mixture of solutions of the two types.

Very few systems are in one of the diagonal forms above, but some systems can be transformed into these forms via coordinate transformations. One such class of systems is those for which the dynamics matrix has distinct (nonrepeating) eigenvalues. In this case there is a matrix $T \in \mathbb{R}^{n \times n}$ such that the matrix TAT^{-1} is in (block) diagonal form, with the block diagonal elements corresponding to the eigenvalues of the original matrix A (see Exercise 4.14). If we choose new coordinates $z = Tx$, then

$$
\frac{dz}{dt} = T\dot{x} = TAx = TAT^{-1}z
$$

and the linear system has a (block) diagonal dynamics matrix. Furthermore, the eigenvalues of the transformed system are the same as the original system since if v is an eigenvector of A, then $w = Tv$ can be shown to be an eigenvector of TAT^{-1}. We can reason about the stability of the original system by noting that $x(t) = T^{-1}z(t)$, and so if the transformed system is stable (or asymptotically stable), then the original system has the same type of stability.

This analysis shows that for linear systems with distinct eigenvalues, the stability of the system can be completely determined by examining the real part of the eigenvalues of the dynamics matrix. For more general systems, we make use of the following theorem, proved in the next chapter:

Theorem 4.1 (Stability of a linear system). *The system*

$$
\frac{dx}{dt} = Ax
$$

is asymptotically stable if and only if all eigenvalues of A all have a strictly negative real part and is unstable if any eigenvalue of A has a strictly positive real part.

Example 4.6 Compartment model
Consider the two-compartment module for drug delivery introduced in Section 3.6.

Using concentrations as state variables and denoting the state vector by x, the system dynamics are given by

$$\frac{dx}{dt} = \begin{bmatrix} -k_0 - k_1 & k_1 \\ k_2 & -k_2 \end{bmatrix} x + \begin{bmatrix} b_0 \\ 0 \end{bmatrix} u, \qquad y = \begin{bmatrix} 0 & 1 \end{bmatrix} x,$$

where the input u is the rate of injection of a drug into compartment 1 and the concentration of the drug in compartment 2 is the measured output y. We wish to design a feedback control law that maintains a constant output given by $y = y_d$.

We choose an output feedback control law of the form

$$u = -k(y - y_d) + u_d,$$

where u_d is the rate of injection required to maintain the desired concentration and k is a feedback gain that should be chosen such that the closed loop system is stable. Substituting the control law into the system, we obtain

$$\frac{dx}{dt} = \begin{bmatrix} -k_0 - k_1 & k_1 - b_0 k \\ k_2 & -k_2 \end{bmatrix} x + \begin{bmatrix} b_0 \\ 0 \end{bmatrix} (u_d + k y_d) =: Ax + Bu_e,$$

$$y = \begin{bmatrix} 0 & 1 \end{bmatrix} x =: Cx.$$

The equilibrium concentration $x_e \in \mathbb{R}^2$ is given by $x_e = -A^{-1} B u_e$ and

$$y_e = -CA^{-1} B u_e = \frac{b_0 k_2}{k_0 k_2 + b_0 k_2 k} (u_d + k y_d).$$

Choosing u_d such that $y_e = y_d$ provides the constant rate of injection required to maintain the desired output. We can now shift coordinates to place the equilibrium point at the origin, which yields (after some algebra)

$$\frac{dz}{dt} = \begin{bmatrix} -k_0 - k_1 & k_1 - b_0 k \\ k_2 & -k_2 \end{bmatrix} z,$$

where $z = x - x_e$. We can now apply the results of Theorem 4.1 to determine the stability of the system. The eigenvalues of the system are given by the roots of the characteristic polynomial

$$\lambda(s) = s^2 + (k_0 + k_1 + k_2)s + (k_0 k_2 + b_0 k_2 k).$$

While the specific form of the roots is messy, it can be shown that the roots are positive as long as the linear term and the constant term are both positive (Exercise 4.16). Hence the system is stable for any $k > 0$. ▽

Stability Analysis via Linear Approximation

An important feature of differential equations is that it is often possible to determine the local stability of an equilibrium point by approximating the system by a linear system. The following example illustrates the basic idea.

Example 4.7 Inverted pendulum

Consider again an inverted pendulum whose open loop dynamics are given by

$$\frac{dx}{dt} = \begin{bmatrix} x_2 \\ \sin x_1 - \gamma x_2 \end{bmatrix},$$

where we have defined the state as $x = (\theta, \dot{\theta})$. We first consider the equilibrium point at $x = (0, 0)$, corresponding to the straight-up position. If we assume that the angle $\theta = x_1$ remains small, then we can replace $\sin x_1$ with x_1 and $\cos x_1$ with 1, which gives the approximate system

$$\frac{dx}{dt} = \begin{bmatrix} x_2 \\ x_1 - \gamma x_2 \end{bmatrix} = \begin{bmatrix} 0 & 1 \\ 1 & -\gamma \end{bmatrix} x. \tag{4.9}$$

Intuitively, this system should behave similarly to the more complicated model as long as x_1 is small. In particular, it can be verified that the equilibrium point $(0, 0)$ is unstable by plotting the phase portrait or computing the eigenvalues of the dynamics matrix in equation (4.9)

We can also approximate the system around the stable equilibrium point at $x = (\pi, 0)$. In this case we have to expand $\sin x_1$ and $\cos x_1$ around $x_1 = \pi$, according to the expansions

$$\sin(\pi + \theta) = -\sin\theta \approx -\theta, \qquad \cos(\pi + \theta) = -\cos(\theta) \approx -1.$$

If we define $z_1 = x_1 - \pi$ and $z_2 = x_2$, the resulting approximate dynamics are given by

$$\frac{dz}{dt} = \begin{bmatrix} z_2 \\ -z_1 - \gamma z_2 \end{bmatrix} = \begin{bmatrix} 0 & 1 \\ -1 & -\gamma \end{bmatrix} z. \tag{4.10}$$

Note that $z = (0, 0)$ is the equilibrium point for this system and that it has the same basic form as the dynamics shown in Figure 4.8. Figure 4.11 shows the phase portraits for the original system and the approximate system around the corresponding equilibrium points. Note that they are very similar, although not exactly the same. It can be shown that if a linear approximation has either asymptotically stable or unstable equilibrium points, then the local stability of the original system must be the same (Theorem 4.3). ∇

More generally, suppose that we have a nonlinear system

$$\frac{dx}{dt} = F(x)$$

that has an equilibrium point at x_e. Computing the Taylor series expansion of the vector field, we can write

$$\frac{dx}{dt} = F(x_e) + \frac{\partial F}{\partial x}\bigg|_{x_e} (x - x_e) + \text{higher-order terms in } (x - x_e).$$

Since $F(x_e) = 0$, we can approximate the system by choosing a new state variable

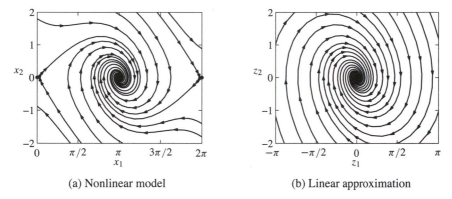

(a) Nonlinear model   (b) Linear approximation

Figure 4.11: Comparison between the phase portraits for the full nonlinear systems (a) and its linear approximation around the origin (b). Notice that near the equilibrium point at the center of the plots, the phase portraits (and hence the dynamics) are almost identical.

$z = x - x_e$ and writing

$$\frac{dz}{dt} = Az, \quad \text{where} \quad A = \left.\frac{\partial F}{\partial x}\right|_{x_e}. \tag{4.11}$$

We call the system (4.11) the *linear approximation* of the original nonlinear system or the *linearization* at x_e.

The fact that a linear model can be used to study the behavior of a nonlinear system near an equilibrium point is a powerful one. Indeed, we can take this even further and use a local linear approximation of a nonlinear system to design a feedback law that keeps the system near its equilibrium point (design of dynamics). Thus, feedback can be used to make sure that solutions remain close to the equilibrium point, which in turn ensures that the linear approximation used to stabilize it is valid.

Linear approximations can also be used to understand the stability of nonequilibrium solutions, as illustrated by the following example.

Example 4.8 Stable limit cycle
Consider the system given by equation (4.6),

$$\frac{dx_1}{dt} = x_2 + x_1(1 - x_1^2 - x_2^2), \qquad \frac{dx_2}{dt} = -x_1 + x_2(1 - x_1^2 - x_2^2),$$

whose phase portrait is shown in Figure 4.5. The differential equation has a periodic solution

$$x_1(t) = x_1(0)\cos t + x_2(0)\sin t, \tag{4.12}$$

with $x_1^2(0) + x_2^2(0) = 1$.

To explore the stability of this solution, we introduce polar coordinates r and φ, which are related to the state variables x_1 and x_2 by

$$x_1 = r\cos\varphi, \qquad x_2 = r\sin\varphi.$$

Differentiation gives the following linear equations for $\dot{r}$ and $\dot{\varphi}$:

$$\dot{x}_1 = \dot{r}\cos\varphi - r\dot{\varphi}\sin\varphi, \qquad \dot{x}_2 = \dot{r}\sin\varphi + r\dot{\varphi}\cos\varphi.$$

Solving this linear system for $\dot{r}$ and $\dot{\varphi}$ gives, after some calculation,

$$\frac{dr}{dt} = r(1 - r^2), \qquad \frac{d\varphi}{dt} = -1.$$

Notice that the equations are decoupled; hence we can analyze the stability of each state separately.

The equation for r has three equilibria: $r = 0, r = 1$ and $r = -1$ (not realizable since r must be positive). We can analyze the stability of these equilibria by linearizing the radial dynamics with $F(r) = r(1 - r^2)$. The corresponding linear dynamics are given by

$$\frac{dr}{dt} = \left.\frac{\partial F}{\partial r}\right|_{r_e} r = (1 - 3r_e^2)r, \qquad r_e = 0,\ 1,$$

where we have abused notation and used r to represent the deviation from the equilibrium point. It follows from the sign of $(1 - 3r_e^2)$ that the equilibrium $r = 0$ is unstable and the equilibrium $r = 1$ is asymptotically stable. Thus for any initial condition $r > 0$ the solution goes to $r = 1$ as time goes to infinity, but if the system starts with $r = 0$, it will remain at the equilibrium for all times. This implies that all solutions to the original system that do not start at $x_1 = x_2 = 0$ will approach the circle $x_1^2 + x_2^2 = 1$ as time increases.

To show the stability of the full solution (4.12), we must investigate the behavior of neighboring solutions with different initial conditions. We have already shown that the radius r will approach that of the solution (4.12) as long as $r(0) > 0$. The equation for the angle φ can be integrated analytically to give $\varphi(t) = -t + \varphi(0)$, which shows that solutions starting at different angles φ will neither converge nor diverge. Thus, the unit circle is *attracting*, but the solution (4.12) is only stable, not asymptotically stable. The behavior of the system is illustrated by the simulation in Figure 4.12. Notice that the solutions approach the circle rapidly, but that there is a constant phase shift between the solutions. ∇

4.4 Lyapunov Stability Analysis

We now return to the study of the full nonlinear system

$$\frac{dx}{dt} = F(x), \quad x \in \mathbb{R}^n. \tag{4.13}$$

Having defined when a solution for a nonlinear dynamical system is stable, we can now ask how to prove that a given solution is stable, asymptotically stable or unstable. For physical systems, one can often argue about stability based on dissipation of energy. The generalization of that technique to arbitrary dynamical systems is based on the use of Lyapunov functions in place of energy.

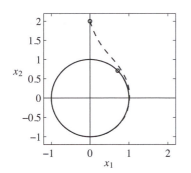 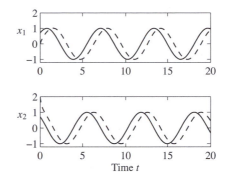

Figure 4.12: Solution curves for a stable limit cycle. The phase portrait on the left shows that the trajectory for the system rapidly converges to the stable limit cycle. The starting points for the trajectories are marked by circles in the phase portrait. The time domain plots on the right show that the states do not converge to the solution but instead maintain a constant phase error.

In this section we will describe techniques for determining the stability of solutions for a nonlinear system (4.13). We will generally be interested in stability of equilibrium points, and it will be convenient to assume that $x_e = 0$ is the equilibrium point of interest. (If not, rewrite the equations in a new set of coordinates $z = x - x_e$.)

Lyapunov Functions

A *Lyapunov function* $V : \mathbb{R}^n \to \mathbb{R}$ is an energy-like function that can be used to determine the stability of a system. Roughly speaking, if we can find a nonnegative function that always decreases along trajectories of the system, we can conclude that the minimum of the function is a stable equilibrium point (locally).

To describe this more formally, we start with a few definitions. We say that a continuous function V is *positive definite* if $V(x) > 0$ for all $x \neq 0$ and $V(0) = 0$. Similarly, a function is *negative definite* if $V(x) < 0$ for all $x \neq 0$ and $V(0) = 0$. We say that a function V is *positive semidefinite* if $V(x) \geq 0$ for all x, but $V(x)$ can be zero at points other than just $x = 0$.

To illustrate the difference between a positive definite function and a positive semidefinite function, suppose that $x \in \mathbb{R}^2$ and let

$$V_1(x) = x_1^2, \qquad V_2(x) = x_1^2 + x_2^2.$$

Both V_1 and V_2 are always nonnegative. However, it is possible for V_1 to be zero even if $x \neq 0$. Specifically, if we set $x = (0, c)$, where $c \in \mathbb{R}$ is any nonzero number, then $V_1(x) = 0$. On the other hand, $V_2(x) = 0$ if and only if $x = (0, 0)$. Thus V_1 is positive semidefinite and V_2 is positive definite.

We can now characterize the stability of an equilibrium point $x_e = 0$ for the system (4.13).

Theorem 4.2 (Lyapunov stability theorem). *Let V be a nonnegative function on*

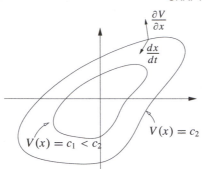

Figure 4.13: Geometric illustration of Lyapunov's stability theorem. The closed contours represent the level sets of the Lyapunov function $V(x) = c$. If dx/dt points inward to these sets at all points along the contour, then the trajectories of the system will always cause $V(x)$ to decrease along the trajectory.

$\mathbb{R}^n$ *and let $\dot{V}$ represent the time derivative of V along trajectories of the system dynamics* (4.13)*:*

$$\dot{V} = \frac{\partial V}{\partial x} \frac{dx}{dt} = \frac{\partial V}{\partial x} F(x).$$

Let $B_r = B_r(0)$ be a ball of radius r around the origin. If there exists $r > 0$ such that V is positive definite and $\dot{V}$ is negative semidefinite for all $x \in B_r$, then $x = 0$ is locally stable in the sense of Lyapunov. If V is positive definite and $\dot{V}$ is negative definite in B_r, then $x = 0$ is locally asymptotically stable.

If V satisfies one of the conditions above, we say that V is a (local) *Lyapunov function* for the system. These results have a nice geometric interpretation. The level curves for a positive definite function are the curves defined by $V(x) = c$, $c > 0$, and for each c this gives a closed contour, as shown in Figure 4.13. The condition that $\dot{V}(x)$ is negative simply means that the vector field points toward lower-level contours. This means that the trajectories move to smaller and smaller values of V and if $\dot{V}$ is negative definite then x must approach 0.

Example 4.9 Scalar nonlinear system
Consider the scalar nonlinear system

$$\frac{dx}{dt} = \frac{2}{1 + x} - x.$$

This system has equilibrium points at $x = 1$ and $x = -2$. We consider the equilibrium point at $x = 1$ and rewrite the dynamics using $z = x - 1$:

$$\frac{dz}{dt} = \frac{2}{2 + z} - z - 1,$$

which has an equilibrium point at $z = 0$. Now consider the candidate Lyapunov function

$$V(z) = \frac{1}{2} z^2,$$

which is globally positive definite. The derivative of V along trajectories of the system is given by

$$\dot{V}(z) = z\dot{z} = \frac{2z}{2+z} - z^2 - z.$$

If we restrict our analysis to an interval B_r, where $r < 2$, then $2 + z > 0$ and we can multiply through by $2 + z$ to obtain

$$2z - (z^2 + z)(2 + z) = -z^3 - 3z^2 = -z^2(z + 3) < 0, \qquad z \in B_r, \, r < 2.$$

It follows that $\dot{V}(z) < 0$ for all $z \in B_r, z \neq 0$, and hence the equilibrium point $x_e = 1$ is locally asymptotically stable. $\qquad\qquad\qquad\qquad\qquad\qquad\qquad \nabla$

A slightly more complicated situation occurs if $\dot{V}$ is negative semidefinite. In this case it is possible that $\dot{V}(x) = 0$ when $x \neq 0$, and hence x could stop decreasing in value. The following example illustrates this case.

Example 4.10 Hanging pendulum
A normalized model for a hanging pendulum is

$$\frac{dx_1}{dt} = x_2, \qquad \frac{dx_2}{dt} = -\sin x_1,$$

where x_1 is the angle between the pendulum and the vertical, with positive x_1 corresponding to counterclockwise rotation. The equation has an equilibrium $x_1 = x_2 = 0$, which corresponds to the pendulum hanging straight down. To explore the stability of this equilibrium we choose the total energy as a Lyapunov function:

$$V(x) = 1 - \cos x_1 + \frac{1}{2}x_2^2 \approx \frac{1}{2}x_1^2 + \frac{1}{2}x_2^2.$$

The Taylor series approximation shows that the function is positive definite for small x. The time derivative of $V(x)$ is

$$\dot{V} = \dot{x}_1 \sin x_1 + \dot{x}_2 x_2 = x_2 \sin x_1 - x_2 \sin x_1 = 0.$$

Since this function is positive semidefinite, it follows from Lyapunov's theorem that the equilibrium is stable but not necessarily asymptotically stable. When perturbed, the pendulum actually moves in a trajectory that corresponds to constant energy. ∇

Lyapunov functions are not always easy to find, and they are not unique. In many cases energy functions can be used as a starting point, as was done in Example 4.10. It turns out that Lyapunov functions can always be found for any stable system (under certain conditions), and hence one knows that if a system is stable, a Lyapunov function exists (and vice versa). Recent results using sum-of-squares methods have provided systematic approaches for finding Lyapunov systems [167]. Sum-of-squares techniques can be applied to a broad variety of systems, including systems whose dynamics are described by polynomial equations, as well as hybrid systems, which can have different models for different regions of state space.

For a linear dynamical system of the form

$$\frac{dx}{dt} = Ax,$$

it is possible to construct Lyapunov functions in a systematic manner. To do so, we consider quadratic functions of the form

$$V(x) = x^T P x,$$

where $P \in \mathbb{R}^{n \times n}$ is a symmetric matrix ($P = P^T$). The condition that V be positive definite is equivalent to the condition that P be a *positive definite matrix*:

$$x^T P x > 0, \quad \text{for all } x \neq 0,$$

which we write as $P > 0$. It can be shown that if P is symmetric, then P is positive definite if and only if all of its eigenvalues are real and positive.

Given a candidate Lyapunov function $V(x) = x^T P x$, we can now compute its derivative along flows of the system:

$$\dot{V} = \frac{\partial V}{\partial x} \frac{dx}{dt} = x^T (A^T P + P A) x =: -x^T Q x.$$

The requirement that $\dot{V}$ be negative definite (for asymptotic stability) becomes a condition that the matrix Q be positive definite. Thus, to find a Lyapunov function for a linear system it is sufficient to choose a $Q > 0$ and solve the *Lyapunov equation*:

$$A^T P + P A = -Q. \tag{4.14}$$

This is a linear equation in the entries of P, and hence it can be solved using linear algebra. It can be shown that the equation always has a solution if all of the eigenvalues of the matrix A are in the left half-plane. Moreover, the solution P is positive definite if Q is positive definite. It is thus always possible to find a quadratic Lyapunov function for a stable linear system. We will defer a proof of this until Chapter 5, where more tools for analysis of linear systems will be developed.

Knowing that we have a direct method to find Lyapunov functions for linear systems, we can now investigate the stability of nonlinear systems. Consider the system

$$\frac{dx}{dt} = F(x) =: Ax + \tilde{F}(x), \tag{4.15}$$

where $F(0) = 0$ and $\tilde{F}(x)$ contains terms that are second order and higher in the elements of x. The function Ax is an approximation of $F(x)$ near the origin, and we can determine the Lyapunov function for the linear approximation and investigate if it is also a Lyapunov function for the full nonlinear system. The following example illustrates the approach.

Example 4.11 Genetic switch
Consider the dynamics of a set of repressors connected together in a cycle, as shown in Figure 4.14a. The normalized dynamics for this system were given in Exercise 2.9:

$$\frac{dz_1}{d\tau} = \frac{\mu}{1 + z_2^n} - z_1, \qquad \frac{dz_2}{d\tau} = \frac{\mu}{1 + z_1^n} - z_2, \tag{4.16}$$

where z_1 and z_2 are scaled versions of the protein concentrations, n and μ are

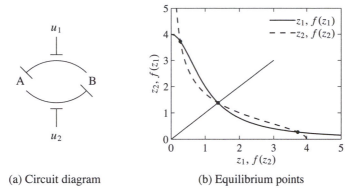

(a) Circuit diagram   (b) Equilibrium points

Figure 4.14: Stability of a genetic switch. The circuit diagram in (a) represents two proteins that are each repressing the production of the other. The inputs u_1 and u_2 interfere with this repression, allowing the circuit dynamics to be modified. The equilibrium points for this circuit can be determined by the intersection of the two curves shown in (b).

parameters that describe the interconnection between the genes and we have set the external inputs u_1 and u_2 to zero.

The equilibrium points for the system are found by equating the time derivatives to zero. We define

$$f(u) = \frac{\mu}{1 + u^n}, \qquad f'(u) = \frac{df}{du} = \frac{-\mu n u^{n-1}}{(1 + u^n)^2},$$

and the equilibrium points are defined as the solutions of the equations

$$z_1 = f(z_2), \qquad z_2 = f(z_1).$$

If we plot the curves $(z_1, f(z_1))$ and $(f(z_2), z_2)$ on a graph, then these equations will have a solution when the curves intersect, as shown in Figure 4.14b. Because of the shape of the curves, it can be shown that there will always be three solutions: one at $z_{1e} = z_{2e}$, one with $z_{1e} < z_{2e}$ and one with $z_{1e} > z_{2e}$. If $\mu \gg 1$, then we can show that the solutions are given approximately by

$$z_{1e} \approx \mu, \quad z_{2e} \approx \frac{1}{\mu^{n-1}}; \qquad z_{1e} = z_{2e}; \qquad z_{1e} \approx \frac{1}{\mu^{n-1}}, \quad z_{2e} \approx \mu. \qquad (4.17)$$

To check the stability of the system, we write $f(u)$ in terms of its Taylor series expansion about u_e:

$$f(u) = f(u_e) + f'(u_e) \cdot (u - u_e) + \frac{1}{2} f''(u_e) \cdot (u - u_e)^2 + \text{higher-order terms},$$

where f' represents the first derivative of the function, and f'' the second. Using these approximations, the dynamics can then be written as

$$\frac{dw}{dt} = \begin{bmatrix} -1 & f'(z_{2e}) \\ f'(z_{1e}) & -1 \end{bmatrix} w + \tilde{F}(w),$$

where $w = z - z_e$ is the shifted state and $\tilde{F}(w)$ represents quadratic and higher-order terms.

We now use equation (4.14) to search for a Lyapunov function. Choosing $Q = I$ and letting $P \in \mathbb{R}^{2 \times 2}$ have elements p_{ij}, we search for a solution of the equation

$$
\begin{bmatrix} -1 & f_2' \\ f_1' & -1 \end{bmatrix} \begin{bmatrix} p_{11} & p_{12} \\ p_{12} & p_{22} \end{bmatrix} + \begin{bmatrix} p_{11} & p_{12} \\ p_{12} & p_{22} \end{bmatrix} \begin{bmatrix} -1 & f_1' \\ f_2' & -1 \end{bmatrix} = \begin{bmatrix} -1 & 0 \\ 0 & -1 \end{bmatrix},
$$

where $f_1' = f'(z_{1e})$ and $f_2' = f'(z_{2e})$. Note that we have set $p_{21} = p_{12}$ to force P to be symmetric. Multiplying out the matrices, we obtain

$$
\begin{bmatrix} -2p_{11} + 2f_2'p_{12} & p_{11}f_1' - 2p_{12} + p_{22}f_2' \\ p_{11}f_1' - 2p_{12} + p_{22}f_2' & -2p_{22} + 2f_1'p_{12} \end{bmatrix} = \begin{bmatrix} -1 & 0 \\ 0 & -1 \end{bmatrix},
$$

which is a set of *linear* equations for the unknowns p_{ij}. We can solve these linear equations to obtain

$$
p_{11} = -\frac{f_1'^2 - f_2'f_1' + 2}{4(f_1'f_2' - 1)}, \qquad p_{12} = -\frac{f_1' + f_2'}{4(f_1'f_2' - 1)}, \qquad p_{22} = -\frac{f_2'^2 - f_1'f_2' + 2}{4(f_1'f_2' - 1)}.
$$

To check that $V(w) = w^T P w$ is a Lyapunov function, we must verify that $V(w)$ is positive definite function or equivalently that $P > 0$. Since P is a 2×2 symmetric matrix, it has two real eigenvalues λ_1 and λ_2 that satisfy

$$
\lambda_1 + \lambda_2 = \text{trace}(P), \qquad \lambda_1 \cdot \lambda_2 = \det(P).
$$

In order for P to be positive definite we must have that λ_1 and λ_2 are positive, and we thus require that

$$
\text{trace}(P) = \frac{f_1'^2 - 2f_2'f_1' + f_2'^2 + 4}{4 - 4f_1'f_2'} > 0, \quad \det(P) = \frac{f_1'^2 - 2f_2'f_1' + f_2'^2 + 4}{16 - 16f_1'f_2'} > 0.
$$

We see that $\text{trace}(P) = 4\det(P)$ and the numerator of the expressions is just $(f_1 - f_2)^2 + 4 > 0$, so it suffices to check the sign of $1 - f_1'f_2'$. In particular, for P to be positive definite, we require that

$$
f'(z_{1e})f'(z_{2e}) < 1.
$$

We can now make use of the expressions for f' defined earlier and evaluate at the approximate locations of the equilibrium points derived in equation (4.17). For the equilibrium points where $z_{1e} \neq z_{2e}$, we can show that

$$
f'(z_{1e})f'(z_{2e}) \approx f'(\mu)f'(\frac{1}{\mu^{n-1}}) = \frac{-\mu n \mu^{n-1}}{(1 + \mu^n)^2} \cdot \frac{-\mu n \mu^{-(n-1)^2}}{1 + \mu^{-n(n-1)}} \approx n^2 \mu^{-n^2 + n}.
$$

Using $n = 2$ and $\mu \approx 200$ from Exercise 2.9, we see that $f'(z_{1e})f'(z_{2e}) \ll 1$ and hence P is a positive definite. This implies that V is a positive definite function and hence a potential Lyapunov function for the system.

To determine if the system (4.16) is stable, we now compute $\dot{V}$ at the equilibrium

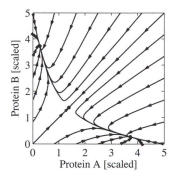
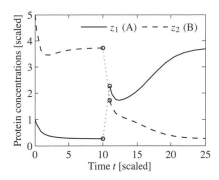

Figure 4.15: Dynamics of a genetic switch. The phase portrait on the left shows that the switch has three equilibrium points, corresponding to protein A having a concentration greater than, equal to or less than protein B. The equilibrium point with equal protein concentrations is unstable, but the other equilibrium points are stable. The simulation on the right shows the time response of the system starting from two different initial conditions. The initial portion of the curve corresponds to initial concentrations $z(0) = (1, 5)$ and converges to the equilibrium where $z_{1e} < z_{2e}$. At time $t = 10$, the concentrations are perturbed by $+2$ in z_1 and -2 in z_2, moving the state into the region of the state space whose solutions converge to the equilibrium point where $z_{2e} < z_{1e}$.

point. By construction,

$$\dot{V} = w^T(PA + A^TP)w + \tilde{F}^T(w)Pw + w^TP\tilde{F}(w)$$
$$= -w^Tw + \tilde{F}^T(w)Pw + w^TP\tilde{F}(w).$$

Since all terms in $\tilde{F}$ are quadratic or higher order in w, it follows that $\tilde{F}^T(w)Pw$ and $w^TP\tilde{F}(w)$ consist of terms that are at least third order in w. Therefore if w is sufficiently close to zero, then the cubic and higher-order terms will be smaller than the quadratic terms. Hence, sufficiently close to $w = 0$, $\dot{V}$ is negative definite, allowing us to conclude that these equilibrium points are both stable.

Figure 4.15 shows the phase portrait and time traces for a system with $\mu = 4$, illustrating the bistable nature of the system. When the initial condition starts with a concentration of protein B greater than that of A, the solution converges to the equilibrium point at (approximately) $(1/\mu^{n-1}, \mu)$. If A is greater than B, then it goes to $(\mu, 1/\mu^{n-1})$. The equilibrium point with $z_{1e} = z_{2e}$ is unstable. ∇

More generally, we can investigate what the linear approximation tells about the stability of a solution to a nonlinear equation. The following theorem gives a partial answer for the case of stability of an equilibrium point.

Theorem 4.3. *Consider the dynamical system (4.15) with $F(0) = 0$ and $\tilde{F}$ such that $\lim \|\tilde{F}(x)\|/\|x\| \to 0$ as $\|x\| \to 0$. If the real parts of all eigenvalues of A are strictly less than zero, then $x_e = 0$ is a locally asymptotically stable equilibrium point of equation (4.15).*

This theorem implies that asymptotic stability of the linear approximation implies *local* asymptotic stability of the original nonlinear system. The theorem is very important for control because it implies that stabilization of a linear approximation of a nonlinear system results in a stable equilibrium for the nonlinear system. The proof of this theorem follows the technique used in Example 4.11. A formal proof can be found in [123].

 ## Krasovski–Lasalle Invariance Principle

For general nonlinear systems, especially those in symbolic form, it can be difficult to find a positive definite function V whose derivative is strictly negative definite. The Krasovski–Lasalle theorem enables us to conclude the asymptotic stability of an equilibrium point under less restrictive conditions, namely, in the case where $\dot{V}$ is negative semidefinite, which is often easier to construct. However, it applies only to time-invariant or periodic systems. This section makes use of some additional concepts from dynamical systems; see Hahn [94] or Khalil [123] for a more detailed description.

We will deal with the time-invariant case and begin by introducing a few more definitions. We denote the solution trajectories of the time-invariant system

$$\frac{dx}{dt} = F(x) \tag{4.18}$$

as $x(t; a)$, which is the solution of equation (4.18) at time t starting from a at $t_0 = 0$. The ω *limit set* of a trajectory $x(t; a)$ is the set of all points $z \in \mathbb{R}^n$ such that there exists a strictly increasing sequence of times t_n such that $x(t_n; a) \to z$ as $n \to \infty$. A set $M \subset \mathbb{R}^n$ is said to be an *invariant set* if for all $b \in M$, we have $x(t; b) \in M$ for all $t \geq 0$. It can be proved that the ω limit set of every trajectory is closed and invariant. We may now state the Krasovski–Lasalle principle.

Theorem 4.4 (Krasovski–Lasalle principle). *Let $V : \mathbb{R}^n \to \mathbb{R}$ be a locally positive definite function such that on the compact set $\Omega_r = \{x \in \mathbb{R}^n : V(x) \leq r\}$ we have $\dot{V}(x) \leq 0$. Define*

$$S = \{x \in \Omega_r : \dot{V}(x) = 0\}.$$

As $t \to \infty$, the trajectory tends to the largest invariant set inside S; i.e., its ω limit set is contained inside the largest invariant set in S. In particular, if S contains no invariant sets other than $x = 0$, then 0 is asymptotically stable.

Proofs are given in [128] and [135].

Lyapunov functions can often be used to design stabilizing controllers, as is illustrated by the following example, which also illustrates how the Krasovski–Lasalle principle can be applied.

Example 4.12 Inverted pendulum
Following the analysis in Example 2.7, an inverted pendulum can be described by the following normalized model:

$$\frac{dx_1}{dt} = x_2, \qquad \frac{dx_2}{dt} = \sin x_1 + u \cos x_1, \tag{4.19}$$

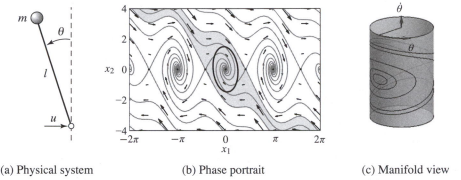

(a) Physical system   (b) Phase portrait   (c) Manifold view

Figure 4.16: Stabilized inverted pendulum. A control law applies a force u at the bottom of the pendulum to stabilize the inverted position (a). The phase portrait (b) shows that the equilibrium point corresponding to the vertical position is stabilized. The shaded region indicates the set of initial conditions that converge to the origin. The ellipse corresponds to a level set of a Lyapunov function $V(x)$ for which $V(x) > 0$ and $\dot{V}(x) < 0$ for all points inside the ellipse. This can be used as an estimate of the region of attraction of the equilibrium point. The actual dynamics of the system evolve on a manifold (c).

where x_1 is the angular deviation from the upright position and u is the (scaled) acceleration of the pivot, as shown in Figure 4.16a. The system has an equilibrium at $x_1 = x_2 = 0$, which corresponds to the pendulum standing upright. This equilibrium is unstable.

To find a stabilizing controller we consider the following candidate for a Lyapunov function:

$$V(x) = (\cos x_1 - 1) + a(1 - \cos^2 x_1) + \frac{1}{2}x_2^2 \approx \left(a - \frac{1}{2}\right)x_1^2 + \frac{1}{2}x_2^2.$$

The Taylor series expansion shows that the function is positive definite near the origin if $a > 0.5$. The time derivative of $V(x)$ is

$$\dot{V} = -\dot{x}_1 \sin x_1 + 2a\dot{x}_1 \sin x_1 \cos x_1 + \dot{x}_2 x_2 = x_2(u + 2a \sin x_1) \cos x_1.$$

Choosing the feedback law

$$u = -2a \sin x_1 - x_2 \cos x_1$$

gives

$$\dot{V} = -x_2^2 \cos^2 x_1.$$

It follows from Lyapunov's theorem that the equilibrium is locally stable. However, since the function is only negative semidefinite, we cannot conclude asymptotic stability using Theorem 4.2. However, note that $\dot{V} = 0$ implies that $x_2 = 0$ or $x_1 = \pi/2 \pm n\pi$.

If we restrict our analysis to a small neighborhood of the origin Ω_r, $r \ll \pi/2$, then we can define

$$S = \{(x_1, x_2) \in \Omega_r : x_2 = 0\}$$

and we can compute the largest invariant set inside S. For a trajectory to remain in this set we must have $x_2 = 0$ for all t and hence $\dot{x}_2(t) = 0$ as well. Using the dynamics of the system (4.19), we see that $x_2(t) = 0$ and $\dot{x}_2(t) = 0$ implies $x_1(t) = 0$ as well. Hence the largest invariant set inside S is $(x_1, x_2) = 0$, and we can use the Krasovski–Lasalle principle to conclude that the origin is locally asymptotically stable. A phase portrait of the closed loop system is shown in Figure 4.16b.

In the analysis and the phase portrait, we have treated the angle of the pendulum $\theta = x_1$ as a real number. In fact, θ is an angle with $\theta = 2\pi$ equivalent to $\theta = 0$. Hence the dynamics of the system actually evolves on a *manifold* (smooth surface) as shown in Figure 4.16c. Analysis of nonlinear dynamical systems on manifolds is more complicated, but uses many of the same basic ideas presented here. ∇

4.5 Parametric and Nonlocal Behavior

Most of the tools that we have explored are focused on the local behavior of a fixed system near an equilibrium point. In this section we briefly introduce some concepts regarding the global behavior of nonlinear systems and the dependence of a system's behavior on parameters in the system model.

Regions of Attraction

To get some insight into the behavior of a nonlinear system we can start by finding the equilibrium points. We can then proceed to analyze the local behavior around the equilibria. The behavior of a system near an equilibrium point is called the *local* behavior of the system.

The solutions of the system can be very different far away from an equilibrium point. This is seen, for example, in the stabilized pendulum in Example 4.12. The inverted equilibrium point is stable, with small oscillations that eventually converge to the origin. But far away from this equilibrium point there are trajectories that converge to other equilibrium points or even cases in which the pendulum swings around the top multiple times, giving very long oscillations that are topologically different from those near the origin.

To better understand the dynamics of the system, we can examine the set of all initial conditions that converge to a given asymptotically stable equilibrium point. This set is called the *region of attraction* for the equilibrium point. An example is shown by the shaded region of the phase portrait in Figure 4.16b. In general, computing regions of attraction is difficult. However, even if we cannot determine the region of attraction, we can often obtain patches around the stable equilibria that are attracting. This gives partial information about the behavior of the system.

One method for approximating the region of attraction is through the use of Lyapunov functions. Suppose that V is a local Lyapunov function for a system around an equilibrium point x_0. Let Ω_r be a set on which $V(x)$ has a value less than r,

$$\Omega_r = \{x \in \mathbb{R}^n : V(x) \leq r\},$$

and suppose that $\dot{V}(x) \leq 0$ for all $x \in \Omega_r$, with equality only at the equilibrium point x_0. Then Ω_r is inside the region of attraction of the equilibrium point. Since this approximation depends on the Lyapunov function and the choice of Lyapunov function is not unique, it can sometimes be a very conservative estimate.

It is sometimes the case that we can find a Lyapunov function V such that V is positive definite and $\dot{V}$ is negative (semi-) definite for all $x \in \mathbb{R}^n$. In many instances it can then be shown that the region of attraction for the equilibrium point is the entire state space, and the equilibrium point is said to be *globally* stable.

Example 4.13 Stabilized inverted pendulum
Consider again the stabilized inverted pendulum from Example 4.12. The Lyapunov function for the system was

$$V(x) = (\cos x_1 - 1) + a(1 - \cos^2 x_1) + \frac{1}{2}x_2^2,$$

and $\dot{V}$ was negative semidefinite for all x and nonzero when $x_1 \neq \pm\pi/2$. Hence for any x such that $|x_2| < \pi/2$, $V(x) > 0$ will be inside the invariant set defined by the level curves of $V(x)$. One of these level sets is shown in Figure 4.16b. $\quad\nabla$

Bifurcations

Another important property of nonlinear systems is how their behavior changes as the parameters governing the dynamics change. We can study this in the context of models by exploring how the location of equilibrium points, their stability, their regions of attraction and other dynamic phenomena, such as limit cycles, vary based on the values of the parameters in the model.

Consider a differential equation of the form

$$\frac{dx}{dt} = F(x, \mu), \quad x \in \mathbb{R}^n, \mu \in \mathbb{R}^k, \tag{4.20}$$

where x is the state and μ is a set of parameters that describe the family of equations. The equilibrium solutions satisfy

$$F(x, \mu) = 0,$$

and as μ is varied, the corresponding solutions $x_e(\mu)$ can also vary. We say that the system (4.20) has a *bifurcation* at $\mu = \mu^*$ if the behavior of the system changes qualitatively at μ^*. This can occur either because of a change in stability type or a change in the number of solutions at a given value of μ.

Example 4.14 Predator–prey
Consider the predator–prey system described in Section 3.7. The dynamics of the system are given by

$$\frac{dH}{dt} = rH\left(1 - \frac{H}{k}\right) - \frac{aHL}{c+H}, \qquad \frac{dL}{dt} = b\frac{aHL}{c+H} - dL, \tag{4.21}$$

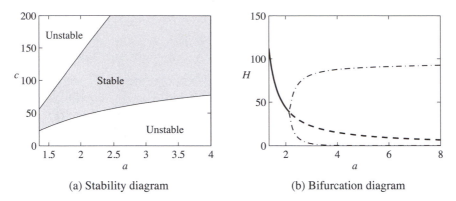

(a) Stability diagram   (b) Bifurcation diagram

Figure 4.17: Bifurcation analysis of the predator–prey system. (a) Parametric stability diagram showing the regions in parameter space for which the system is stable. (b) Bifurcation diagram showing the location and stability of the equilibrium point as a function of a. The solid line represents a stable equilibrium point, and the dashed line represents an unstable equilibrium point. The dashed-dotted lines indicate the upper and lower bounds for the limit cycle at that parameter value (computed via simulation). The nominal values of the parameters in the model are $a = 3.2, b = 0.6, c = 50, d = 0.56, k = 125$ and $r = 1.6$.

where H and L are the numbers of hares (prey) and lynxes (predators) and $a, b,$ c, d, k and r are parameters that model a given predator–prey system (described in more detail in Section 3.7). The system has an equilibrium point at $H_e > 0$ and $L_e > 0$ that can be found numerically.

To explore how the parameters of the model affect the behavior of the system, we choose to focus on two specific parameters of interest: a, the interaction coefficient between the populations and c, a parameter affecting the prey consumption rate. Figure 4.17a is a numerically computed *parametric stability diagram* showing the regions in the chosen parameter space for which the equilibrium point is stable (leaving the other parameters at their nominal values). We see from this figure that for certain combinations of a and c we get a stable equilibrium point, while at other values this equilibrium point is unstable.

Figure 4.17b is a numerically computed *bifurcation diagram* for the system. In this plot, we choose one parameter to vary (a) and then plot the equilibrium value of one of the states (H) on the vertical axis. The remaining parameters are set to their nominal values. A solid line indicates that the equilibrium point is stable; a dashed line indicates that the equilibrium point is unstable. Note that the stability in the bifurcation diagram matches that in the parametric stability diagram for $c = 50$ (the nominal value) and a varying from 1.35 to 4. For the predator–prey system, when the equilibrium point is unstable, the solution converges to a stable limit cycle. The amplitude of this limit cycle is shown by the dashed-dotted line in Figure 4.17b.

∇

A particular form of bifurcation that is very common when controlling linear systems is that the equilibrium remains fixed but the stability of the equilibrium

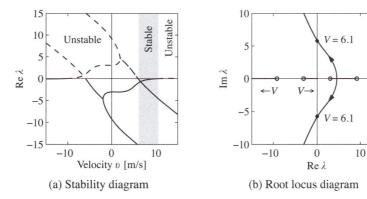

(a) Stability diagram   (b) Root locus diagram

Figure 4.18: Stability plots for a bicycle moving at constant velocity. The plot in (a) shows the real part of the system eigenvalues as a function of the bicycle velocity v. The system is stable when all eigenvalues have negative real part (shaded region). The plot in (b) shows the locus of eigenvalues on the complex plane as the velocity v is varied and gives a different view of the stability of the system. This type of plot is called a *root locus diagram*.

changes as the parameters are varied. In such a case it is revealing to plot the eigenvalues of the system as a function of the parameters. Such plots are called *root locus diagrams* because they give the locus of the eigenvalues when parameters change. Bifurcations occur when parameter values are such that there are eigenvalues with zero real part. Computing environments such as LabVIEW, MATLAB and Mathematica have tools for plotting root loci.

Example 4.15 Root locus diagram for a bicycle model
Consider the linear bicycle model given by equation (3.7) in Section 3.2. Introducing the state variables $x_1 = \varphi$, $x_2 = \delta$, $x_3 = \dot{\varphi}$ and $x_4 = \dot{\delta}$ and setting the steering torque $T = 0$, the equations can be written as

$$\frac{dx}{dt} = \begin{bmatrix} 0 & I \\ -M^{-1}(K_0 + K_2 v_0^2) & -M^{-1}C v_0 \end{bmatrix} x =: Ax,$$

where I is a 2×2 identity matrix and v_0 is the velocity of the bicycle. Figure 4.18a shows the real parts of the eigenvalues as a function of velocity. Figure 4.18b shows the dependence of the eigenvalues of A on the velocity v_0. The figures show that the bicycle is unstable for low velocities because two eigenvalues are in the right half-plane. As the velocity increases, these eigenvalues move into the left half-plane, indicating that the bicycle becomes self-stabilizing. As the velocity is increased further, there is an eigenvalue close to the origin that moves into the right half-plane, making the bicycle unstable again. However, this eigenvalue is small and so it can easily be stabilized by a rider. Figure 4.18a shows that the bicycle is self-stabilizing for velocities between 6 and 10 m/s.   ∇

Parametric stability diagrams and bifurcation diagrams can provide valuable insights into the dynamics of a nonlinear system. It is usually necessary to carefully choose the parameters that one plots, including combining the natural parameters

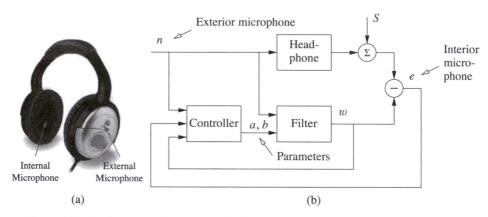

Figure 4.19: Headphones with noise cancellation. Noise is sensed by the exterior microphone
(a) and sent to a filter in such a way that it cancels the noise that penetrates the head phone
(b). The filter parameters a and b are adjusted by the controller. S represents the input signal
to the headphones.

of the system to eliminate extra parameters when possible. Computer programs
such as AUTO, LOCBIF and XPPAUT provide numerical algorithms for producing
stability and bifurcation diagrams.

Design of Nonlinear Dynamics Using Feedback

In most of the text we will rely on linear approximations to design feedback laws
that stabilize an equilibrium point and provide a desired level of performance.
However, for some classes of problems the feedback controller must be nonlinear to
accomplish its function. By making use of Lyapunov functions we can often design
a nonlinear control law that provides stable behavior, as we saw in Example 4.12.

One way to systematically design a nonlinear controller is to begin with a
candidate Lyapunov function $V(x)$ and a control system $\dot{x} = f(x, u)$. We say
that $V(x)$ is a *control Lyapunov function* if for every x there exists a u such that
$\dot{V}(x) = \frac{\partial V}{\partial x} f(x, u) < 0$. In this case, it may be possible to find a function $\alpha(x)$
such that $u = \alpha(x)$ stabilizes the system. The following example illustrates the
approach.

Example 4.16 Noise cancellation
Noise cancellation is used in consumer electronics and in industrial systems to
reduce the effects of noise and vibrations. The idea is to locally reduce the effect of
noise by generating opposing signals. A pair of headphones with noise cancellation
such as those shown in Figure 4.19a is a typical example. A schematic diagram of
the system is shown in Figure 4.19b. The system has two microphones, one outside
the headphones that picks up exterior noise n and another inside the headphones that
picks up the signal e, which is a combination of the desired signal and the external
noise that penetrates the headphone. The signal from the exterior microphone is
filtered and sent to the headphones in such a way that it cancels the external noise

that penetrates into the headphones. The parameters of the filter are adjusted by a feedback mechanism to make the noise signal in the internal microphone as small as possible. The feedback is inherently nonlinear because it acts by changing the parameters of the filter.

To analyze the system we assume for simplicity that the propagation of external noise into the headphones is modeled by a first-order dynamical system described by

$$\frac{dz}{dt} = a_0 z + b_0 n, \tag{4.22}$$

where z is the sound level and the parameters $a_0 < 0$ and b_0 are not known. Assume that the filter is a dynamical system of the same type:

$$\frac{dw}{dt} = aw + bn.$$

We wish to find a controller that updates a and b so that they converge to the (unknown) parameters a_0 and b_0. Introduce $x_1 = e = w - z$, $x_2 = a - a_0$ and $x_3 = b - b_0$; then

$$\frac{dx_1}{dt} = a_0(w - z) + (a - a_0)w + (b - b_0)n = a_0 x_1 + x_2 w + x_3 n. \tag{4.23}$$

We will achieve noise cancellation if we can find a feedback law for changing the parameters a and b so that the error e goes to zero. To do this we choose

$$V(x_1, x_2, x_3) = \frac{1}{2}\left(\alpha x_1^2 + x_2^2 + x_3^2\right)$$

as a candidate Lyapunov function for (4.23). The derivative of V is

$$\dot{V} = \alpha x_1 \dot{x}_1 + x_2 \dot{x}_2 + x_3 \dot{x}_3 = \alpha a_0 x_1^2 + x_2(\dot{x}_2 + \alpha w x_1) + x_3(\dot{x}_3 + \alpha n x_1).$$

Choosing

$$\dot{x}_2 = -\alpha w x_1 = -\alpha w e, \qquad \dot{x}_3 = -\alpha n x_1 = -\alpha n e, \tag{4.24}$$

we find that $\dot{V} = \alpha a_0 x_1^2 < 0$, and it follows that the quadratic function will decrease as long as $e = x_1 = w - z \neq 0$. The nonlinear feedback (4.24) thus attempts to change the parameters so that the error between the signal and the noise is small. Notice that feedback law (4.24) does not use the model (4.22) explicitly.

A simulation of the system is shown in Figure 4.20. In the simulation we have represented the signal as a pure sinusoid and the noise as broad band noise. The figure shows the dramatic improvement with noise cancellation. The sinusoidal signal is not visible without noise cancellation. The filter parameters change quickly from their initial values $a = b = 0$. Filters of higher order with more coefficients are used in practice.   ∇

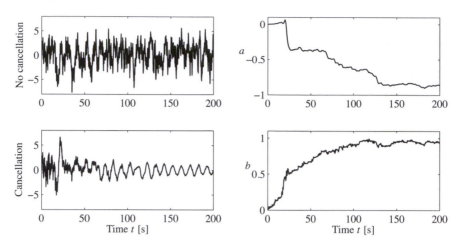

Figure 4.20: Simulation of noise cancellation. The top left figure shows the headphone signal without noise cancellation, and the bottom left figure shows the signal with noise cancellation. The right figures show the parameters a and b of the filter.

4.6 Further Reading

The field of dynamical systems has a rich literature that characterizes the possible features of dynamical systems and describes how parametric changes in the dynamics can lead to topological changes in behavior. Readable introductions to dynamical systems are given by Strogatz [188] and the highly illustrated text by Abraham and Shaw [2]. More technical treatments include Andronov, Vitt and Khaikin [8], Guckenheimer and Holmes [91] and Wiggins [201]. For students with a strong interest in mechanics, the texts by Arnold [13] and Marsden and Ratiu [147] provide an elegant approach using tools from differential geometry. Finally, good treatments of dynamical systems methods in biology are given by Wilson [203] and Ellner and Guckenheimer [70]. There is a large literature on Lyapunov stability theory, including the classic texts by Malkin [144], Hahn [94] and Krasovski [128]. We highly recommend the comprehensive treatment by Khalil [123].

Exercises

4.1 (Time-invariant systems) Show that if we have a solution of the differential equation (4.1) given by $x(t)$ with initial condition $x(t_0) = x_0$, then $\tilde{x}(\tau) = x(t - t_0)$ is a solution of the differential equation

$$\frac{d\tilde{x}}{d\tau} = F(\tilde{x})$$

with initial condition $\tilde{x}(0) = x_0$, where $\tau = t - t_0$.

4.2 (Flow in a tank) A cylindrical tank has cross section A m^2, effective outlet area a m^2 and inflow q_{in} m^3/s. An energy balance shows that the outlet velocity is

$v = \sqrt{2gh}$ m/s, where g m/s^2 is the acceleration of gravity and h is the distance between the outlet and the water level in the tank (in meters). Show that the system can be modeled by

$$\frac{dh}{dt} = -\frac{a}{A}\sqrt{2gh} - \frac{1}{A}q_{in}, \qquad q_{out} = a\sqrt{2gh}.$$

Use the parameters $A = 0.2, a = 0.01$. Simulate the system when the inflow is zero and the initial level is $h = 0.2$. Do you expect any difficulties in the simulation?

4.3 (Cruise control) Consider the cruise control system described in Section 3.1. Generate a phase portrait for the closed loop system on flat ground ($\theta = 0$), in third gear, using a PI controller (with $k_p = 0.5$ and $k_i = 0.1$), $m = 1000$ kg and desired speed 20 m/s. Your system model should include the effects of saturating the input between 0 and 1.

4.4 (Lyapunov functions) Consider the second-order system

$$\frac{dx_1}{dt} = -ax_1, \qquad \frac{dx_2}{dt} = -bx_1 - cx_2,$$

where $a, b, c > 0$. Investigate whether the functions

$$V_1(x) = \frac{1}{2}x_1^2 + \frac{1}{2}x_2^2, \qquad V_2(x) = \frac{1}{2}x_1^2 + \frac{1}{2}\left(x_2 + \frac{b}{c-a}x_1\right)^2$$

are Lyapunov functions for the system and give any conditions that must hold.

4.5 (Damped spring–mass system) Consider a damped spring–mass system with dynamics

$$m\ddot{q} + c\dot{q} + kq = 0.$$

A natural candidate for a Lyapunov function is the total energy of the system, given by

$$V = \frac{1}{2}m\dot{q}^2 + \frac{1}{2}kq^2.$$

Use the Krasovski–Lasalle theorem to show that the system is asymptotically stable.

4.6 (Electric generator) The following simple model for an electric generator connected to a strong power grid was given in Exercise 2.7:

$$J\frac{d^2\varphi}{dt^2} = P_m - P_e = P_m - \frac{EV}{X}\sin\varphi.$$

The parameter

$$a = \frac{P_{max}}{P_m} = \frac{EV}{XP_m} \tag{4.25}$$

is the ratio between the maximum deliverable power $P_{max} = EV/X$ and the mechanical power P_m.

(a) Consider a as a bifurcation parameter and discuss how the equilibria depend on a.

(b) For $a > 1$, show that there is a center at $\varphi_0 = \arcsin(1/a)$ and a saddle at $\varphi = \pi - \varphi_0$.

(c) Show that if $P_m/J = 1$ there is a solution through the saddle that satisfies

$$\frac{1}{2}\left(\frac{d\varphi}{dt}\right)^2 - \varphi + \varphi_0 - a\cos\varphi - \sqrt{a^2 - 1} = 0. \tag{4.26}$$

Use simulation to show that the stability region is the interior of the area enclosed by this solution. Investigate what happens if the system is in equilibrium with a value of a that is slightly larger than 1 and a suddenly decreases, corresponding to the reactance of the line suddenly increasing.

4.7 (Lyapunov equation) Show that Lyapunov equation (4.14) always has a solution if all of the eigenvalues of A are in the left half-plane. (Hint: Use the fact that the Lyapunov equation is linear in P and start with the case where A has distinct eigenvalues.)

4.8 (Congestion control) Consider the congestion control problem described in Section 3.4. Confirm that the equilibrium point for the system is given by equation (3.21) and compute the stability of this equilibrium point using a linear approximation.

4.9 (Swinging up a pendulum) Consider the inverted pendulum, discussed in Example 4.4, that is described by

$$\ddot{\theta} = \sin\theta + u\cos\theta,$$

where θ is the angle between the pendulum and the vertical and the control signal u is the acceleration of the pivot. Using the energy function

$$V(\theta, \dot{\theta}) = \cos\theta - 1 + \frac{1}{2}\dot{\theta}^2,$$

show that the state feedback $u = k(V_0 - V)\dot{\theta}\cos\theta$ causes the pendulum to "swing up" to the upright position.

4.10 (Root locus diagram) Consider the linear system

$$\frac{dx}{dt} = \begin{bmatrix} 0 & 1 \\ 0 & -3 \end{bmatrix} x + \begin{bmatrix} -1 \\ 4 \end{bmatrix} u, \qquad y = \begin{bmatrix} 1 & 0 \end{bmatrix} x,$$

with the feedback $u = -ky$. Plot the location of the eigenvalues as a function the parameter k.

4.11 (Discrete-time Lyapunov function) Consider a nonlinear discrete-time system with dynamics $x[k+1] = f(x[k])$ and equilibrium point $x_e = 0$. Suppose there exists a smooth, positive definite function $V : \mathbb{R}^n \to \mathbb{R}$ such that $V(f(x)) - V(x) < 0$ for $x \neq 0$ and $V(0) = 0$. Show that $x_e = 0$ is (locally) asymptotically stable.

4.12 (Operational amplifier oscillator) An op amp circuit for an oscillator was shown in Exercise 3.5. The oscillatory solution for that linear circuit was stable but not asymptotically stable. A schematic of a modified circuit that has nonlinear elements is shown in the figure below.

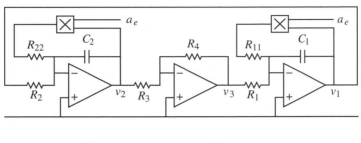

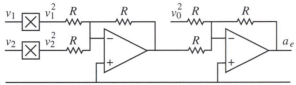

The modification is obtained by making a feedback around each operational amplifier that has capacitors using multipliers. The signal $a_e = v_1^2 + v_2^2 - v_0^2$ is the amplitude error. Show that the system is modeled by

$$\frac{dv_1}{dt} = \frac{R_4}{R_1 R_3 C_1} v_2 + \frac{1}{R_{11} C_1} v_1 (v_0^2 - v_1^2 - v_2^2),$$

$$\frac{dv_2}{dt} = -\frac{1}{R_2 C_2} v_1 + \frac{1}{R_{22} C_2} v_2 (v_0^2 - v_1^2 - v_2^2).$$

Show that the circuit gives an oscillation with a stable limit cycle with amplitude v_0. (Hint: Use the results of Example 4.8.)

4.13 (Self-activating genetic circuit) Consider the dynamics of a genetic circuit that implements *self-activation*: the protein produced by the gene is an activator for the protein, thus stimulating its own production through positive feedback. Using the models presented in Example 2.13, the dynamics for the system can be written as

$$\frac{dm}{dt} = \frac{\alpha p^2}{1 + k p^2} + \alpha_0 - \gamma m, \qquad \frac{dp}{dt} = \beta m - \delta p, \qquad (4.27)$$

for $p, m \geq 0$. Find the equilibrium points for the system and analyze the local stability of each using Lyapunov analysis.

4.14 (Diagonal systems) Let $A \in \mathbb{R}^{n \times n}$ be a square matrix with real eigenvalues $\lambda_1, \ldots, \lambda_n$ and corresponding eigenvectors $v_1, \ldots, v_n$.

(a) Show that if the eigenvalues are distinct ($\lambda_i \neq \lambda_j$ for $i \neq j$), then $v_i \neq v_j$ for $i \neq j$.

(b) Show that the eigenvectors form a basis for $\mathbb{R}^n$ so that any vector x can be written as $x = \sum \alpha_i v_i$ for $\alpha_i \in \mathbb{R}$.

(c) Let $T = \begin{bmatrix} v_1 & v_2 & \cdots & v_n \end{bmatrix}$ and show that $T^{-1} A T$ is a diagonal matrix of the form (4.8).

(d) Show that if some of the λ_i are complex numbers, then A can be written as

$$A = \begin{bmatrix} \Lambda_1 & & 0 \\ 0 & \ddots & \\ 0 & & \Lambda_k \end{bmatrix} \qquad \text{where} \qquad \Lambda_i = \lambda \in \mathbb{R} \quad \text{or} \quad \Lambda_i = \begin{bmatrix} \sigma & \omega \\ -\omega & \sigma \end{bmatrix}.$$

in an appropriate set of coordinates.

This form of the dynamics of a linear system is often referred to as *modal form*.

4.15 (Furuta pendulum) The Furuta pendulum, an inverted pendulum on a rotating arm, is shown to the left in the figure below.

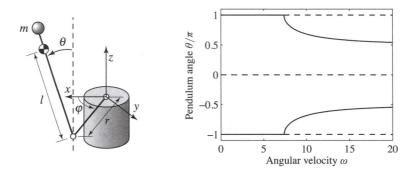

Consider the situation when the pendulum arm is spinning with constant rate. The system has multiple equilibrium points that depend on the angular velocity ω, as shown in the bifurcation diagram on the right.

The equations of motion for the system are given by

$$J_p\ddot{\theta} - J_p\omega_0^2 \sin\theta \cos\theta - m_p gl \sin\theta = 0,$$

where J_p is the moment of inertia of the pendulum with respect to its pivot, m_p is the pendulum mass, l is the distance between the pivot and the center of mass of the pendulum and ω_0 is the the rate of rotation of the arm.

(a) Determine the equilibria for the system and the condition(s) for stability of each equilibrium point (in terms of ω).

(b) Consider the angular velocity as a bifurcation parameter and verify the bifurcation diagram given above. This is an example of a *pitchfork bifurcation*.

4.16 (Routh-Hurwitz criterion) Consider a linear differential equation with the characteristic polynomial

$$\lambda(s) = s^2 + a_1 s + a_2, \qquad \lambda(s) = s^3 + a_1 s^2 + a_2 s + a_3.$$

Show that the system is asymptotically stable if and only if all the coefficients a_i are positive and if $a_1 a_2 > a_3$. This is a special case of a more general set of criteria known as the Routh-Hurwitz criterion.

Chapter Five
Linear Systems

Few physical elements display truly linear characteristics. For example the relation between force on a spring and displacement of the spring is always nonlinear to some degree. The relation between current through a resistor and voltage drop across it also deviates from a straight-line relation. However, if in each case the relation is reasonably *linear, then it will be found that the system behavior will be very close to that obtained by assuming an ideal, linear physical element, and the analytical simplification is so enormous that we make linear assumptions wherever we can possibly do so in good conscience.*

Robert H. Cannon, *Dynamics of Physical Systems*, 1967 [49].

In Chapters 2–4 we considered the construction and analysis of differential equation models for dynamical systems. In this chapter we specialize our results to the case of linear, time-invariant input/output systems. Two central concepts are the matrix exponential and the convolution equation, through which we can completely characterize the behavior of a linear system. We also describe some properties of the input/output response and show how to approximate a nonlinear system by a linear one.

5.1 Basic Definitions

We have seen several instances of linear differential equations in the examples in the previous chapters, including the spring–mass system (damped oscillator) and the operational amplifier in the presence of small (nonsaturating) input signals. More generally, many dynamical systems can be modeled accurately by linear differential equations. Electrical circuits are one example of a broad class of systems for which linear models can be used effectively. Linear models are also broadly applicable in mechanical engineering, for example, as models of small deviations from equilibria in solid and fluid mechanics. Signal-processing systems, including digital filters of the sort used in CD and MP3 players, are another source of good examples, although these are often best modeled in discrete time (as described in more detail in the exercises).

In many cases, we *create* systems with a linear input/output response through the use of feedback. Indeed, it was the desire for linear behavior that led Harold S. Black to the invention of the negative feedback amplifier. Almost all modern signal processing systems, whether analog or digital, use feedback to produce linear or near-linear input/output characteristics. For these systems, it is often useful to represent the input/output characteristics as linear, ignoring the internal details required to get that linear response.

For other systems nonlinearities cannot be ignored, especially if one cares about the global behavior of the system. The predator–prey problem is one example of this: to capture the oscillatory behavior of the interdependent populations we must include the nonlinear coupling terms. Other examples include switching behavior and generating periodic motion for locomotion. However, if we care about what happens near an equilibrium point, it often suffices to approximate the nonlinear dynamics by their local linearization, as we already explored briefly in Section 4.3. The linearization is essentially an approximation of the nonlinear dynamics around the desired operating point.

Linearity

We now proceed to define linearity of input/output systems more formally. Consider a state space system of the form

$$\frac{dx}{dt} = f(x, u), \qquad y = h(x, u), \tag{5.1}$$

where $x \in \mathbb{R}^n$, $u \in \mathbb{R}^p$ and $y \in \mathbb{R}^q$. As in the previous chapters, we will usually restrict ourselves to the single-input, single-output case by taking $p = q = 1$. We also assume that all functions are smooth and that for a reasonable class of inputs (e.g., piecewise continuous functions of time) the solutions of equation (5.1) exist for all time.

It will be convenient to assume that the origin $x = 0$, $u = 0$ is an equilibrium point for this system ($\dot{x} = 0$) and that $h(0, 0) = 0$. Indeed, we can do so without loss of generality. To see this, suppose that $(x_e, u_e) \neq (0, 0)$ is an equilibrium point of the system with output $y_e = h(x_e, u_e)$. Then we can define a new set of states, inputs and outputs,

$$\tilde{x} = x - x_e, \qquad \tilde{u} = u - u_e, \qquad \tilde{y} = y - y_e,$$

and rewrite the equations of motion in terms of these variables:

$$\frac{d}{dt}\tilde{x} = f(\tilde{x} + x_e, \tilde{u} + u_e) =: \tilde{f}(\tilde{x}, \tilde{u}),$$
$$\tilde{y} = h(\tilde{x} + x_e, \tilde{u} + u_e) - y_e =: \tilde{h}(\tilde{x}, \tilde{u}).$$

In the new set of variables, the origin is an equilibrium point with output 0, and hence we can carry out our analysis in this set of variables. Once we have obtained our answers in this new set of variables, we simply "translate" them back to the original coordinates using $x = \tilde{x} + x_e$, $u = \tilde{u} + u_e$ and $y = \tilde{y} + y_e$.

Returning to the original equations (5.1), now assuming without loss of generality that the origin is the equilibrium point of interest, we write the output $y(t)$ corresponding to the initial condition $x(0) = x_0$ and input $u(t)$ as $y(t; x_0, u)$. Using this notation, a system is said to be a *linear input/output system* if the following

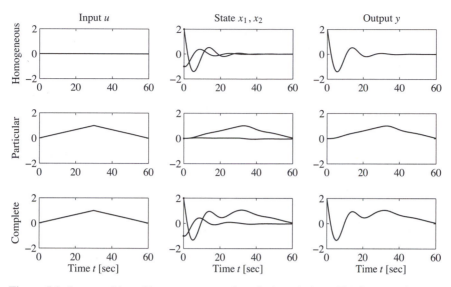

Figure 5.1: Superposition of homogeneous and particular solutions. The first row shows the input, state and output corresponding to the initial condition response. The second row shows the same variables corresponding to zero initial condition but nonzero input. The third row is the complete solution, which is the sum of the two individual solutions.

conditions are satisfied:

$$\text{(i)} \quad y(t; \alpha x_1 + \beta x_2, 0) = \alpha y(t; x_1, 0) + \beta y(t; x_2, 0),$$

$$\text{(ii)} \quad y(t; \alpha x_0, \delta u) = \alpha y(t; x_0, 0) + \delta y(t; 0, u), \tag{5.2}$$

$$\text{(iii)} \quad y(t; 0, \delta u_1 + \gamma u_2) = \delta y(t; 0, u_1) + \gamma y(t; 0, u_2).$$

Thus, we define a system to be linear if the outputs are jointly linear in the initial condition response ($u = 0$) and the forced response ($x(0) = 0$). Property (iii) is a statement of the *principle of superposition*: the response of a linear system to the sum of two inputs u_1 and u_2 is the sum of the outputs y_1 and y_2 corresponding to the individual inputs.

The general form of a linear state space system is

$$\frac{dx}{dt} = Ax + Bu, \qquad y = Cx + Du, \tag{5.3}$$

where $A \in \mathbb{R}^{n \times n}$, $B \in \mathbb{R}^{n \times p}$, $C \in \mathbb{R}^{q \times n}$ and $D \in \mathbb{R}^{q \times p}$. In the special case of a single-input, single-output system, B is a column vector, C is a row vector and D is scalar. Equation (5.3) is a system of linear first-order differential equations with input u, state x and output y. It is easy to show that given solutions $x_1(t)$ and $x_2(t)$ for this set of equations, they satisfy the linearity conditions.

We define $x_h(t)$ to be the solution with zero input (the *homogeneous solution*) and the solution $x_p(t)$ to be the solution with zero initial condition (a *particular solution*). Figure 5.1 illustrates how these two individual solutions can be superimposed to form the complete solution.

It is also possible to show that if a finite-dimensional dynamical system is input/output linear in the sense we have described, it can always be represented by a state space equation of the form (5.3) through an appropriate choice of state variables. In Section 5.2 we will give an explicit solution of equation (5.3), but we illustrate the basic form through a simple example.

Example 5.1 Scalar system
Consider the first-order differential equation

$$\frac{dx}{dt} = ax + u, \qquad y = x,$$

with $x(0) = x_0$. Let $u_1 = A \sin \omega_1 t$ and $u_2 = B \cos \omega_2 t$. The homogeneous solution is $x_h(t) = e^{at} x_0$, and two particular solutions with $x(0) = 0$ are

$$x_{p1}(t) = -A \frac{-\omega_1 e^{at} + \omega_1 \cos \omega_1 t + a \sin \omega_1 t}{a^2 + \omega_1^2},$$

$$x_{p2}(t) = B \frac{a e^{at} - a \cos \omega_2 t + \omega_2 \sin \omega_2 t}{a^2 + \omega_2^2}.$$

Suppose that we now choose $x(0) = \alpha x_0$ and $u = u_1 + u_2$. Then the resulting solution is the weighted sum of the individual solutions:

$$
\begin{aligned}
x(t) = e^{at} & \left(\alpha x_0 + \frac{A\omega_1}{a^2 + \omega_1^2} + \frac{Ba}{a^2 + \omega_2^2} \right) \\
& - A \frac{\omega_1 \cos \omega_1 t + a \sin \omega_1 t}{a^2 + \omega_1^2} + B \frac{-a \cos \omega_2 t + \omega_2 \sin \omega_2 t}{a^2 + \omega_2^2}.
\end{aligned}
\tag{5.4}
$$

To see this, substitute equation (5.4) into the differential equation. Thus, the properties of a linear system are satisfied. ∇

Time Invariance

Time invariance is an important concept that is used to describe a system whose properties do not change with time. More precisely, for a time-invariant system if the input $u(t)$ gives output $y(t)$, then if we shift the time at which the input is applied by a constant amount a, $u(t + a)$ gives the output $y(t + a)$. Systems that are linear and time-invariant, often called *LTI systems*, have the interesting property that their response to an arbitrary input is completely characterized by their response to step inputs or their response to short "impulses."

To explore the consequences of time invariance, we first compute the response to a piecewise constant input. Assume that the system is initially at rest and consider the piecewise constant input shown in Figure 5.2a. The input has jumps at times t_k, and its values after the jumps are $u(t_k)$. The input can be viewed as a combination of steps: the first step at time t_0 has amplitude $u(t_0)$, the second step at time t_1 has amplitude $u(t_1) - u(t_0)$, etc.

Assuming that the system is initially at an equilibrium point (so that the initial condition response is zero), the response to the input can be obtained by superim-

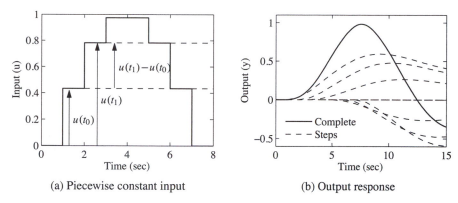

(a) Piecewise constant input   (b) Output response

Figure 5.2: Response to piecewise constant inputs. A piecewise constant signal can be represented as a sum of step signals (a), and the resulting output is the sum of the individual outputs (b).

posing the responses to a combination of step inputs. Let $H(t)$ be the response to a unit step applied at time 0. The response to the first step is then $H(t - t_0)u(t_0)$, the response to the second step is $H(t - t_1)\big(u(t_1) - u(t_0)\big)$, and we find that the complete response is given by

$$
\begin{aligned}
y(t) &= H(t - t_0)u(t_0) + H(t - t_1)\big(u(t_1) - u(t_0)\big) + \cdots \\
&= \big(H(t - t_0) - H(t - t_1)\big)u(t_0) + \big(H(t - t_1) - H(t - t_2)\big)u(t_1) + \cdots \\
&= \sum_{n=0}^{t_n < t} \infty\big(H(t - t_n) - H(t - t_{n+1})\big)u(t_n) \\
&= \sum_{n=0}^{t_n < t} \frac{H(t - t_n) - H(t - t_{n+1})}{t_{n+1} - t_n}u(t_n)\big(t_{n+1} - t_n\big).
\end{aligned}
$$

An example of this computation is shown in Figure 5.2b.

The response to a continuous input signal is obtained by taking the limit as $t_{n+1} - t_n \to 0$, which gives

$$
y(t) = \int_0^t H'(t - \tau)u(\tau)d\tau, \tag{5.5}
$$

where H' is the derivative of the step response, also called the *impulse response*. The response of a linear time-invariant system to any input can thus be computed from the step response. Notice that the output depends only on the input since we assumed the system was initially at rest, $x(0) = 0$. We will derive equation (5.5) in a slightly different way in the Section 5.3.

5.2 The Matrix Exponential

Equation (5.5) shows that the output of a linear system can be written as an integral over the inputs $u(t)$. In this section and the next we derive a more general version of this formula, which includes nonzero initial conditions. We begin by exploring the initial condition response using the matrix exponential.

Initial Condition Response

Although we have shown that the solution of a linear set of differential equations defines a linear input/output system, we have not fully computed the solution of the system. We begin by considering the homogeneous response corresponding to the system

$$\frac{dx}{dt} = Ax. \tag{5.6}$$

For the *scalar* differential equation

$$\frac{dx}{dt} = ax, \qquad x \in \mathbb{R}, \, a \in \mathbb{R},$$

the solution is given by the exponential

$$x(t) = e^{at}x(0).$$

We wish to generalize this to the vector case, where A becomes a matrix. We define the *matrix exponential* as the infinite series

$$e^X = I + X + \frac{1}{2}X^2 + \frac{1}{3!}X^3 + \cdots = \sum_{k=0}^{\infty} \frac{1}{k!}X^k, \tag{5.7}$$

where $X \in \mathbb{R}^{n \times n}$ is a square matrix and I is the $n \times n$ identity matrix. We make use of the notation

$$X^0 = I, \qquad X^2 = XX, \qquad X^n = X^{n-1}X,$$

which defines what we mean by the "power" of a matrix. Equation (5.7) is easy to remember since it is just the Taylor series for the scalar exponential, applied to the matrix X. It can be shown that the series in equation (5.7) converges for any matrix $X \in \mathbb{R}^{n \times n}$ in the same way that the normal exponential is defined for any scalar $a \in \mathbb{R}$.

Replacing X in equation (5.7) by At, where $t \in \mathbb{R}$, we find that

$$e^{At} = I + At + \frac{1}{2}A^2t^2 + \frac{1}{3!}A^3t^3 + \cdots = \sum_{k=0}^{\infty} \frac{1}{k!}A^k t^k,$$

and differentiating this expression with respect to t gives

$$\frac{d}{dt}e^{At} = A + A^2t + \frac{1}{2}A^3t^2 + \cdots = A\sum_{k=0}^{\infty} \frac{1}{k!}A^k t^k = Ae^{At}. \tag{5.8}$$

Multiplying by $x(0)$ from the right, we find that $x(t) = e^{At}x(0)$ is the solution to the differential equation (5.6) with initial condition $x(0)$. We summarize this important result as a proposition.

Proposition 5.1. *The solution to the homogeneous system of differential equations* (5.6) *is given by*

$$x(t) = e^{At}x(0).$$

Notice that the form of the solution is exactly the same as for scalar equations, but we must put the vector $x(0)$ on the right of the matrix e^{At}.

The form of the solution immediately allows us to see that the solution is linear in the initial condition. In particular, if $x_{h1}(t)$ is the solution to equation (5.6) with initial condition $x(0) = x_{01}$ and $x_{h2}(t)$ with initial condition $x(0) = x_{02}$, then the solution with initial condition $x(0) = \alpha x_{01} + \beta x_{02}$ is given by

$$x(t) = e^{At}(\alpha x_{01} + \beta x_{02}) = (\alpha e^{At} x_{01} + \beta e^{At} x_{02}) = \alpha x_{h1}(t) + \beta x_{h2}(t).$$

Similarly, we see that the corresponding output is given by

$$y(t) = Cx(t) = \alpha y_{h1}(t) + \beta y_{h2}(t),$$

where $y_{h1}(t)$ and $y_{h2}(t)$ are the outputs corresponding to $x_{h1}(t)$ and $x_{h2}(t)$.

We illustrate computation of the matrix exponential by two examples.

Example 5.2 Double integrator
A very simple linear system that is useful in understanding basic concepts is the second-order system given by

$$\ddot{q} = u, \qquad y = q.$$

This system is called a *double integrator* because the input u is integrated twice to determine the output y.

In state space form, we write $x = (q, \dot{q})$ and

$$\frac{dx}{dt} = \begin{bmatrix} 0 & 1 \\ 0 & 0 \end{bmatrix} x + \begin{bmatrix} 0 \\ 1 \end{bmatrix} u.$$

The dynamics matrix of a double integrator is

$$A = \begin{bmatrix} 0 & 1 \\ 0 & 0 \end{bmatrix},$$

and we find by direct calculation that $A^2 = 0$ and hence

$$e^{At} = \begin{bmatrix} 1 & t \\ 0 & 1 \end{bmatrix}.$$

Thus the homogeneous solution ($u = 0$) for the double integrator is given by

$$x(t) = \begin{bmatrix} 1 & t \\ 0 & 1 \end{bmatrix} \begin{bmatrix} x_1(0) \\ x_2(0) \end{bmatrix} = \begin{bmatrix} x_1(0) + tx_2(0) \\ x_2(0) \end{bmatrix},$$

$$y(t) = x_1(0) + tx_2(0).$$

∇

Example 5.3 Undamped oscillator

A simple model for an oscillator, such as the spring–mass system with zero damping, is

$$\ddot{q} + \omega_0^2 q = u.$$

Putting the system into state space form, the dynamics matrix for this system can be written as

$$A = \begin{bmatrix} 0 & \omega_0 \\ -\omega_0 & 0 \end{bmatrix} \quad \text{and} \quad e^{At} = \begin{bmatrix} \cos \omega_0 t & \sin \omega_0 t \\ -\sin \omega_0 t & \cos \omega_0 t \end{bmatrix}.$$

This expression for e^{At} can be verified by differentiation:

$$\frac{d}{dt} e^{At} = \begin{bmatrix} -\omega_0 \sin \omega_0 t & \omega_0 \cos \omega_0 t \\ -\omega_0 \cos \omega_0 t & -\omega_0 \sin \omega_0 t \end{bmatrix}$$

$$= \begin{bmatrix} 0 & \omega_0 \\ -\omega_0 & 0 \end{bmatrix} \begin{bmatrix} \cos \omega_0 t & \sin \omega_0 t \\ -\sin \omega_0 t & \cos \omega_0 t \end{bmatrix} = A e^{At}.$$

The solution is then given by

$$x(t) = e^{At} x(0) = \begin{bmatrix} \cos \omega_0 t & \sin \omega_0 t \\ -\sin \omega_0 t & \cos \omega_0 t \end{bmatrix} \begin{bmatrix} x_1(0) \\ x_2(0) \end{bmatrix}.$$

If the system has damping,

$$\ddot{q} + 2\zeta \omega_0 q + \omega_0^2 q = u,$$

the solution is more complicated, but the matrix exponential can be shown to be

$$e^{-\omega_0 \zeta t} \begin{bmatrix} \dfrac{\zeta e^{i\omega_d t} - \zeta e^{-i\omega_d t}}{2\sqrt{\zeta^2 - 1}} + \dfrac{e^{i\omega_d t} + e^{-i\omega_d t}}{2} & \dfrac{e^{i\omega_d t} - e^{-i\omega_d t}}{2\sqrt{\zeta^2 - 1}} \\ \dfrac{e^{-i\omega_d t} - e^{i\omega_d t}}{2\sqrt{\zeta^2 - 1}} & \dfrac{\zeta e^{-i\omega_d t} - \zeta e^{i\omega_d t}}{2\sqrt{\zeta^2 - 1}} + \dfrac{e^{i\omega_d t} + e^{-i\omega_d t}}{2} \end{bmatrix},$$

where $\omega_d = \omega_0 \sqrt{\zeta^2 - 1}$. Note that ω_d and $\sqrt{\zeta^2 - 1}$ can be either real or complex, but the combinations of terms will always yield a real value for the entries in the matrix exponential. ∇

An important class of linear systems are those that can be converted into diagonal form. Suppose that we are given a system

$$\frac{dx}{dt} = Ax$$

such that all the eigenvalues of A are distinct. It can be shown (Exercise 4.14) that we can find an invertible matrix T such that $T A T^{-1}$ is diagonal. If we choose a set of coordinates $z = Tx$, then in the new coordinates the dynamics become

$$\frac{dz}{dt} = T\frac{dx}{dt} = TAx = TAT^{-1}z.$$

By construction of T, this system will be diagonal.

Now consider a diagonal matrix A and the corresponding kth power of At, which is also diagonal:

$$A = \begin{bmatrix} \lambda_1 & & & 0 \\ & \lambda_2 & & \\ & & \ddots & \\ 0 & & & \lambda_n \end{bmatrix}, \qquad (At)^k = \begin{bmatrix} \lambda_1^k t^k & & & 0 \\ & \lambda_2^k t^k & & \\ & & \ddots & \\ 0 & & & \lambda_n^k t^k \end{bmatrix},$$

It follows from the series expansion that the matrix exponential is given by

$$e^{At} = \begin{bmatrix} e^{\lambda_1 t} & & & 0 \\ & e^{\lambda_2 t} & & \\ & & \ddots & \\ 0 & & & e^{\lambda_n t} \end{bmatrix}.$$

A similar expansion can be done in the case where the eigenvalues are complex, using a block diagonal matrix, similar to what was done in Section 4.3.

Jordan Form

Some matrices with equal eigenvalues cannot be transformed to diagonal form. They can, however, be transformed to a closely related form, called the *Jordan form*, in which the dynamics matrix has the eigenvalues along the diagonal. When there are equal eigenvalues, there may be 1's appearing in the superdiagonal indicating that there is coupling between the states.

More specifically, we define a matrix to be in Jordan form if it can be written as

$$J = \begin{bmatrix} J_1 & 0 & \cdots & 0 & 0 \\ 0 & J_2 & 0 & 0 & 0 \\ \vdots & & \ddots & \ddots & \vdots \\ 0 & 0 & & J_{k-1} & 0 \\ 0 & 0 & \cdots & 0 & J_k \end{bmatrix}, \quad \text{where} \quad J_i = \begin{bmatrix} \lambda_i & 1 & 0 & \cdots & 0 \\ 0 & \lambda_i & 1 & & 0 \\ \vdots & & \ddots & \ddots & \vdots \\ 0 & 0 & & \lambda_i & 1 \\ 0 & 0 & \cdots & 0 & \lambda_i \end{bmatrix}.$$

$$(5.9)$$

Each matrix J_i is called a *Jordan block*, and λ_i for that block corresponds to an eigenvalue of J. A first-order Jordan block can be represented as a system consisting of an integrator with feedback λ. A Jordan block of higher order can be represented as series connections of such systems, as illustrated in Figure 5.3.

Theorem 5.2 (Jordan decomposition). *Any matrix $A \in \mathbb{R}^{n \times n}$ can be transformed into Jordan form with the eigenvalues of A determining λ_i in the Jordan form.*

Proof. See any standard text on linear algebra, such as Strang [187]. The special case where the eigenvalues are distinct is examined in Exercise 4.14.   □

Converting a matrix into Jordan form can be complicated, although MATLAB can do this conversion for numerical matrices using the `jordan` function. The structure of the resulting Jordan form is particularly interesting since there is no

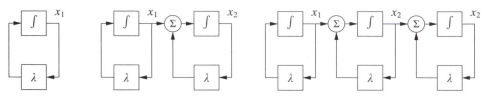

Figure 5.3: Representations of linear systems where the dynamics matrices are Jordan blocks. A first-order Jordan block can be represented as an integrator with feedback λ, as shown on the left. Second- and third-order Jordan blocks can be represented as series connections of integrators with feedback, as shown on the right.

requirement that the individual λ_i's be unique, and hence for a given eigenvalue we can have one or more Jordan blocks of different sizes.

Once a matrix is in Jordan form, the exponential of the matrix can be computed in terms of the Jordan blocks:

$$
e^J = \begin{bmatrix} e^{J_1} & 0 & \cdots & 0 \\ 0 & e^{J_2} & & \vdots \\ \vdots & & \ddots & 0 \\ 0 & \cdots & 0 & e^{J_k}. \end{bmatrix}. \tag{5.10}
$$

This follows from the block diagonal form of J. The exponentials of the Jordan blocks can in turn be written as

$$
e^{J_i t} = \begin{bmatrix} 1 & t & \frac{t^2}{2!} & \cdots & \frac{t^{n-1}}{(n-1)!} \\ 0 & 1 & t & \cdots & \frac{t^{n-2}}{(n-2)!} \\ \vdots & & 1 & \ddots & \vdots \\ & & & \ddots & t \\ 0 & \cdots & & 0 & 1 \end{bmatrix} e^{\lambda_i t}. \tag{5.11}
$$

When there are multiple eigenvalues, the invariant subspaces associated with each eigenvalue correspond to the Jordan blocks of the matrix A. Note that λ may be complex, in which case the transformation T that converts a matrix into Jordan form will also be complex. When λ has a nonzero imaginary component, the solutions will have oscillatory components since

$$
e^{\sigma + i\omega t} = e^{\sigma t}(\cos \omega t + i \sin \omega t).
$$

We can now use these results to prove Theorem 4.1, which states that the equilibrium point $x_e = 0$ of a linear system is asymptotically stable if and only if Re $\lambda_i < 0$.

Proof of Theorem 4.1. Let $T \in \mathbb{C}^{n \times n}$ be an invertible matrix that transforms A into Jordan form, $J = TAT^{-1}$. Using coordinates $z = Tx$, we can write the solution $z(t)$ as

$$
z(t) = e^{Jt} z(0).
$$

Since any solution $x(t)$ can be written in terms of a solution $z(t)$ with $z(0) = Tx(0)$, it follows that it is sufficient to prove the theorem in the transformed coordinates.

The solution $z(t)$ can be written in terms of the elements of the matrix exponential. From equation (5.11) these elements all decay to zero for arbitrary $z(0)$ if and only if $\operatorname{Re} \lambda_i < 0$. Furthermore, if any λ_i has positive real part, then there exists an initial condition $z(0)$ such that the corresponding solution increases without bound. Since we can scale this initial condition to be arbitrarily small, it follows that the equilibrium point is unstable if any eigenvalue has positive real part.   □

The existence of a canonical form allows us to prove many properties of linear systems by changing to a set of coordinates in which the A matrix is in Jordan form. We illustrate this in the following proposition, which follows along the same lines as the proof of Theorem 4.1.

Proposition 5.3. *Suppose that the system*

$$\frac{dx}{dt} = Ax$$

has no eigenvalues with strictly positive real part and one or more eigenvalues with zero real part. Then the system is stable if and only if the Jordan blocks corresponding to each eigenvalue with zero real part are scalar (1×1) blocks.

Proof. See Exercise 5.6b.   □

The following example illustrates the use of the Jordan form.

Example 5.4 Linear model of a vectored thrust aircraft
Consider the dynamics of a vectored thrust aircraft such as that described in Example 2.9. Suppose that we choose $u_1 = u_2 = 0$ so that the dynamics of the system become

$$\frac{dz}{dt} = \begin{bmatrix} z_4 \\ z_5 \\ z_6 \\ -g \sin z_3 - \frac{c}{m} z_4 \\ -g(\cos z_3 - 1) - \frac{c}{m} z_5 \\ 0 \end{bmatrix}, \tag{5.12}$$

where $z = (x, y, \theta, \dot{x}, \dot{y}, \dot{\theta})$. The equilibrium points for the system are given by setting the velocities $\dot{x}$, $\dot{y}$ and $\dot{\theta}$ to zero and choosing the remaining variables to satisfy

$$\begin{aligned} -g \sin z_{3,e} &= 0 \\ -g(\cos z_{3,e} - 1) &= 0 \end{aligned} \qquad \Longrightarrow \qquad z_{3,e} = \theta_e = 0.$$

This corresponds to the upright orientation for the aircraft. Note that x_e and y_e are not specified. This is because we can translate the system to a new (upright) position and still obtain an equilibrium point.

(a) Mode 1   (b) Mode 2

Figure 5.4: Modes of vibration for a system consisting of two masses connected by springs. In (a) the masses move left and right in synchronization in (b) they move toward or against each other.

To compute the stability of the equilibrium point, we compute the linearization using equation (4.11):

$$A = \left. \frac{\partial F}{\partial z} \right|_{z_e} = \begin{bmatrix} 0 & 0 & 0 & 1 & 0 & 0 \\ 0 & 0 & 0 & 0 & 1 & 0 \\ 0 & 0 & 0 & 0 & 0 & 1 \\ 0 & 0 & -g & -c/m & 0 & 0 \\ 0 & 0 & 0 & 0 & -c/m & 0 \\ 0 & 0 & 0 & 0 & 0 & 0 \end{bmatrix}.$$

The eigenvalues of the system can be computed as

$$\lambda(A) = \{0, 0, 0, 0, -c/m, -c/m\}.$$

We see that the linearized system is not asymptotically stable since not all of the eigenvalues have strictly negative real part.

To determine whether the system is stable in the sense of Lyapunov, we must make use of the Jordan form. It can be shown that the Jordan form of A is given by

$$J = \left[\begin{array}{cccc|c|c} 0 & 0 & 0 & 0 & 0 & 0 \\ \hline 0 & 0 & 1 & 0 & 0 & 0 \\ 0 & 0 & 0 & 1 & 0 & 0 \\ 0 & 0 & 0 & 0 & 0 & 0 \\ \hline 0 & 0 & 0 & 0 & -c/m & 0 \\ \hline 0 & 0 & 0 & 0 & 0 & -c/m \end{array} \right].$$

Since the second Jordan block has eigenvalue 0 and is not a simple eigenvalue, the linearization is unstable.   ∇

Eigenvalues and Modes

The eigenvalues and eigenvectors of a system provide a description of the types of behavior the system can exhibit. For oscillatory systems, the term *mode* is often used to describe the vibration patterns that can occur. Figure 5.4 illustrates the modes for a system consisting of two masses connected by springs. One pattern is when both masses oscillate left and right in unison, and another is when the masses move toward and away from each other.

The initial condition response of a linear system can be written in terms of a matrix exponential involving the dynamics matrix A. The properties of the matrix A

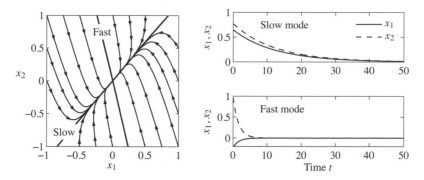

Figure 5.5: The notion of modes for a second-order system with real eigenvalues. The left figure shows the phase portrait and the modes corresponding to solutions that start on the eigenvectors (bold lines). The corresponding time functions are shown on the right.

therefore determine the resulting behavior of the system. Given a matrix $A \in \mathbb{R}^{n \times n}$, recall that v is an eigenvector of A with eigenvalue λ if

$$Av = \lambda v.$$

In general λ and v may be complex-valued, although if A is real-valued, then for any eigenvalue λ its complex conjugate λ^* will also be an eigenvalue (with v^* as the corresponding eigenvector).

Suppose first that λ and v are a real-valued eigenvalue/eigenvector pair for A. If we look at the solution of the differential equation for $x(0) = v$, it follows from the definition of the matrix exponential that

$$e^{At}v = \left(I + At + \frac{1}{2}A^2t^2 + \cdots\right)v = v + \lambda tv + \frac{\lambda^2 t^2}{2}v + \cdots = e^{\lambda t}v.$$

The solution thus lies in the subspace spanned by the eigenvector. The eigenvalue λ describes how the solution varies in time, and this solution is often called a *mode* of the system. (In the literature, the term "mode" is also often used to refer to the eigenvalue rather than the solution.)

If we look at the individual elements of the vectors x and v, it follows that

$$\frac{x_i(t)}{x_j(t)} = \frac{e^{\lambda t}v_i}{e^{\lambda t}v_j} = \frac{v_i}{v_j},$$

and hence the ratios of the components of the state x are constants for a (real) mode. The eigenvector thus gives the "shape" of the solution and is also called a *mode shape* of the system. Figure 5.5 illustrates the modes for a second-order system consisting of a fast mode and a slow mode. Notice that the state variables have the same sign for the slow mode and different signs for the fast mode.

The situation is more complicated when the eigenvalues of A are complex. Since A has real elements, the eigenvalues and the eigenvectors are complex conjugates

$\lambda = \sigma \pm i\omega$ and $v = u \pm iw$, which implies that

$$u = \frac{v + v^*}{2}, \qquad w = \frac{v - v^*}{2i}.$$

Making use of the matrix exponential, we have

$$e^{At}v = e^{\lambda t}(u + iw) = e^{\sigma t}\big((u \cos \omega t - w \sin \omega t) + i(u \sin \omega t + w \cos \omega t)\big),$$

from which it follows that

$$e^{At}u = \frac{1}{2}\big(e^{At}v + e^{At}v^*\big) = ue^{\sigma t}\cos \omega t - we^{\sigma t}\sin \omega t,$$

$$e^{At}w = \frac{1}{2i}\big(e^{At}v - e^{At}v^*\big) = ue^{\sigma t}\sin \omega t + we^{\sigma t}\cos \omega t.$$

A solution with initial conditions in the subspace spanned by the real part u and imaginary part w of the eigenvector will thus remain in that subspace. The solution will be a logarithmic spiral characterized by σ and ω. We again call the solution corresponding to λ a mode of the system, and v the mode shape.

If a matrix A has n distinct eigenvalues $\lambda_1, \ldots, \lambda_n$, then the initial condition response can be written as a linear combination of the modes. To see this, suppose for simplicity that we have all real eigenvalues with corresponding unit eigenvectors $v_1, \ldots, v_n$. From linear algebra, these eigenvectors are linearly independent, and we can write the initial condition $x(0)$ as

$$x(0) = \alpha_1 v_1 + \alpha_2 v_2 + \cdots + \alpha_n v_n.$$

Using linearity, the initial condition response can be written as

$$x(t) = \alpha_1 e^{\lambda_1 t} v_1 + \alpha_2 e^{\lambda_2 t} v_2 + \cdots + \alpha_n e^{\lambda_n t} v_n.$$

Thus, the response is a linear combination of the modes of the system, with the amplitude of the individual modes growing or decaying as $e^{\lambda_i t}$. The case for distinct complex eigenvalues follows similarly (the case for nondistinct eigenvalues is more subtle and requires making use of the Jordan form discussed in the previous section).

Example 5.5 Coupled spring–mass system
Consider the spring–mass system shown in Figure 5.4, but with the addition of dampers on each mass. The equations of motion of the system are

$$m\ddot{q}_1 = -2kq_1 - c\dot{q}_1 + kq_2, \qquad m\ddot{q}_2 = kq_1 - 2kq_2 - c\dot{q}_2.$$

In state space form, we define the state to be $x = (q_1, q_2, \dot{q}_1, \dot{q}_2)$, and we can rewrite the equations as

$$\frac{dx}{dt} = \begin{bmatrix} 0 & 0 & 1 & 0 \\ 0 & 0 & 0 & 1 \\ -\dfrac{2k}{m} & \dfrac{k}{m} & -\dfrac{c}{m} & 0 \\ \dfrac{k}{m} & -\dfrac{2k}{m} & 0 & -\dfrac{c}{m} \end{bmatrix} x.$$

We now define a transformation $z = Tx$ that puts this system into a simpler form. Let $z_1 = \frac{1}{2}(q_1 + q_2)$, $z_2 = \dot{z}_1$, $z_3 = \frac{1}{2}(q_1 - q_2)$ and $z_4 = \dot{z}_3$, so that

$$z = Tx = \frac{1}{2}\begin{bmatrix} 1 & 1 & 0 & 0 \\ 0 & 0 & 1 & 1 \\ 1 & -1 & 0 & 0 \\ 0 & 0 & 1 & -1 \end{bmatrix} x.$$

In the new coordinates, the dynamics become

$$\frac{dz}{dt} = \begin{bmatrix} 0 & 1 & 0 & 0 \\ -\dfrac{k}{m} & -\dfrac{c}{m} & 0 & 0 \\ 0 & 0 & 0 & 1 \\ 0 & 0 & -\dfrac{3k}{m} & -\dfrac{c}{m} \end{bmatrix} z,$$

and we see that the system is in block diagonal (or *modal*) form.

In the z coordinates, the states z_1 and z_2 parameterize one mode with eigenvalues $\lambda \approx c/(2\sqrt{km}) \pm i\sqrt{k/m}$, and the states z_3 and z_4 another mode with $\lambda \approx c/(2\sqrt{3km}) \pm i\sqrt{3k/m}$. From the form of the transformation T we see that these modes correspond exactly to the modes in Figure 5.4, in which q_1 and q_2 move either toward or against each other. The real and imaginary parts of the eigenvalues give the decay rates σ and frequencies ω for each mode.   ∇

5.3 Input/Output Response

In the previous section we saw how to compute the initial condition response using the matrix exponential. In this section we derive the convolution equation, which includes the inputs and outputs as well.

The Convolution Equation

We return to the general input/output case in equation (5.3), repeated here:

$$\frac{dx}{dt} = Ax + Bu, \qquad y = Cx + Du. \tag{5.13}$$

Using the matrix exponential, the solution to equation (5.13) can be written as follows.

Theorem 5.4. *The solution to the linear differential equation (5.13) is given by*

$$x(t) = e^{At}x(0) + \int_0^t e^{A(t-\tau)} Bu(\tau)d\tau. \tag{5.14}$$

Proof. To prove this, we differentiate both sides and use the property (5.8) of the matrix exponential. This gives

$$\frac{dx}{dt} = Ae^{At}x(0) + \int_0^t Ae^{A(t-\tau)} Bu(\tau)d\tau + Bu(t) = Ax + Bu,$$

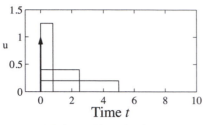

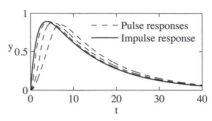

(a) Pulse and impulse functions   (b) Pulse and impulse responses

Figure 5.6: Pulse response and impulse response. (a) The rectangles show pulses of width 5, 2.5 and 0.8, each with total area equal to 1. The arrow denotes an impulse $\delta(t)$ defined by equation (5.17). The corresponding pulse responses for a linear system with eigenvalues $\lambda = \{-0.08, -0.62\}$ are shown in (b) as dashed lines. The solid line is the true impulse response, which is well approximated by a pulse of duration 0.8.

which proves the result. Notice that the calculation is essentially the same as for proving the result for a first-order equation.   □

It follows from equations (5.13) and (5.14) that the input/output relation for a linear system is given by

$$y(t) = Ce^{At}x(0) + \int_0^t Ce^{A(t-\tau)}Bu(\tau)d\tau + Du(t). \qquad (5.15)$$

It is easy to see from this equation that the output is jointly linear in both the initial conditions and the input, which follows from the linearity of matrix/vector multiplication and integration.

Equation (5.15) is called the *convolution equation*, and it represents the general form of the solution of a system of coupled linear differential equations. We see immediately that the dynamics of the system, as characterized by the matrix A, play a critical role in both the stability and performance of the system. Indeed, the matrix exponential describes *both* what happens when we perturb the initial condition and how the system responds to inputs.

 Another interpretation of the convolution equation can be given using the concept of the *impulse response* of a system. Consider the application of an input signal $u(t)$ given by the following equation:

$$u(t) = p_\epsilon(t) = \begin{cases} 0 & t < 0 \\ 1/\epsilon & 0 \le t < \epsilon \\ 0 & t \ge \epsilon. \end{cases} \qquad (5.16)$$

This signal is a *pulse* of duration ϵ and amplitude $1/\epsilon$, as illustrated in Figure 5.6a. We define an *impulse* $\delta(t)$ to be the limit of this signal as $\epsilon \to 0$:

$$\delta(t) = \lim_{\epsilon \to 0} p_\epsilon(t). \qquad (5.17)$$

This signal, sometimes called a *delta function,* is not physically achievable but provides a convenient abstraction in understanding the response of a system. Note that the integral of an impulse is 1:

$$\int_0^t \delta(\tau)\,d\tau = \int_0^t \lim_{\epsilon \to 0} p_\epsilon(t)\,d\tau = \lim_{\epsilon \to 0} \int_0^t p_\epsilon(t)\,d\tau$$

$$= \lim_{\epsilon \to 0} \int_0^\epsilon 1/\epsilon\,d\tau = 1 \qquad t > 0.$$

In particular, the integral of an impulse over an arbitrarily short period of time is identically 1.

We define the *impulse response* of a system $h(t)$ to be the output corresponding to having an impulse as its input:

$$h(t) = \int_0^t C e^{A(t-\tau)} B\delta(\tau)\,d\tau = C e^{At} B, \tag{5.18}$$

where the second equality follows from the fact that $\delta(t)$ is zero everywhere except the origin and its integral is identically 1. We can now write the convolution equation in terms of the initial condition response, the convolution of the impulse response and the input signal, and the direct term:

$$y(t) = C e^{At} x(0) + \int_0^t h(t-\tau) u(\tau)\,d\tau + Du(t). \tag{5.19}$$

One interpretation of this equation, explored in Exercise 5.2, is that the response of the linear system is the superposition of the response to an infinite set of shifted impulses whose magnitudes are given by the input $u(t)$. This is essentially the argument used in analyzing Figure 5.2 and deriving equation (5.5). Note that the second term in equation (5.19) is identical to equation (5.5), and it can be shown that the impulse response is formally equivalent to the derivative of the step response.

The use of pulses as approximations of the impulse function also provides a mechanism for identifying the dynamics of a system from data. Figure 5.6b shows the pulse responses of a system for different pulse widths. Notice that the pulse responses approach the impulse response as the pulse width goes to zero. As a general rule, if the fastest eigenvalue of a stable system has real part $-\sigma_{max}$, then a pulse of length ϵ will provide a good estimate of the impulse response if $\epsilon\sigma_{max} \ll 1$. Note that for Figure 5.6, a pulse width of $\epsilon = 1$ s gives $\epsilon\sigma_{max} = 0.62$ and the pulse response is already close to the impulse response.

Coordinate Invariance

The components of the input vector u and the output vector y are given by the chosen inputs and outputs of a model, but the state variables depend on the coordinate frame chosen to represent the state. This choice of coordinates affects the values of the matrices A, B and C that are used in the model. (The direct term D is not affected since it maps inputs to outputs.) We now investigate some of the consequences of changing coordinate systems.

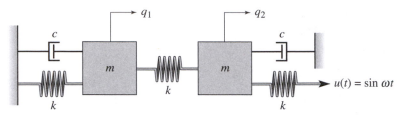

Figure 5.7: Coupled spring mass system. Each mass is connected to two springs with stiffness k and a viscous damper with damping coefficient c. The mass on the right is drive through a spring connected to a sinusoidally varying attachment.

Introduce new coordinates z by the transformation $z = Tx$, where T is an invertible matrix. It follows from equation (5.3) that

$$\frac{dz}{dt} = T(Ax + Bu) = TAT^{-1}z + TBu =: \tilde{A}z + \tilde{B}u,$$

$$y = Cx + Du = CT^{-1}z + Du =: \tilde{C}z + Du.$$

The transformed system has the same form as equation (5.3), but the matrices A, B and C are different:

$$\tilde{A} = TAT^{-1}, \qquad \tilde{B} = TB, \qquad \tilde{C} = CT^{-1}. \tag{5.20}$$

There are often special choices of coordinate systems that allow us to see a particular property of the system, hence coordinate transformations can be used to gain new insight into the dynamics.

We can also compare the solution of the system in transformed coordinates to that in the original state coordinates. We make use of an important property of the exponential map,

$$e^{TST^{-1}} = Te^{S}T^{-1},$$

which can be verified by substitution in the definition of the matrix exponential. Using this property, it is easy to show that

$$x(t) = T^{-1}z(t) = T^{-1}e^{\tilde{A}t}Tx(0) + T^{-1}\int_0^t e^{\tilde{A}(t-\tau)}\tilde{B}u(\tau)\,d\tau.$$

From this form of the equation, we see that if it is possible to transform A into a form $\tilde{A}$ for which the matrix exponential is easy to compute, we can use that computation to solve the general convolution equation for the untransformed state x by simple matrix multiplications. This technique is illustrated in the following example.

Example 5.6 Coupled spring–mass system
Consider the coupled spring–mass system shown in Figure 5.7. The input to this system is the sinusoidal motion of the end of the rightmost spring, and the output is the position of each mass, q_1 and q_2. The equations of motion are given by

$$m\ddot{q}_1 = -2kq_1 - c\dot{q}_1 + kq_2, \qquad m\ddot{q}_2 = kq_1 - 2kq_2 - c\dot{q}_2 + ku.$$

In state space form, we define the state to be $x = (q_1, q_2, \dot{q}_1, \dot{q}_2)$, and we can rewrite the equations as

$$\frac{dx}{dt} = \begin{bmatrix} 0 & 0 & 1 & 0 \\ 0 & 0 & 0 & 1 \\ -\dfrac{2k}{m} & \dfrac{k}{m} & -\dfrac{c}{m} & 0 \\ \dfrac{k}{m} & -\dfrac{2k}{m} & 0 & -\dfrac{c}{m} \end{bmatrix} x + \begin{bmatrix} 0 \\ 0 \\ 0 \\ \dfrac{k}{m} \end{bmatrix} u.$$

This is a coupled set of four differential equations and is quite complicated to solve in analytical form.

The dynamics matrix is the same as in Example 5.5, and we can use the coordinate transformation defined there to put the system in modal form:

$$\frac{dz}{dt} = \begin{bmatrix} 0 & 1 & 0 & 0 \\ -\dfrac{k}{m} & -\dfrac{c}{m} & 0 & 0 \\ 0 & 0 & 0 & 1 \\ 0 & 0 & -\dfrac{3k}{m} & -\dfrac{c}{m} \end{bmatrix} z + \begin{bmatrix} 0 \\ \dfrac{k}{2m} \\ 0 \\ -\dfrac{k}{2m} \end{bmatrix} u.$$

Note that the resulting matrix equations are block diagonal and hence decoupled. We can solve for the solutions by computing the solutions of two sets of second-order systems represented by the states (z_1, z_2) and (z_3, z_4). Indeed, the functional form of each set of equations is identical to that of a single spring–mass system. (The explicit solution is derived in Section 6.3.)

Once we have solved the two sets of independent second-order equations, we can recover the dynamics in the original coordinates by inverting the state transformation and writing $x = T^{-1}z$. We can also determine the stability of the system by looking at the stability of the independent second-order systems. ▽

Steady-State Response

Given a linear input/output system

$$\frac{dx}{dt} = Ax + Bu, \qquad y = Cx + Du, \tag{5.21}$$

the general form of the solution to equation (5.21) is given by the convolution equation:

$$y(t) = Ce^{At}x(0) + \int_0^t Ce^{A(t-\tau)}Bu(\tau)d\tau + Du(t).$$

We see from the form of this equation that the solution consists of an initial condition response and an input response.

The input response, corresponding to the last two terms in the equation above, itself consists of two components—the *transient response* and the *steady-state*

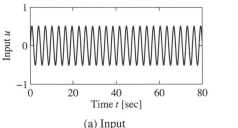

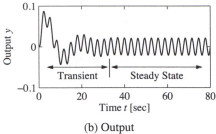

(a) Input   (b) Output

Figure 5.8: Transient versus steady-state response. The input to a linear system is shown in (a), and the corresponding output with $x(0) = 0$ is shown in (b). The output signal initially undergoes a transient before settling into its steady-state behavior.

response. The transient response occurs in the first period of time after the input is applied and reflects the mismatch between the initial condition and the steady-state solution. The steady-state response is the portion of the output response that reflects the long-term behavior of the system under the given inputs. For inputs that are periodic the steady-state response will often be periodic, and for constant inputs the response will often be constant. An example of the transient and the steady-state response for a periodic input is shown in Figure 5.8.

A particularly common form of input is a *step input*, which represents an abrupt change in input from one value to another. A *unit step* (sometimes called the Heaviside step function) is defined as

$$u = S(t) = \begin{cases} 0 & t = 0 \\ 1 & t > 0. \end{cases}$$

The *step response* of the system (5.21) is defined as the output $y(t)$ starting from zero initial condition (or the appropriate equilibrium point) and given a step input. We note that the step input is discontinuous and hence is not practically implementable. However, it is a convenient abstraction that is widely used in studying input/output systems.

We can compute the step response to a linear system using the convolution equation. Setting $x(0) = 0$ and using the definition of the step input above, we have

$$y(t) = \int_0^t Ce^{A(t-\tau)}Bu(\tau)d\tau + Du(t) = C\int_0^t e^{A(t-\tau)}Bd\tau + D$$

$$= C\int_0^t e^{A\sigma}Bd\sigma + D = C\left(A^{-1}e^{A\sigma}B\right)\Big|_{\sigma=0}^{\sigma=t} + D$$

$$= CA^{-1}e^{At}B - CA^{-1}B + D.$$

If A has eigenvalues with negative real part (implying that the origin is a stable

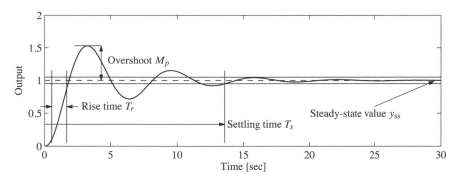

Figure 5.9: Sample step response. The rise time, overshoot, settling time and steady-state value give the key performance properties of the signal.

equilibrium point in the absence of any input), then we can rewrite the solution as

$$y(t) = \underbrace{C A^{-1} e^{At} B}_{\text{transient}} + \underbrace{D - C A^{-1} B}_{\text{steady-state}}, \qquad t > 0. \tag{5.22}$$

The first term is the transient response and decays to zero as $t \to \infty$. The second term is the steady-state response and represents the value of the output for large time.

A sample step response is shown in Figure 5.9. Several terms are used when referring to a step response. The *steady-state value* y_{ss} of a step response is the final level of the output, assuming it converges. The *rise time* T_r is the amount of time required for the signal to go from 10% of its final value to 90% of its final value. It is possible to define other limits as well, but in this book we shall use these percentages unless otherwise indicated. The *overshoot* M_p is the percentage of the final value by which the signal initially rises above the final value. This usually assumes that future values of the signal do not overshoot the final value by more than this initial transient, otherwise the term can be ambiguous. Finally, the *settling time* T_s is the amount of time required for the signal to stay within 2% of its final value for all future times. The settling time is also sometimes defined as reaching 1% or 5% of the final value (see Exercise 5.7). In general these performance measures can depend on the amplitude of the input step, but for linear systems the last three quantities defined above are independent of the size of the step.

Example 5.7 Compartment model
Consider the compartment model illustrated in Figure 5.10 and described in more detail in Section 3.6. Assume that a drug is administered by constant infusion in compartment V_1 and that the drug has its effect in compartment V_2. To assess how quickly the concentration in the compartment reaches steady state we compute the step response, which is shown in Figure 5.10b. The step response is quite slow, with a settling time of 39 min. It is possible to obtain the steady-state concentration much faster by having a faster injection rate initially, as shown in Figure 5.10c. The response of the system in this case can be computed by combining two step

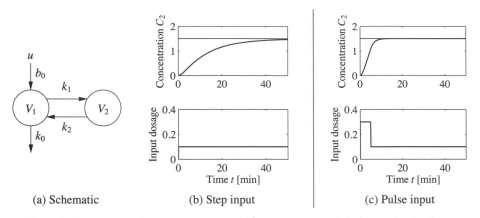

(a) Schematic   (b) Step input   (c) Pulse input

Figure 5.10: Response of a compartment model to a constant drug infusion. A simple diagram of the system is shown in (a). The step response (b) shows the rate of concentration buildup in compartment 2. In (c) a pulse of initial concentration is used to speed up the response.

responses (Exercise 5.3).   ▽

Another common input signal to a linear system is a sinusoid (or a combination of sinusoids). The *frequency response* of an input/output system measures the way in which the system responds to a sinusoidal excitation on one of its inputs. As we have already seen for scalar systems, the particular solution associated with a sinusoidal excitation is itself a sinusoid at the same frequency. Hence we can compare the magnitude and phase of the output sinusoid to the input. More generally, if a system has a sinusoidal output response at the same frequency as the input forcing, we can speak of the frequency response of the system.

To see this in more detail, we must evaluate the convolution equation (5.15) for $u = \cos \omega t$. This turns out to be a very messy calculation, but we can make use of the fact that the system is linear to simplify the derivation. In particular, we note that

$$\cos \omega t = \frac{1}{2}\left(e^{i\omega t} + e^{-i\omega t}\right).$$

Since the system is linear, it suffices to compute the response of the system to the complex input $u(t) = e^{st}$ and we can then reconstruct the input to a sinusoid by averaging the responses corresponding to $s = i\omega$ and $s = -i\omega$.

Applying the convolution equation to the input $u = e^{st}$ we have

$$y(t) = Ce^{At}x(0) + \int_0^t Ce^{A(t-\tau)}Be^{s\tau}\,d\tau + De^{st}$$

$$= Ce^{At}x(0) + Ce^{At}\int_0^t e^{(sI-A)\tau}B\,d\tau + De^{st}.$$

If we assume that none of the eigenvalues of A are equal to $s = \pm i\omega$, then the matrix $sI - A$ is invertible, and we can write

$$
\begin{aligned}
y(t) &= C e^{At} x(0) + C e^{At} \left((sI - A)^{-1} e^{(sI-A)\tau} B \right) \Big|_0^t + D e^{st} \\
&= C e^{At} x(0) + C e^{At} (sI - A)^{-1} \left(e^{(sI-A)t} - I \right) B + D e^{st} \\
&= C e^{At} x(0) + C (sI - A)^{-1} e^{st} B - C e^{At} (sI - A)^{-1} B + D e^{st},
\end{aligned}
$$

and we obtain

$$
y(t) = \underbrace{C e^{At} \left(x(0) - (sI - A)^{-1} B \right)}_{\text{transient}} + \underbrace{\left(C (sI - A)^{-1} B + D \right) e^{st}}_{\text{steady-state}}. \tag{5.23}
$$

Notice that once again the solution consists of both a transient component and a steady-state component. The transient component decays to zero if the system is asymptotically stable and the steady-state component is proportional to the (complex) input $u = e^{st}$.

We can simplify the form of the solution slightly further by rewriting the steady-state response as

$$
y_{ss}(t) = M e^{i\theta} e^{st} = M e^{(st + i\theta)},
$$

where

$$
M e^{i\theta} = C (sI - A)^{-1} B + D \tag{5.24}
$$

and M and θ represent the magnitude and phase of the complex number $C(sI - A)^{-1} B + D$. When $s = i\omega$, we say that M is the *gain* and θ is the *phase* of the system at a given forcing frequency ω. Using linearity and combining the solutions for $s = +i\omega$ and $s = -i\omega$, we can show that if we have an input $u = A_u \sin(\omega t + \psi)$ and an output $y = A_y \sin(\omega t + \varphi)$, then

$$
\text{gain}(\omega) = \frac{A_y}{A_u} = M, \qquad \text{phase}(\omega) = \varphi - \psi = \theta.
$$

The steady-state solution for a sinusoid $u = \cos \omega t$ is now given by

$$
y_{ss}(t) = M \cos(\omega t + \theta).
$$

If the phase θ is positive, we say that the output *leads* the input, otherwise we say it *lags* the input.

A sample sinusoidal response is illustrated in Figure 5.11a. The dashed line shows the input sinusoid, which has amplitude 1. The output sinusoid is shown as a solid line and has a different amplitude plus a shifted phase. The gain is the ratio of the amplitudes of the sinusoids, which can be determined by measuring the height of the peaks. The phase is determined by comparing the ratio of the time between zero crossings of the input and output to the overall period of the sinusoid:

$$
\theta = -2\pi \cdot \frac{\Delta T}{T}.
$$

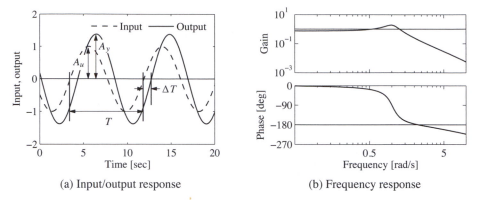

(a) Input/output response   (b) Frequency response

Figure 5.11: Response of a linear system to a sinusoid. (a) A sinusoidal input of magnitude A_u (dashed) gives a sinusoidal output of magnitude A_y (solid), delayed by ΔT seconds. (b) Frequency response, showing gain and phase. The gain is given by the ratio of the output amplitude to the input amplitude, $M = A_y/A_u$. The phase lag is given by $\theta = -2\pi\,\Delta T/T$; it is negative for the case shown because the output lags the input.

A convenient way to view the frequency response is to plot how the gain and phase in equation (5.24) depend on ω (through $s = i\omega$). Figure 5.11b shows an example of this type of representation.

Example 5.8 Active band-pass filter
Consider the op amp circuit shown in Figure 5.12a. We can derive the dynamics of the system by writing the *nodal equations*, which state that the sum of the currents at any node must be zero. Assuming that $v_- = v_+ = 0$, as we did in Section 3.3, we have

$$0 = \frac{v_1 - v_2}{R_1} - C_1 \frac{dv_2}{dt}, \qquad 0 = C_1 \frac{dv_2}{dt} + \frac{v_3}{R_2} + C_2 \frac{dv_3}{dt}.$$

Choosing v_2 and v_3 as our states and using these equations, we obtain

$$\frac{dv_2}{dt} = \frac{v_1 - v_2}{R_1 C_1}, \qquad \frac{dv_3}{dt} = \frac{-v_3}{R_2 C_2} - \frac{v_1 - v_2}{R_1 C_2}.$$

Rewriting these in linear state space form, we obtain

$$\frac{dx}{dt} = \begin{bmatrix} -\dfrac{1}{R_1 C_1} & 0 \\[2ex] \dfrac{1}{R_1 C_2} & -\dfrac{1}{R_2 C_2} \end{bmatrix} x + \begin{bmatrix} \dfrac{1}{R_1 C_1} \\[2ex] -\dfrac{1}{R_1 C_2} \end{bmatrix} u, \qquad y = \begin{bmatrix} 0 & 1 \end{bmatrix} x, \qquad (5.25)$$

where $x = (v_2, v_3)$, $u = v_1$ and $y = v_3$.

The frequency response for the system can be computed using equation (5.24):

$$Me^{j\theta} = C(sI - A)^{-1}B + D = -\frac{R_2}{R_1} \frac{R_1 C_1 s}{(1 + R_1 C_1 s)(1 + R_2 C_2 s)}, \qquad s = i\omega.$$

The magnitude and phase are plotted in Figure 5.12b for $R_1 = 100\ \Omega$, $R_2 = 5\ \text{k}\Omega$ and $C_1 = C_2 = 100\ \mu\text{F}$. We see that the circuit passes through signals with

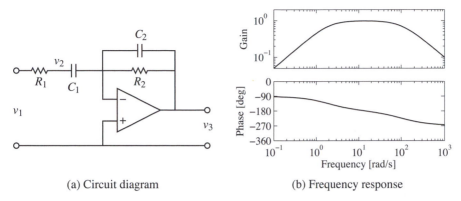

(a) Circuit diagram   (b) Frequency response

Figure 5.12: Active band-pass filter. The circuit diagram (a) shows an op amp with two RC filters arranged to provide a band-pass filter. The plot in (b) shows the gain and phase of the filter as a function of frequency. Note that the phase starts at -90° due to the negative gain of the operational amplifier.

frequencies at about 10 rad/s, but attenuates frequencies below 5 rad/s and above 50 rad/s. At 0.1 rad/s the input signal is attenuated by 20× (0.05). This type of circuit is called a *band-pass filter* since it passes through signals in the band of frequencies between 5 and 50 rad/s.   ▽

As in the case of the step response, a number of standard properties are defined for frequency responses. The gain of a system at $\omega = 0$ is called the *zero frequency gain* and corresponds to the ratio between a constant input and the steady output:

$$M_0 = -CA^{-1}B + D.$$

The zero frequency gain is well defined only if A is invertible (and, in particular, if it does not have eigenvalues at 0). It is also important to note that the zero frequency gain is a relevant quantity only when a system is stable about the corresponding equilibrium point. So, if we apply a constant input $u = r$, then the corresponding equilibrium point $x_e = -A^{-1}Br$ must be stable in order to talk about the zero frequency gain. (In electrical engineering, the zero frequency gain is often called the *DC gain*. DC stands for direct current and reflects the common separation of signals in electrical engineering into a direct current (zero frequency) term and an alternating current (AC) term.)

The *bandwidth* ω_b of a system is the frequency range over which the gain has decreased by no more than a factor of $1/\sqrt{2}$ from its reference value. For systems with nonzero, finite zero frequency gain, the bandwidth is the frequency where the gain has decreased by $1/\sqrt{2}$ from the zero frequency gain. For systems that attenuate low frequencies but pass through high frequencies, the reference gain is taken as the high-frequency gain. For a system such as the band-pass filter in Example 5.8, bandwidth is defined as the range of frequencies where the gain is larger than $1/\sqrt{2}$ of the gain at the center of the band. (For Example 5.8 this would give a bandwidth of approximately 50 rad/s.)

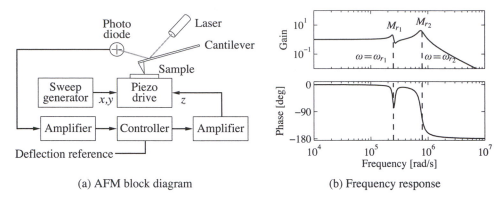

(a) AFM block diagram   (b) Frequency response

Figure 5.13: AFM frequency response. (a) A block diagram for the vertical dynamics of an atomic force microscope in contact mode. The plot in (b) shows the gain and phase for the piezo stack. The response contains two frequency peaks at resonances of the system, along with an antiresonance at $\omega = 268$ krad/s. The combination of a resonant peak followed by an antiresonance is common for systems with multiple lightly damped modes.

Another important property of the frequency response is the *resonant peak* M_r, the largest value of the frequency response, and the *peak frequency* ω_{mr}, the frequency where the maximum occurs. These two properties describe the frequency of the sinusoidal input that produces the largest possible output and the gain at the frequency.

Example 5.9 Atomic force microscope in contact mode
Consider the model for the vertical dynamics of the atomic force microscope in contact mode, discussed in Section 3.5. The basic dynamics are given by equation (3.23). The piezo stack can be modeled by a second-order system with undamped natural frequency ω_3 and damping ratio ζ_3. The dynamics are then described by the linear system

$$\frac{dx}{dt} = \begin{bmatrix} 0 & 1 & 0 & 0 \\ -k_2/(m_1+m_2) & -c_2/(m_1+m_2) & 1/m_2 & 0 \\ 0 & 0 & 0 & \omega_3 \\ 0 & 0 & -\omega_3 & -2\zeta_3\omega_3 \end{bmatrix} x + \begin{bmatrix} 0 \\ 0 \\ 0 \\ \omega_3 \end{bmatrix} u,$$

$$y = \frac{m_2}{m_1+m_2} \begin{bmatrix} \dfrac{m_1 k_2}{m_1+m_2} & \dfrac{m_1 c_2}{m_1+m_2} & 1 & 0 \end{bmatrix} x,$$

where the input signal is the drive signal to the amplifier and the output is the elongation of the piezo. The frequency response of the system is shown in Figure 5.13b. The zero frequency gain of the system is $M_0 = 1$. There are two resonant poles with peaks $M_{r1} = 2.12$ at $\omega_{mr1} = 238$ krad/s and $M_{r2} = 4.29$ at $\omega_{mr2} = 746$ krad/s. The bandwidth of the system, defined as the lowest frequency where the gain is $\sqrt{2}$ less than the zero frequency gain, is $\omega_b = 292$ krad/s. There is also a dip in the gain $M_d = 0.556$ for $\omega_{md} = 268$ krad/s. This dip, called an *antiresonance*, is associated with a dip in the phase and limits the performance when the system is controlled by simple controllers, as we will see in Chapter 10. ∇

Sampling

It is often convenient to use both differential and difference equations in modeling and control. For linear systems it is straightforward to transform from one to the other. Consider the general linear system described by equation (5.13) and assume that the control signal is constant over a sampling interval of constant length h. It follows from equation (5.14) of Theorem 5.4 that

$$x(t + h) = e^{Ah}x(t) + \int_t^{t+h} e^{A(t+h-\tau)} Bu(k)\, d\tau = \Phi x(t) + \Gamma u(t), \qquad (5.26)$$

where we have assumed that the discontinuous control signal is continuous from the right. The behavior of the system at the sampling times $t = kh$ is described by the difference equation

$$x[k + 1] = \Phi x[k] + \Gamma u[k], \qquad y[k] = Cx[k] + Du[k]. \qquad (5.27)$$

Notice that the difference equation (5.27) is an exact representation of the behavior of the system at the sampling instants. Similar expressions can also be obtained if the control signal is linear over the sampling interval.

The transformation from (5.26) to (5.27) is called *sampling*. The relations between the system matrices in the continuous and sampled representations are as follows:

$$\Phi = e^{Ah}, \quad \Gamma = \left(\int_0^h e^{As}\, ds\right) B; \qquad A = \frac{1}{h}\log \Phi, \quad B = \left(\int_0^h e^{As}\, ds\right)^{-1}\Gamma. \tag{5.28}$$

Notice that if A is invertible, we have

$$\Gamma = A^{-1}\left(e^{Ah} - I\right)B.$$

All continuous-time systems can be sampled to obtain a discrete-time version, but there are discrete-time systems that do not have a continuous-time equivalent. The precise condition is that the matrix Φ cannot have real eigenvalues on the negative real axis.

Example 5.10 IBM Lotus server

In Example 2.4 we described how the dynamics of an IBM Lotus server were obtained as the discrete-time system

$$y[k + 1] = ay[k] + bu[k],$$

where $a = 0.43$, $b = 0.47$ and the sampling period is $h = 60$ s. A differential equation model is needed if we would like to design control systems based on continuous-time theory. Such a model is obtained by applying equation (5.28); hence

$$A = \frac{\log a}{h} = -0.0141, \qquad B = \left(\int_0^h e^{At}\, dt\right)^{-1} b = 0.0116,$$

and we find that the difference equation can be interpreted as a sampled version of the ordinary differential equation

$$\frac{dx}{dt} = -0.0141x + 0.0116u.$$

$$\nabla$$

5.4 Linearization

As described at the beginning of the chapter, a common source of linear system models is through the approximation of a nonlinear system by a linear one. These approximations are aimed at studying the local behavior of a system, where the nonlinear effects are expected to be small. In this section we discuss how to locally approximate a system by its linearization and what can be said about the approximation in terms of stability. We begin with an illustration of the basic concept using the cruise control example from Chapter 3.

Example 5.11 Cruise control
The dynamics for the cruise control system were derived in Section 3.1 and have the form

$$m\frac{dv}{dt} = \alpha_n u T(\alpha_n v) - mgC_r \, \text{sgn}(v) - \frac{1}{2}\rho C_v A v^2 - mg\sin\theta, \qquad (5.29)$$

where the first term on the right-hand side of the equation is the force generated by the engine and the remaining three terms are the rolling friction, aerodynamic drag and gravitational disturbance force. There is an equilibrium (v_e, u_e) when the force applied by the engine balances the disturbance forces.

To explore the behavior of the system near the equilibrium we will linearize the system. A Taylor series expansion of equation (5.29) around the equilibrium gives

$$\frac{d(v - v_e)}{dt} = a(v - v_e) - b_g(\theta - \theta_e) + b(u - u_e) + \text{higher order terms}, \quad (5.30)$$

where

$$a = \frac{u_e \alpha_n^2 T'(\alpha_n v_e) - \rho C_v A v_e}{m}, \qquad b_g = g\cos\theta_e, \qquad b = \frac{\alpha_n T(\alpha_n v_e)}{m}. \quad (5.31)$$

Notice that the term corresponding to rolling friction disappears if $v \neq 0$. For a car in fourth gear with $v_e = 25$ m/s, $\theta_e = 0$ and the numerical values for the car from Section 3.1, the equilibrium value for the throttle is $u_e = 0.1687$ and the parameters are $a = -0.0101$, $b = 1.32$ and $c = 9.8$. This linear model describes how small perturbations in the velocity about the nominal speed evolve in time.

Figure 5.14 shows a simulation of a cruise controller with linear and nonlinear models; the differences between the linear and nonlinear models are small, and hence the linearized model provides a reasonable approximation. ∇

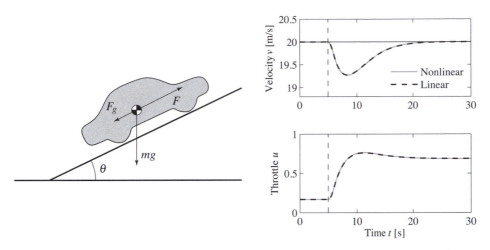

Figure 5.14: Simulated response of a vehicle with PI cruise control as it climbs a hill with a slope of 4°. The solid line is the simulation based on a nonlinear model, and the dashed line shows the corresponding simulation using a linear model. The controller gains are $k_p = 0.5$ and $k_i = 0.1$.

Jacobian Linearization Around an Equilibrium Point

To proceed more formally, consider a single-input, single-output nonlinear system

$$\frac{dx}{dt} = f(x, u), \qquad x \in \mathbb{R}^n, u \in \mathbb{R},$$

$$y = h(x, u), \qquad y \in \mathbb{R}, \tag{5.32}$$

with an equilibrium point at $x = x_e, u = u_e$. Without loss of generality we can assume that $x_e = 0$ and $u_e = 0$, although initially we will consider the general case to make the shift of coordinates explicit.

To study the *local* behavior of the system around the equilibrium point (x_e, u_e), we suppose that $x - x_e$ and $u - u_e$ are both small, so that nonlinear perturbations around this equilibrium point can be ignored compared with the (lower-order) linear terms. This is roughly the same type of argument that is used when we do small-angle approximations, replacing $\sin \theta$ with θ and $\cos \theta$ with 1 for θ near zero.

As we did in Chapter 4, we define a new set of state variables z, as well as inputs v and outputs w:

$$z = x - x_e, \qquad v = u - u_e, \qquad w = y - h(x_e, u_e).$$

These variables are all close to zero when we are near the equilibrium point, and so in these variables the nonlinear terms can be thought of as the higher-order terms in a Taylor series expansion of the relevant vector fields (assuming for now that these exist).

Formally, the *Jacobian linearization* of the nonlinear system (5.32) is

$$\frac{dz}{dt} = Az + Bv, \qquad w = Cz + Dv, \tag{5.33}$$

where

$$A = \frac{\partial f}{\partial x}\bigg|_{(x_e, u_e)}, \quad B = \frac{\partial f}{\partial u}\bigg|_{(x_e, u_e)}, \quad C = \frac{\partial h}{\partial x}\bigg|_{(x_e, u_e)}, \quad D = \frac{\partial h}{\partial u}\bigg|_{(x_e, u_e)}. \quad (5.34)$$

The system (5.33) approximates the original system (5.32) when we are near the equilibrium point about which the system was linearized. Using Theorem 4.3, if the linearization is asymptotically stable, then the equilibrium point x_e is locally asymptotically stable for the full nonlinear system.

It is important to note that we can define the linearization of a system only near an equilibrium point. To see this, consider a polynomial system

$$\frac{dx}{dt} = a_0 + a_1 x + a_2 x^2 + a_3 x^3 + u,$$

where $a_0 \neq 0$. A set of equilibrium points for this system is given by $(x_e, u_e) = (x_e, -a_0 - a_1 x_e - a_2 x_e^2 - a_3 x_e^3)$, and we can linearize around any of them. Suppose that we try to linearize around the origin of the system $x = 0, u = 0$. If we drop the higher-order terms in x, then we get

$$\frac{dx}{dt} = a_0 + a_1 x + u,$$

which is *not* the Jacobian linearization if $a_0 \neq 0$. The constant term must be kept, and it is not present in (5.33). Furthermore, even if we kept the constant term in the approximate model, the system would quickly move away from this point (since it is "driven" by the constant term a_0), and hence the approximation could soon fail to hold.

Software for modeling and simulation frequently has facilities for performing linearization symbolically or numerically. The MATLAB command `trim` finds the equilibrium, and `linmod` extracts linear state space models from a SIMULINK system around an operating point.

Example 5.12 Vehicle steering
Consider the vehicle steering system introduced in Example 2.8. The nonlinear equations of motion for the system are given by equations (2.23)–(2.25) and can be written as

$$\frac{d}{dt}\begin{bmatrix} x \\ y \\ \theta \end{bmatrix} = \begin{bmatrix} v \cos(\alpha(\delta) + \theta) \\ v \sin(\alpha(\delta) + \theta) \\ \frac{v_0}{b} \tan\delta \end{bmatrix}, \quad \alpha(\delta) = \arctan\left(\frac{a \tan\delta}{b}\right),$$

where x, y and θ are the position and orientation of the center of mass of the vehicle, v_0 is the velocity of the rear wheel, b is the distance between the front and rear wheels and δ is the angle of the front wheel. The function $\alpha(\delta)$ is the angle between the velocity vector and the vehicle's length axis.

We are interested in the motion of the vehicle about a straight-line path ($\theta = \theta_0$) with fixed velocity $v_0 \neq 0$. To find the relevant equilibrium point, we first set $\dot\theta = 0$ and we see that we must have $\delta = 0$, corresponding to the steering wheel being straight. This also yields $\alpha = 0$. Looking at the first two equations in the dynamics, we see that the motion in the xy direction is by definition *not* at equilibrium since $\dot{\xi}^2 + \dot{\eta}^2 = v_0^2 \neq 0$. Therefore we cannot formally linearize the full model.

Suppose instead that we are concerned with the lateral deviation of the vehicle from a straight line. For simplicity, we let $\theta_e = 0$, which corresponds to driving along the x axis. We can then focus on the equations of motion in the y and θ directions. With some abuse of notation we introduce the state $x = (y, \theta)$ and $u = \delta$. The system is then in standard form with

$$
f(x, u) = \begin{bmatrix} v \sin(\alpha(u) + x_2) \\ \dfrac{v_0}{b} \tan u \end{bmatrix}, \quad \alpha(u) = \arctan\left(\dfrac{a \tan u}{b}\right), \quad h(x, u) = x_1.
$$

The equilibrium point of interest is given by $x = (0, 0)$ and $u = 0$. To compute the linearized model around this equilibrium point, we make use of the formulas (5.34). A straightforward calculation yields

$$
A = \left.\frac{\partial f}{\partial x}\right|_{\substack{x=0 \\ u=0}} = \begin{bmatrix} 0 & v_0 \\ 0 & 0 \end{bmatrix}, \qquad B = \left.\frac{\partial f}{\partial u}\right|_{\substack{x=0 \\ u=0}} = \begin{bmatrix} av_0/b \\ v_0/b \end{bmatrix},
$$

$$
C = \left.\frac{\partial h}{\partial x}\right|_{\substack{x=0 \\ u=0}} = \begin{bmatrix} 1 & 0 \end{bmatrix}, \qquad D = \left.\frac{\partial h}{\partial u}\right|_{\substack{x=0 \\ u=0}} = 0,
$$

and the linearized system

$$
\frac{dx}{dt} = Ax + Bu, \qquad y = Cx + Du \tag{5.35}
$$

thus provides an approximation to the original nonlinear dynamics.

The linearized model can be simplified further by introducing normalized variables, as discussed in Section 2.3. For this system, we choose the wheel base b as the length unit and the unit as the time required to travel a wheel base. The normalized state is thus $z = (x_1/b, x_2)$, and the new time variable is $\tau = v_0 t/b$. The model (5.35) then becomes

$$
\frac{dz}{d\tau} = \begin{bmatrix} z_2 + \gamma u \\ u \end{bmatrix} = \begin{bmatrix} 0 & 1 \\ 0 & 0 \end{bmatrix} z + \begin{bmatrix} \gamma \\ 1 \end{bmatrix} u, \qquad y = \begin{bmatrix} 1 & 0 \end{bmatrix} z, \tag{5.36}
$$

where $\gamma = a/b$. The normalized linear model for vehicle steering with nonslipping wheels is thus a linear system with only one parameter. ∇

Feedback Linearization

Another type of linearization is the use of feedback to convert the dynamics of a nonlinear system into those of a linear one. We illustrate the basic idea with an example.

Example 5.13 Cruise control
Consider again the cruise control system from Example 5.11, whose dynamics are

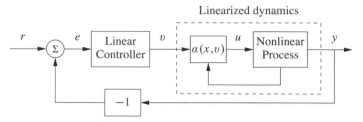

Figure 5.15: Feedback linearization. A nonlinear feedback of the form $u = \alpha(x, v)$ is used to modify the dynamics of a nonlinear process so that the response from the input v to the output y is linear. A linear controller can then be used to regulate the system's dynamics.

given in equation (5.29):

$$m\frac{dv}{dt} = \alpha_n u T(\alpha_n v) - mgC_r \, \text{sgn}(v) - \frac{1}{2}\rho C_d A v^2 - mg \sin \theta.$$

If we choose u as a feedback law of the form

$$u = \frac{1}{\alpha_n T(\alpha_n v)}\left(u' + mgC_r \, \text{sgn}(v) + \frac{1}{2}\rho C_v A v^2\right), \tag{5.37}$$

then the resulting dynamics become

$$m\frac{dv}{dt} = u' + d, \tag{5.38}$$

where $d = -mg \sin \theta$ is the disturbance force due the slope of the road. If we now define a feedback law for u' (such as a proportional-integral-derivative [PID] controller), we can use equation (5.37) to compute the final input that should be commanded.

Equation (5.38) is a linear differential equation. We have essentially "inverted" the nonlinearity through the use of the feedback law (5.37). This requires that we have an accurate measurement of the vehicle velocity v as well as an accurate model of the torque characteristics of the engine, gear ratios, drag and friction characteristics and mass of the car. While such a model is not generally available (remembering that the parameter values can change), if we design a good feedback law for u', then we can achieve robustness to these uncertainties. ∇

More generally, we say that a system of the form

$$\frac{dx}{dt} = f(x, u), \qquad y = h(x),$$

is *feedback linearizable* if we can find a control law $u = \alpha(x, v)$ such that the resulting closed loop system is input/output linear with input v and output y, as shown in Figure 5.15. To fully characterize such systems is beyond the scope of this text, but we note that in addition to changes in the input, the general theory also allows for (nonlinear) changes in the states that are used to describe the system, keeping only the input and output variables fixed. More details of this process can be found in the textbooks by Isidori [106] and Khalil [123].

One case that comes up relatively frequently, and is hence worth special mention,
is the set of mechanical systems of the form

$$M(q)\ddot{q} + C(q, \dot{q}) = B(q)u.$$

Here $q \in \mathbb{R}^n$ is the configuration of the mechanical system, $M(q) \in \mathbb{R}^{n \times n}$ is
the configuration-dependent inertia matrix, $C(q, \dot{q}) \in \mathbb{R}^n$ represents the Coriolis
forces and additional nonlinear forces (such as stiffness and friction) and $B(q) \in$
$\mathbb{R}^{n \times p}$ is the input matrix. If $p = n$, then we have the same number of inputs and
configuration variables, and if we further have that $B(q)$ is an invertible matrix for
all configurations q, then we can choose

$$u = B^{-1}(q)\big(M(q)v - C(q, \dot{q})\big). \tag{5.39}$$

The resulting dynamics become

$$M(q)\ddot{q} = M(q)v \qquad \Longrightarrow \qquad \ddot{q} = v,$$

which is a linear system. We can now use the tools of linear system theory to
analyze and design control laws for the linearized system, remembering to apply
equation (5.39) to obtain the actual input that will be applied to the system.

This type of control is common in robotics, where it goes by the name of
computed torque, and in aircraft flight control, where it is called *dynamic inversion*. Some modeling tools like Modelica can generate the code for the inverse
model automatically. One caution is that feedback linearization can often cancel
out beneficial terms in the natural dynamics, and hence it must be used with care.
Extensions that do not require complete cancellation of nonlinearities are discussed
in Khalil [123] and Krstić et al. [129].

5.5 Further Reading

The majority of the material in this chapter is classical and can be found in most
books on dynamics and control theory, including early works on control such as
James, Nichols and Phillips [110] and more recent textbooks such as Dorf and
Bishop [61], Franklin, Powell and Emami-Naeini [79] and Ogata [162]. An excellent presentation of linear systems based on the matrix exponential is given in the
book by Brockett [44], a more comprehensive treatment is given by Rugh [171] and
an elegant mathematical treatment is given in Sontag [182]. Material on feedback
linearization can be found in books on nonlinear control theory such as Isidori [106]
and Khalil [123]. The idea of characterizing dynamics by considering the responses
to step inputs is due to Heaviside, he also introduced an operator calculus to analyze
linear systems. The unit step is therefore also called the *Heaviside step function*.
Analysis of linear systems was simplified significantly, but Heaviside's work was
heavily criticized because of lack of mathematical rigor, as described in the biog-
raphy by Nahin [157]. The difficulties were cleared up later by the mathematician
Laurent Schwartz who developed *distribution theory* in the late 1940s. In engineering, linear systems have traditionally been analyzed using Laplace transforms as

described in Gardner and Barnes [81]. Use of the matrix exponential started with developments of control theory in the 1960s, strongly stimulated by a textbook by Zadeh and Desoer [207]. Use of matrix techniques expanded rapidly when the powerful methods of numeric linear algebra were packaged in programs like LabVIEW, MATLAB and Mathematica.

Exercises

5.1 (Response to the derivative of a signal) Show that if $y(t)$ is the output of a linear system corresponding to input $u(t)$, then the output corresponding to an input $\dot{u}(t)$ is given by $\dot{y}(t)$. (Hint: Use the definition of the derivative: $\dot{y}(t) = \lim_{\epsilon \to 0} \big(y(t + \epsilon) - y(t)\big)/\epsilon$.)

 5.2 (Impulse response and convolution) Show that a signal $u(t)$ can be decomposed in terms of the impulse function $\delta(t)$ as

$$u(t) = \int_0^t \delta(t - \tau)u(\tau)\, d\tau$$

and use this decomposition plus the principle of superposition to show that the response of a linear system to an input $u(t)$ (assuming a zero initial condition) can be written as

$$y(t) = \int_0^t h(t - \tau)u(\tau)\, d\tau,$$

where $h(t)$ is the impulse response of the system.

5.3 (Pulse response for a compartment model) Consider the compartment model given in Example 5.7. Compute the step response for the system and compare it with Figure 5.10b. Use the principle of superposition to compute the response to the 5 s pulse input shown in Figure 5.10c. Use the parameter values $k_0 = 0.1$, $k_1 = 0.1, k_2 = 0.5$ and $b_0 = 1.5$.

5.4 (Matrix exponential for second-order system) Assume that $\zeta < 1$ and let $\omega_d = \omega_0\sqrt{1 - \zeta^2}$. Show that

$$\exp \begin{bmatrix} -\zeta\omega_0 & \omega_d \\ -\omega_d & -\zeta\omega_0 \end{bmatrix} t = \begin{bmatrix} e^{-\zeta\omega_0 t}\cos\omega_d t & e^{-\zeta\omega_0 t}\sin\omega_d t \\ -e^{-\zeta\omega_0 t}\sin\omega_d t & e^{-\zeta\omega_0 t}\cos\omega_d t \end{bmatrix}.$$

5.5 (Lyapunov function for a linear system) Consider a linear system $\dot{x} = Ax$ with $\operatorname{Re}\lambda_j < 0$ for all eigenvalues λ_j of the matrix A. Show that the matrix

$$P = \int_0^\infty e^{A^T \tau} Q e^{A\tau}\, d\tau$$

defines a Lyapunov function of the form $V(x) = x^T P x$.

5.6 (Nondiagonal Jordan form) Consider a linear system with a Jordan form that is non-diagonal.

(a) Prove Proposition 5.3 by showing that if the system contains a real eigenvalue $\lambda = 0$ with a nontrivial Jordan block, then there exists an initial condition with a solution that grows in time.

(b) Extend this argument to the case of complex eigenvalues with Re $\lambda = 0$ by using the block Jordan form

$$J_i = \begin{bmatrix} 0 & \omega & 1 & 0 \\ -\omega & 0 & 0 & 1 \\ 0 & 0 & 0 & \omega \\ 0 & 0 & -\omega & 0 \end{bmatrix}.$$

5.7 (Rise time for a first-order system) Consider a first-order system of the form

$$\tau \frac{dx}{dt} = -x + u, \qquad y = x.$$

We say that the parameter τ is the *time constant* for the system since the zero input system approaches the origin as $e^{-t/\tau}$. For a first-order system of this form, show that the rise time for a step response of the system is approximately 2τ, and that $1\%, 2\%$, and 5% settling times approximately corresponds to $4.6\tau, 4\tau$ and 3τ.

5.8 (Discrete-time systems) Consider a linear discrete-time system of the form

$$x[k+1] = Ax[k] + Bu[k], \qquad y[k] = Cx[k] + Du[k].$$

(a) Show that the general form of the output of a discrete-time linear system is given by the discrete-time convolution equation:

$$y[k] = C A^k x[0] + \sum_{j=0}^{k-1} C A^{k-j-1} Bu[j] + Du[k].$$

(b) Show that a discrete-time linear system is asymptotically stable if and only if all the eigenvalues of A have a magnitude strictly less than 1.

(c) Let $u[k] = \sin(\omega k)$ represent an oscillatory input with frequency $\omega < \pi$ (to avoid "aliasing"). Show that the steady-state component of the response has gain M and phase θ, where

$$Me^{i\theta} = C(e^{i\omega}I - A)^{-1}B + D.$$

(d) Show that if we have a nonlinear discrete-time system

$$x[k] = f(x[k], u[k]), \qquad x[k] \in \mathbb{R}^n, \ u \in \mathbb{R},$$
$$y[k] = h(x[k], u[k]), \qquad y \in \mathbb{R},$$

then we can linearize the system around an equilibrium point (x_e, u_e) by defining the matrices A, B, C and D as in equation (5.34).

5.9 (Keynesian economics) Consider the following simple Keynesian macroeco-
nomic model in the form of a linear discrete-time system discussed in Exercise 5.8:

$$
\begin{bmatrix} C[t+1] \\ I[t+1] \end{bmatrix} = \begin{bmatrix} a & a \\ ab-b & ab \end{bmatrix} \begin{bmatrix} C[t] \\ I[t] \end{bmatrix} + \begin{bmatrix} a \\ ab \end{bmatrix} G[t],
$$

$$
Y[t] = C[t] + I[t] + G[t].
$$

Determine the eigenvalues of the dynamics matrix. When are the magnitudes of
the eigenvalues less than 1? Assume that the system is in equilibrium with constant
values capital spending C, investment I and government expenditure G. Explore
what happens when government expenditure increases by 10%. Use the values
$a = 0.25$ and $b = 0.5$.

5.10 Consider a scalar system

$$
\frac{dx}{dt} = 1 - x^3 + u.
$$

Compute the equilibrium points for the unforced system ($u = 0$) and use a Taylor
series expansion around the equilibrium point to compute the linearization. Verify
that this agrees with the linearization in equation (5.33).

5.11 (Transcriptional regulation) Consider the dynamics of a genetic circuit that implements *self-repression*: the protein produced by a gene is a repressor for that gene,
thus restricting its own production. Using the models presented in Example 2.13,
the dynamics for the system can be written as

$$
\frac{dm}{dt} = \frac{\alpha}{1 + kp^2} + \alpha_0 - \gamma m - u, \qquad \frac{dp}{dt} = \beta m - \delta p, \qquad (5.40)
$$

where u is a disturbance term that affects RNA transcription and $m, p \geq 0$. Find
the equilibrium points for the system and use the linearized dynamics around each
equilibrium point to determine the local stability of the equilibrium point and the
step response of the system to a disturbance.

Chapter Six
State Feedback

Intuitively, the state may be regarded as a kind of information storage or memory or accumulation of past causes. We must, of course, demand that the set of internal states Σ be sufficiently rich to carry all information about the past history of Σ to predict the effect of the past upon the future. We do not insist, however, that the state is the least *such information although this is often a convenient assumption.*

R. E. Kalman, P. L. Falb and M. A. Arbib, *Topics in Mathematical System Theory*, 1969 [117].

This chapter describes how the feedback of a system's state can be used to shape the local behavior of a system. The concept of reachability is introduced and used to investigate how to design the dynamics of a system through assignment of its eigenvalues. In particular, it will be shown that under certain conditions it is possible to assign the system eigenvalues arbitrarily by appropriate feedback of the system state.

6.1 Reachability

One of the fundamental properties of a control system is what set of points in the state space can be reached through the choice of a control input. It turns out that the property of reachability is also fundamental in understanding the extent to which feedback can be used to design the dynamics of a system.

Definition of Reachability

We begin by disregarding the output measurements of the system and focusing on the evolution of the state, given by

$$\frac{dx}{dt} = Ax + Bu, \qquad (6.1)$$

where $x \in \mathbb{R}^n$, $u \in \mathbb{R}$, A is an $n \times n$ matrix and B a column vector. A fundamental question is whether it is possible to find control signals so that any point in the state space can be reached through some choice of input. To study this, we define the *reachable set* $\mathcal{R}(x_0, \leq T)$ as the set of all points x_f such that there exists an input $u(t), 0 \leq t \leq T$ that steers the system from $x(0) = x_0$ to $x(T) = x_f$, as illustrated in Figure 6.1a.

Definition 6.1 (Reachability). A linear system is *reachable* if for any $x_0, x_f \in \mathbb{R}^n$ there exists a $T > 0$ and $u : [0, T] \to \mathbb{R}$ such that the corresponding solution satisfies $x(0) = x_0$ and $x(T) = x_f$.

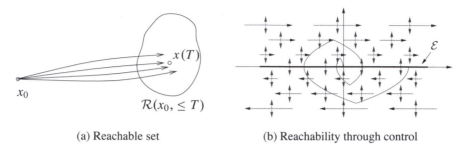

(a) Reachable set   (b) Reachability through control

Figure 6.1: The reachable set for a control system. The set $\mathcal{R}(x_0, \leq T)$ shown in (a) is the set of points reachable from x_0 in time less than T. The phase portrait in (b) shows the dynamics for a double integrator, with the natural dynamics drawn as horizontal arrows and the control inputs drawn as vertical arrows. The set of achievable equilibrium points is the x axis. By setting the control inputs as a function of the state, it is possible to steer the system to the origin, as shown on the sample path.

The definition of reachability addresses whether it is possible to reach all points in the state space in a *transient* fashion. In many applications, the set of points that we are most interested in reaching is the set of equilibrium points of the system (since we can remain at those points once we get there). The set of all possible equilibria for constant controls is given by

$$\mathcal{E} = \{x_e : Ax_e + Bu_e = 0 \text{ for some } u_e \in \mathbb{R}\}.$$

This means that possible equilibria lie in a one- (or possibly higher) dimensional subspace. If the matrix A is invertible, this subspace is spanned by $A^{-1}B$.

The following example provides some insight into the possibilities.

Example 6.1 Double integrator
Consider a linear system consisting of a double integrator whose dynamics are given by

$$\frac{dx_1}{dt} = x_2, \qquad \frac{dx_2}{dt} = u.$$

Figure 6.1b shows a phase portrait of the system. The open loop dynamics ($u = 0$) are shown as horizontal arrows pointed to the right for $x_2 > 0$ and to the left for $x_2 < 0$. The control input is represented by a double-headed arrow in the vertical direction, corresponding to our ability to set the value of $\dot{x}_2$. The set of equilibrium points $\mathcal{E}$ corresponds to the x_1 axis, with $u_e = 0$.

Suppose first that we wish to reach the origin from an initial condition $(a, 0)$. We can directly move the state up and down in the phase plane, but we must rely on the natural dynamics to control the motion to the left and right. If $a > 0$, we can move the origin by first setting $u < 0$, which will cause x_2 to become negative. Once $x_2 < 0$, the value of x_1 will begin to decrease and we will move to the left. After a while, we can set u_2 to be positive, moving x_2 back toward zero and slowing the motion in the x_1 direction. If we bring $x_2 > 0$, we can move the system state in the opposite direction.

Figure 6.1b shows a sample trajectory bringing the system to the origin. Note that if we steer the system to an equilibrium point, it is possible to remain there indefinitely (since $\dot{x}_1 = 0$ when $x_2 = 0$), but if we go to any other point in the state space, we can pass through the point only in a transient fashion. ∇

To find general conditions under which a linear system is reachable, we will first give a heuristic argument based on formal calculations with impulse functions. We note that if we can reach all points in the state space through some choice of input, then we can also reach all equilibrium points.

Testing for Reachability

When the initial state is zero, the response of the system to an input $u(t)$ is given by

$$x(t) = \int_0^t e^{A(t-\tau)} B u(\tau) \, d\tau. \tag{6.2}$$

If we choose the input to be a impulse function $\delta(t)$ as defined in Section 5.3, the state becomes

$$x_\delta = \int_0^t e^{A(t-\tau)} B \delta(\tau) \, d\tau = \frac{dx_S}{dt} = e^{At} B.$$

(Note that the state changes instantaneously in response to the impulse.) We can find the response to the derivative of an impulse function by taking the derivative of the impulse response (Exercise 5.1):

$$x_{\dot{\delta}} = \frac{dx_\delta}{dt} = A e^{At} B.$$

Continuing this process and using the linearity of the system, the input

$$u(t) = \alpha_1 \delta(t) + \alpha_2 \dot{\delta}(t) + \alpha_3 \ddot{\delta}(t) + \cdots + \alpha_n \delta^{(n-1)}(t)$$

gives the state

$$x(t) = \alpha_1 e^{At} B + \alpha_2 A e^{At} B + \alpha_3 A^2 e^{At} B + \cdots + \alpha_n A^{n-1} e^{At} B.$$

Taking the limit as t goes to zero through positive values, we get

$$\lim_{t \to 0+} x(t) = \alpha_1 B + \alpha_2 AB + \alpha_3 A^2 B + \cdots + \alpha_n A^{n-1} B.$$

On the right is a linear combination of the columns of the matrix

$$W_r = \begin{bmatrix} B & AB & \cdots & A^{n-1}B \end{bmatrix}. \tag{6.3}$$

To reach an arbitrary point in the state space, we thus require that there are n linear independent columns of the matrix W_r. The matrix W_r is called the *reachability matrix*.

An input consisting of a sum of impulse functions and their derivatives is a very violent signal. To see that an arbitrary point can be reached with smoother signals we can make use of the convolution equation. Assuming that the initial condition is zero, the state of a linear system is given by

$$x(t) = \int_0^t e^{A(t-\tau)} Bu(\tau)d\tau = \int_0^t e^{A\tau} Bu(t-\tau)d\tau.$$

It follows from the theory of matrix functions, specifically the Cayley–Hamilton theorem (see Exercise 6.10), that

$$e^{A\tau} = I\alpha_0(\tau) + A\alpha_1(\tau) + \cdots + A^{n-1}\alpha_{n-1}(\tau),$$

where $\alpha_i(\tau)$ are scalar functions, and we find that

$$x(t) = B \int_0^t \alpha_0(\tau)u(t-\tau)\,d\tau + AB \int_0^t \alpha_1(\tau)u(t-\tau)\,d\tau$$

$$+ \cdots + A^{n-1}B \int_0^t \alpha_{n-1}(\tau)u(t-\tau)\,d\tau.$$

Again we observe that the right-hand side is a linear combination of the columns of the reachability matrix W_r given by equation (6.3). This basic approach leads to the following theorem.

Theorem 6.1 (Reachability rank condition). *A linear system is reachable if and only if the reachability matrix W_r is invertible.*

The formal proof of this theorem is beyond the scope of this text but follows along the lines of the sketch above and can be found in most books on linear control theory, such as Callier and Desoer [48] or Lewis [136]. We illustrate the concept of reachability with the following example.

Example 6.2 Balance system
Consider the balance system introduced in Example 2.1 and shown in Figure 6.2. Recall that this system is a model for a class of examples in which the center of mass is balanced above a pivot point. One example is the Segway Personal Transporter shown in Figure 6.2a, about which a natural question to ask is whether we can move from one stationary point to another by appropriate application of forces through the wheels.

The nonlinear equations of motion for the system are given in equation (2.9) and repeated here:

$$
\begin{aligned}
(M+m)\ddot{p} - ml\cos\theta\,\ddot{\theta} &= -c\dot{p} - ml\sin\theta\,\dot{\theta}^2 + F, \\
(J+ml^2)\ddot{\theta} - ml\cos\theta\,\ddot{p} &= -\gamma\dot{\theta} + mgl\sin\theta.
\end{aligned}
\tag{6.4}
$$

For simplicity, we take $c = \gamma = 0$. Linearizing around the equilibrium point $x_e = (p, 0, 0, 0)$, the dynamics matrix and the control matrix are

$$
A = \begin{bmatrix} 0 & 0 & 1 & 0 \\ 0 & 0 & 0 & 1 \\ 0 & m^2l^2g/\mu & 0 & 0 \\ 0 & M_t mgl/\mu & 0 & 0 \end{bmatrix}, \qquad
B = \begin{bmatrix} 0 \\ 0 \\ J_t/\mu \\ lm/\mu \end{bmatrix},
$$

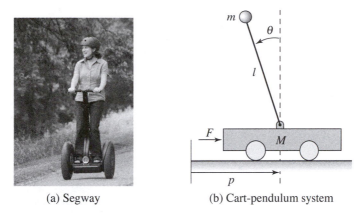

(a) Segway (b) Cart-pendulum system

Figure 6.2: Balance system. The Segway Personal Transporter shown in (a) is an example of a balance system that uses torque applied to the wheels to keep the rider upright. A simplified diagram for a balance system is shown in (b). The system consists of a mass m on a rod of length l connected by a pivot to a cart with mass M.

where $\mu = M_t J_t - m^2 l^2$, $M_t = M + m$ and $J_t = J + ml^2$. The reachability matrix is

$$
W_r = \begin{bmatrix}
0 & J_t/\mu & 0 & gl^3 m^3/\mu^2 \\
0 & lm/\mu & 0 & gl^2 m^2 (m + M)/\mu^2 \\
J_t/\mu & 0 & gl^3 m^3/\mu^2 & 0 \\
lm/\mu & 0 & gl^2 m^2 (m + M)/\mu^2 & 0
\end{bmatrix}.
\tag{6.5}
$$

The determinant of this matrix is

$$
\det(W_r) = \frac{g^2 l^4 m^4}{(\mu)^4} \neq 0,
$$

and we can conclude that the system is reachable. This implies that we can move the system from any initial state to any final state and, in particular, that we can always find an input to bring the system from an initial state to an equilibrium point.

$$\nabla$$

It is useful to have an intuitive understanding of the mechanisms that make a system unreachable. An example of such a system is given in Figure 6.3. The system consists of two identical systems with the same input. Clearly, we cannot separately cause the first and the second systems to do something different since they have the same input. Hence we cannot reach arbitrary states, and so the system is not reachable (Exercise 6.3).

More subtle mechanisms for nonreachability can also occur. For example, if there is a linear combination of states that always remains constant, then the system is not reachable. To see this, suppose that there exists a row vector H such that

$$
0 = \frac{d}{dt} Hx = H(Ax + Bu), \quad \text{for all } u.
$$

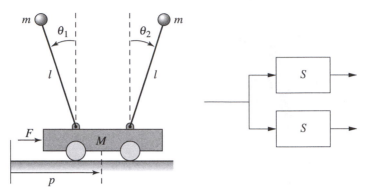

Figure 6.3: An unreachable system. The cart–pendulum system shown on the left has a single input that affects two pendula of equal length and mass. Since the forces affecting the two pendula are the same and their dynamics are identical, it is not possible to arbitrarily control the state of the system. The figure on the right is a block diagram representation of this situation.

Then H is in the left null space of both A and B and it follows that

$$HW_r = H \begin{bmatrix} B & AB & \cdots & A^{n-1}B \end{bmatrix} = 0.$$

Hence the reachability matrix is not full rank. In this case, if we have an initial condition x_0 and we wish to reach a state x_f for which $Hx_0 \neq Hx_f$, then since $Hx(t)$ is constant, no input u can move from x_0 to x_f.

Reachable Canonical Form

As we have already seen in previous chapters, it is often convenient to change coordinates and write the dynamics of the system in the transformed coordinates $z = Tx$. One application of a change of coordinates is to convert a system into a canonical form in which it is easy to perform certain types of analysis.

A linear state space system is in *reachable canonical form* if its dynamics are given by

$$\frac{dz}{dt} = \begin{bmatrix} -a_1 & -a_2 & -a_3 & \cdots & -a_n \\ 1 & 0 & 0 & \cdots & 0 \\ 0 & 1 & 0 & \cdots & 0 \\ \vdots & & \ddots & \ddots & \vdots \\ 0 & & & 1 & 0 \end{bmatrix} z + \begin{bmatrix} 1 \\ 0 \\ 0 \\ \vdots \\ 0 \end{bmatrix} u, \tag{6.6}$$

$$y = \begin{bmatrix} b_1 & b_2 & b_3 & \cdots & b_n \end{bmatrix} z + du.$$

A block diagram for a system in reachable canonical form is shown in Figure 6.4. We see that the coefficients that appear in the A and B matrices show up directly in the block diagram. Furthermore, the output of the system is a simple linear combination of the outputs of the integration blocks.

The characteristic polynomial for a system in reachable canonical form is given

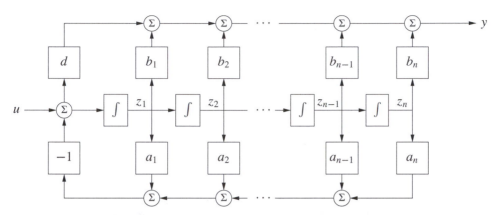

Figure 6.4: Block diagram for a system in reachable canonical form. The individual states of the system are represented by a chain of integrators whose input depends on the weighted values of the states. The output is given by an appropriate combination of the system input and other states.

by

$$\lambda(s) = s^n + a_1 s^{n-1} + \cdots + a_{n-1}s + a_n. \tag{6.7}$$

The reachability matrix also has a relatively simple structure:

$$W_r = \begin{bmatrix} B & AB & \cdots & A^{n-1}B \end{bmatrix} = \begin{bmatrix} 1 & -a_1 & a_1^2 - a_2 & \cdots & * \\ 0 & 1 & -a_1 & \cdots & * \\ \vdots & \vdots & \ddots & \ddots & \vdots \\ 0 & 0 & 0 & 1 & * \\ 0 & 0 & 0 & \cdots & 1 \end{bmatrix},$$

where $*$ indicates a possibly nonzero term. This matrix is full rank since no column can be written as a linear combination of the others because of the triangular structure of the matrix.

We now consider the problem of changing coordinates such that the dynamics of a system can be written in reachable canonical form. Let A, B represent the dynamics of a given system and $\tilde{A}$, $\tilde{B}$ be the dynamics in reachable canonical form. Suppose that we wish to transform the original system into reachable canonical form using a coordinate transformation $z = Tx$. As shown in the last chapter, the dynamics matrix and the control matrix for the transformed system are

$$\tilde{A} = TAT^{-1}, \qquad \tilde{B} = TB.$$

The reachability matrix for the transformed system then becomes

$$\tilde{W}_r = \begin{bmatrix} \tilde{B} & \tilde{A}\tilde{B} & \cdots & \tilde{A}^{n-1}\tilde{B} \end{bmatrix}.$$

Transforming each element individually, we have

$$\tilde{A}\tilde{B} = TAT^{-1}TB = TAB,$$
$$\tilde{A}^2\tilde{B} = (TAT^{-1})^2TB = TAT^{-1}TAT^{-1}TB = TA^2B,$$
$$\vdots$$
$$\tilde{A}^n\tilde{B} = TA^nB,$$

and hence the reachability matrix for the transformed system is

$$\tilde{W}_r = T \begin{bmatrix} B & AB & \cdots & A^{n-1}B \end{bmatrix} = TW_r. \tag{6.8}$$

Since W_r is invertible, we can thus solve for the transformation T that takes the system into reachable canonical form:

$$T = \tilde{W}_r W_r^{-1}.$$

The following example illustrates the approach.

Example 6.3 Transformation to reachable form
Consider a simple two-dimensional system of the form

$$\frac{dx}{dt} = \begin{bmatrix} \alpha & \omega \\ -\omega & \alpha \end{bmatrix} x + \begin{bmatrix} 0 \\ 1 \end{bmatrix} u.$$

We wish to find the transformation that converts the system into reachable canonical form:

$$\tilde{A} = \begin{bmatrix} -a_1 & -a_2 \\ 1 & 0 \end{bmatrix}, \qquad \tilde{B} = \begin{bmatrix} 1 \\ 0 \end{bmatrix}.$$

The coefficients a_1 and a_2 can be determined from the characteristic polynomial for the original system:

$$\lambda(s) = \det(sI - A) = s^2 - 2\alpha s + (\alpha^2 + \omega^2) \quad \Longrightarrow \quad \begin{aligned} a_1 &= -2\alpha, \\ a_2 &= \alpha^2 + \omega^2. \end{aligned}$$

The reachability matrix for each system is

$$W_r = \begin{bmatrix} 0 & \omega \\ 1 & \alpha \end{bmatrix}, \qquad \tilde{W}_r = \begin{bmatrix} 1 & -a_1 \\ 0 & 1 \end{bmatrix}.$$

The transformation T becomes

$$T = \tilde{W}_r W_r^{-1} = \begin{bmatrix} -(a_1 + \alpha)/\omega & 1 \\ 1/\omega & 0 \end{bmatrix} = \begin{bmatrix} \alpha/\omega & 1 \\ 1/\omega & 0 \end{bmatrix},$$

and hence the coordinates

$$\begin{bmatrix} z_1 \\ z_2 \end{bmatrix} = Tx = \begin{bmatrix} \alpha x_1/\omega + x_2 \\ x_1/\omega \end{bmatrix}$$

put the system in reachable canonical form. ∇

We summarize the results of this section in the following theorem.

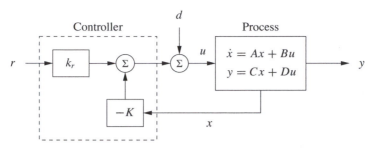

Figure 6.5: A feedback control system with state feedback. The controller uses the system state x and the reference input r to command the process through its input u. We model disturbances via the additive input d.

Theorem 6.2 (Reachable canonical form). *Let A and B be the dynamics and control matrices for a reachable system. Then there exists a transformation $z = Tx$ such that in the transformed coordinates the dynamics and control matrices are in reachable canonical form (6.6) and the characteristic polynomial for A is given by*

$$\det(sI - A) = s^n + a_1 s^{n-1} + \cdots + a_{n-1}s + a_n.$$

One important implication of this theorem is that for any reachable system, we can assume without loss of generality that the coordinates are chosen such that the system is in reachable canonical form. This is particularly useful for proofs, as we shall see later in this chapter. However, for high-order systems, small changes in the coefficients a_i can give large changes in the eigenvalues. Hence, the reachable canonical form is not always well conditioned and must be used with some care.

6.2 Stabilization by State Feedback

The state of a dynamical system is a collection of variables that permits prediction of the future development of a system. We now explore the idea of designing the dynamics of a system through feedback of the state. We will assume that the system to be controlled is described by a linear state model and has a single input (for simplicity). The feedback control law will be developed step by step using a single idea: the positioning of closed loop eigenvalues in desired locations.

State Space Controller Structure

Figure 6.5 is a diagram of a typical control system using state feedback. The full system consists of the process dynamics, which we take to be linear, the controller elements K and k_r, the reference input (or command signal) r and process disturbances d. The goal of the feedback controller is to regulate the output of the system y such that it tracks the reference input in the presence of disturbances and also uncertainty in the process dynamics.

An important element of the control design is the performance specification. The simplest performance specification is that of stability: in the absence of any disturbances, we would like the equilibrium point of the system to be asymptotically stable. More sophisticated performance specifications typically involve giving desired properties of the step or frequency response of the system, such as specifying the desired rise time, overshoot and settling time of the step response. Finally, we are often concerned with the disturbance attenuation properties of the system: to what extent can we experience disturbance inputs d and still hold the output y near the desired value?

Consider a system described by the linear differential equation

$$\frac{dx}{dt} = Ax + Bu, \qquad y = Cx + Du, \tag{6.9}$$

where we have ignored the disturbance signal d for now. Our goal is to drive the output y to a given reference value r and hold it there. Notice that it may not be possible to maintain all equilibria; see Exercise 6.8.

We begin by assuming that all components of the state vector are measured. Since the state at time t contains all the information necessary to predict the future behavior of the system, the most general time-invariant control law is a function of the state and the reference input:

$$u = \alpha(x, r).$$

If the feedback is restricted to be linear, it can be written as

$$u = -Kx + k_r r, \tag{6.10}$$

where r is the reference value, assumed for now to be a constant.

This control law corresponds to the structure shown in Figure 6.5. The negative sign is a convention to indicate that negative feedback is the normal situation. The closed loop system obtained when the feedback (6.10) is applied to the system (6.9) is given by

$$\frac{dx}{dt} = (A - BK)x + Bk_r r. \tag{6.11}$$

We attempt to determine the feedback gain K so that the closed loop system has the characteristic polynomial

$$p(s) = s^n + p_1 s^{n-1} + \cdots + p_{n-1} s + p_n. \tag{6.12}$$

This control problem is called the *eigenvalue assignment problem* or *pole placement problem* (we will define poles more formally in Chapter 8).

Note that k_r does not affect the stability of the system (which is determined by the eigenvalues of $A - BK$) but does affect the steady-state solution. In particular, the equilibrium point and steady-state output for the closed loop system are given by

$$x_e = -(A - BK)^{-1} Bk_r r, \qquad y_e = Cx_e + Du_e,$$

hence k_r should be chosen such that $y_e = r$ (the desired output value). Since k_r is a scalar, we can easily solve to show that if $D = 0$ (the most common case),

$$k_r = -1/\left(C(A - BK)^{-1}B\right). \tag{6.13}$$

Notice that k_r is exactly the inverse of the zero frequency gain of the closed loop system. The solution for $D \neq 0$ is left as an exercise.

Using the gains K and k_r, we are thus able to design the dynamics of the closed loop system to satisfy our goal. To illustrate how to construct such a state feedback control law, we begin with a few examples that provide some basic intuition and insights.

Example 6.4 Vehicle steering
In Example 5.12 we derived a normalized linear model for vehicle steering. The dynamics describing the lateral deviation were given by

$$A = \begin{bmatrix} 0 & 1 \\ 0 & 0 \end{bmatrix}, \qquad B = \begin{bmatrix} \gamma \\ 1 \end{bmatrix},$$

$$C = \begin{bmatrix} 1 & 0 \end{bmatrix}, \qquad D = 0.$$

The reachability matrix for the system is thus

$$W_r = \begin{bmatrix} B & AB \end{bmatrix} = \begin{bmatrix} \gamma & 1 \\ 1 & 0 \end{bmatrix}.$$

The system is reachable since $\det W_r = -1 \neq 0$.

We now want to design a controller that stabilizes the dynamics and tracks a given reference value r of the lateral position of the vehicle. To do this we introduce the feedback

$$u = -Kx + k_r r = -k_1 x_1 - k_2 x_2 + k_r r,$$

and the closed loop system becomes

$$\frac{dx}{dt} = (A - BK)x + Bk_r r = \begin{bmatrix} -\gamma k_1 & 1 - \gamma k_2 \\ -k_1 & -k_2 \end{bmatrix} x + \begin{bmatrix} \gamma k_r \\ k_r \end{bmatrix} r,$$

$$y = Cx + Du = \begin{bmatrix} 1 & 0 \end{bmatrix} x.$$

(6.14)

The closed loop system has the characteristic polynomial

$$\det (sI - A + BK) = \det \begin{bmatrix} s + \gamma k_1 & \gamma k_2 - 1 \\ k_1 & s + k_2 \end{bmatrix} = s^2 + (\gamma k_1 + k_2)s + k_1.$$

Suppose that we would like to use feedback to design the dynamics of the system to have the characteristic polynomial

$$p(s) = s^2 + 2\zeta_c \omega_c s + \omega_c^2.$$

Comparing this polynomial with the characteristic polynomial of the closed loop system, we see that the feedback gains should be chosen as

$$k_1 = \omega_c^2, \qquad k_2 = 2\zeta_c \omega_c - \gamma \omega_c^2.$$

Equation (6.13) gives $k_r = k_1 = \omega_c^2$, and the control law can be written as

$$u = k_1(r - x_1) - k_2 x_2 = \omega_c^2(r - x_1) - (2\zeta_c \omega_c - \gamma \omega_c^2)x_2.$$

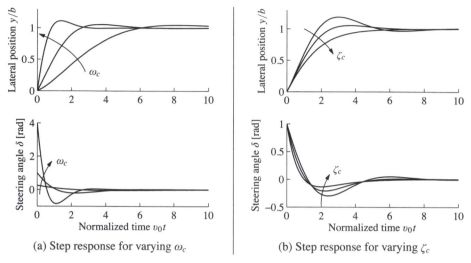

(a) Step response for varying ω_c (b) Step response for varying ζ_c

Figure 6.6: State feedback control of a steering system. Step responses obtained with controllers designed with $\zeta_c = 0.7$ and $\omega_c = 0.5$, 1 and 2 [rad/s] are shown in (a). Notice that response speed increases with increasing ω_c, but that large ω_c also give large initial control actions. Step responses obtained with a controller designed with $\omega_c = 1$ and $\zeta_c = 0.5, 0.7$ and 1 are shown in (b).

The step responses for the closed loop system for different values of the design parameters are shown in Figure 6.6. The effect of ω_c is shown in Figure 6.6a, which shows that the response speed increases with increasing ω_c. The responses for $\omega_c = 0.5$ and 1 have reasonable overshoot. The settling time is about 15 car lengths for $\omega_c = 0.5$ (beyond the end of the plot) and decreases to about 6 car lengths for $\omega_c = 1$. The control signal δ is large initially and goes to zero as time increases because the closed loop dynamics have an integrator. The initial value of the control signal is $u(0) = k_r = \omega_c^2 r$, and thus the achievable response time is limited by the available actuator signal. Notice in particular the dramatic increase in control signal when ω_c changes from 1 to 2. The effect of ζ_c is shown in Figure 6.6b. The response speed and the overshoot increase with decreasing damping. Using these plots, we conclude that reasonable values of the design parameters are to have ω_c in the range of 0.5 to 1 and $\zeta_c \approx 0.7$. ∇

The example of the vehicle steering system illustrates how state feedback can be used to set the eigenvalues of a closed loop system to arbitrary values.

State Feedback for Systems in Reachable Canonical Form

The reachable canonical form has the property that the parameters of the system are the coefficients of the characteristic polynomial. It is therefore natural to consider systems in this form when solving the eigenvalue assignment problem.

Consider a system in reachable canonical form, i.e,

$$\frac{dz}{dt} = \tilde{A}z + \tilde{B}u = \begin{bmatrix} -a_1 & -a_2 & -a_3 & \cdots & -a_n \\ 1 & 0 & 0 & \cdots & 0 \\ 0 & 1 & 0 & \cdots & 0 \\ \vdots & & \ddots & \ddots & \vdots \\ 0 & & & 1 & 0 \end{bmatrix} z + \begin{bmatrix} 1 \\ 0 \\ \vdots \\ 0 \\ 0 \end{bmatrix} u \qquad (6.15)$$

$$y = \tilde{C}z = \begin{bmatrix} b_1 & b_2 & \cdots & b_n \end{bmatrix} z.$$

It follows from (6.7) that the open loop system has the characteristic polynomial

$$\det(sI - A) = s^n + a_1 s^{n-1} + \cdots + a_{n-1}s + a_n.$$

Before making a formal analysis we can gain some insight by investigating the block diagram of the system shown in Figure 6.4. The characteristic polynomial is given by the parameters a_k in the figure. Notice that the parameter a_k can be changed by feedback from state z_k to the input u. It is thus straightforward to change the coefficients of the characteristic polynomial by state feedback.

Returning to equations, introducing the control law

$$u = -\tilde{K}z + k_r r = -\tilde{k}_1 z_1 - \tilde{k}_2 z_2 - \cdots - \tilde{k}_n z_n + k_r r, \qquad (6.16)$$

the closed loop system becomes

$$\frac{dz}{dt} = \begin{bmatrix} -a_1 - \tilde{k}_1 & -a_2 - \tilde{k}_2 & -a_3 - \tilde{k}_3 & \cdots & -a_n - \tilde{k}_n \\ 1 & 0 & 0 & \cdots & 0 \\ 0 & 1 & 0 & \cdots & 0 \\ \vdots & & \ddots & \ddots & \vdots \\ 0 & & & 1 & 0 \end{bmatrix} z + \begin{bmatrix} k_r \\ 0 \\ 0 \\ \vdots \\ 0 \end{bmatrix} r,$$

$$y = \begin{bmatrix} b_n & \cdots & b_2 & b_1 \end{bmatrix} z.$$

$$(6.17)$$

The feedback changes the elements of the first row of the A matrix, which corresponds to the parameters of the characteristic polynomial. The closed loop system thus has the characteristic polynomial

$$s^n + (a_l + \tilde{k}_1)s^{n-1} + (a_2 + \tilde{k}_2)s^{n-2} + \cdots + (a_{n-1} + \tilde{k}_{n-1})s + a_n + \tilde{k}_n.$$

Requiring this polynomial to be equal to the desired closed loop polynomial

$$p(s) = s^n + p_1 s^{n-1} + \cdots + p_{n-1}s + p_n,$$

we find that the controller gains should be chosen as

$$\tilde{k}_1 = p_1 - a_1, \quad \tilde{k}_2 = p_2 - a_2, \quad \cdots \quad \tilde{k}_n = p_n - a_n.$$

This feedback simply replaces the parameters a_i in the system (6.15) by p_i. The feedback gain for a system in reachable canonical form is thus

$$\tilde{K} = \begin{bmatrix} p_1 - a_1 & p_2 - a_2 & \cdots & p_n - a_n \end{bmatrix}. \qquad (6.18)$$

To have zero frequency gain equal to unity, the parameter k_r should be chosen as

$$k_r = \frac{a_n + \tilde{k}_n}{b_n} = \frac{p_n}{b_n}. \tag{6.19}$$

Notice that it is essential to know the precise values of parameters a_n and b_n in order to obtain the correct zero frequency gain. The zero frequency gain is thus obtained by precise calibration. This is very different from obtaining the correct steady-state value by integral action, which we shall see in later sections.

Eigenvalue Assignment

We have seen through the examples how feedback can be used to design the dynamics of a system through assignment of its eigenvalues. To solve the problem in the general case, we simply change coordinates so that the system is in reachable canonical form. Consider the system

$$\frac{dx}{dt} = Ax + Bu, \qquad y = Cx + Du. \tag{6.20}$$

We can change the coordinates by a linear transformation $z = Tx$ so that the transformed system is in reachable canonical form (6.15). For such a system the feedback is given by equation (6.16), where the coefficients are given by equation (6.18). Transforming back to the original coordinates gives the feedback

$$u = -\tilde{K}z + k_r r = -\tilde{K}Tx + k_r r.$$

The results obtained can be summarized as follows.

Theorem 6.3 (Eigenvalue assignment by state feedback). *Consider the system given by equation (6.20), with one input and one output. Let $\lambda(s) = s^n + a_1 s^{n-1} + \cdots + a_{n-1}s + a_n$ be the characteristic polynomial of A. If the system is reachable, then there exists a feedback*

$$u = -Kx + k_r r$$

that gives a closed loop system with the characteristic polynomial

$$p(s) = s^n + p_1 s^{n-1} + \cdots + p_{n-1}s + p_n$$

and unity zero frequency gain between r and y. The feedback gain is given by

$$K = \tilde{K}T = \begin{bmatrix} p_1 - a_1 & p_2 - a_2 & \cdots & p_n - a_n \end{bmatrix} \tilde{W}_r W_r^{-1}, \tag{6.21}$$

where a_i are the coefficients of the characteristic polynomial of the matrix A and the matrices W_r and $\tilde{W}_r$ are given by

$$W_r = \begin{bmatrix} B & AB & \cdots & A^{n-1}B \end{bmatrix}, \quad \tilde{W}_r = \begin{bmatrix} 1 & a_1 & a_2 & \cdots & a_{n-1} \\ 0 & 1 & a_1 & \cdots & a_{n-2} \\ \vdots & & \ddots & \ddots & \vdots \\ 0 & 0 & \cdots & 1 & a_1 \\ 0 & 0 & 0 & \cdots & 1 \end{bmatrix}^{-1}.$$

The reference gain is given by

$$k_r = -1/\left(C(A - BK)^{-1}B\right).$$

For simple problems, the eigenvalue assignment problem can be solved by introducing the elements k_i of K as unknown variables. We then compute the characteristic polynomial

$$\lambda(s) = \det(sI - A + BK)$$

and equate coefficients of equal powers of s to the coefficients of the desired characteristic polynomial

$$p(s) = s^n + p_1 s^{n-1} + \cdots + p_{n-1}s + p_n.$$

This gives a system of linear equations to determine k_i. The equations can always be solved if the system is reachable, exactly as we did in Example 6.4.

Equation (6.21), which is called Ackermann's formula [3, 4], can be used for numeric computations. It is implemented in the MATLAB function `acker`. The MATLAB function `place` is preferable for systems of high order because it is better conditioned numerically.

Example 6.5 Predator–prey
Consider the problem of regulating the population of an ecosystem by modulating the food supply. We use the predator–prey model introduced in Section 3.7. The dynamics for the system are given by

$$\frac{dH}{dt} = (r + u)H\left(1 - \frac{H}{k}\right) - \frac{aHL}{c + H}, \quad H \geq 0,$$

$$\frac{dL}{dt} = b\frac{aHL}{c + H} - dL, \quad L \geq 0.$$

We choose the following nominal parameters for the system, which correspond to the values used in previous simulations:

$$a = 3.2, \quad b = 0.6, \quad c = 50,$$
$$d = 0.56, \quad k = 125 \quad r = 1.6.$$

We take the parameter r, corresponding to the growth rate for hares, as the input to the system, which we might modulate by controlling a food source for the hares. This is reflected in our model by the term $(r + u)$ in the first equation. We choose the number of lynxes as the output of our system.

To control this system, we first linearize the system around the equilibrium point of the system (H_e, L_e), which can be determined numerically to be $x_e \approx$

(20.6, 29.5). This yields a linear dynamical system

$$
\frac{d}{dt}\begin{bmatrix} z_1 \\ z_2 \end{bmatrix} = \begin{bmatrix} 0.13 & -0.93 \\ 0.57 & 0 \end{bmatrix} \begin{bmatrix} z_1 \\ z_2 \end{bmatrix} + \begin{bmatrix} 17.2 \\ 0 \end{bmatrix} v, \quad w = \begin{bmatrix} 0 & 1 \end{bmatrix} \begin{bmatrix} z_1 \\ z_2 \end{bmatrix},
$$

where $z_1 = L - L_e$, $z_2 = H - H_e$ and $v = u$. It is easy to check that the system is reachable around the equilibrium $(z, v) = (0, 0)$, and hence we can assign the eigenvalues of the system using state feedback.

Determining the eigenvalues of the closed loop system requires balancing the ability to modulate the input against the natural dynamics of the system. This can be done by the process of trial and error or by using some of the more systematic techniques discussed in the remainder of the text. For now, we simply choose the desired closed loop eigenvalues to be at $\lambda = \{-0.1, -0.2\}$. We can then solve for the feedback gains using the techniques described earlier, which results in

$$
K = \begin{bmatrix} 0.025 & -0.052 \end{bmatrix}.
$$

Finally, we solve for the reference gain k_r, using equation (6.13) to obtain $k_r = 0.002$.

Putting these steps together, our control law becomes

$$
v = -Kz + k_r r.
$$

In order to implement the control law, we must rewrite it using the original coordinates for the system, yielding

$$
u = u_e - K(x - x_e) + k_r (r - y_e)
$$

$$
= \begin{bmatrix} 0.025 & -0.052 \end{bmatrix} \begin{bmatrix} H - 20.6 \\ L - 29.5 \end{bmatrix} + 0.002 (r - 29.5).
$$

This rule tells us how much we should modulate r_h as a function of the current number of lynxes and hares in the ecosystem. Figure 6.7a shows a simulation of the resulting closed loop system using the parameters defined above and starting with an initial population of 15 hares and 20 lynxes. Note that the system quickly stabilizes the population of lynxes at the reference value ($L = 30$). A phase portrait of the system is given in Figure 6.7b, showing how other initial conditions converge to the stabilized equilibrium population. Notice that the dynamics are very different from the natural dynamics (shown in Figure 3.20).   ∇

The results of this section show that we can use state feedback to design the dynamics of a system, under the strong assumption that we can measure all of the states. We shall address the availability of the states in the next chapter, when we consider output feedback and state estimation. In addition, Theorem 6.3, which states that the eigenvalues can be assigned to arbitrary locations, is also highly idealized and assumes that the dynamics of the process are known to high precision. The robustness of state feedback combined with state estimators is considered in Chapter 12 after we have developed the requisite tools.

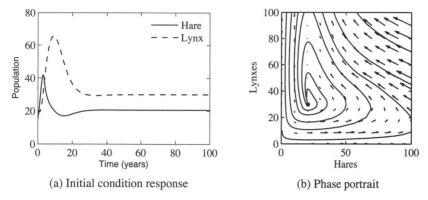

(a) Initial condition response   (b) Phase portrait

Figure 6.7: Simulation results for the controlled predator–prey system. The population of lynxes and hares as a function of time is shown in (a), and a phase portrait for the controlled system is shown in (b). Feedback is used to make the population stable at $H_e = 20.6$ and $L_e = 20$.

6.3 State Feedback Design

The location of the eigenvalues determines the behavior of the closed loop dynamics, and hence where we place the eigenvalues is the main design decision to be made. As with all other feedback design problems, there are trade-offs among the magnitude of the control inputs, the robustness of the system to perturbations and the closed loop performance of the system. In this section we examine some of these trade-offs starting with the special case of second-order systems.

Second-Order Systems

One class of systems that occurs frequently in the analysis and design of feedback systems is second-order linear differential equations. Because of their ubiquitous nature, it is useful to apply the concepts of this chapter to that specific class of systems and build more intuition about the relationship between stability and performance.

The canonical second-order system is a differential equation of the form

$$\ddot{q} + 2\zeta\omega_0\dot{q} + \omega_0^2 q = k\omega_0^2 u, \qquad y = q. \tag{6.22}$$

In state space form, this system can be represented as

$$\frac{dx}{dt} = \begin{bmatrix} 0 & \omega_0 \\ -\omega_0 & -2\zeta\omega_0 \end{bmatrix} x + \begin{bmatrix} 0 \\ k\omega_0 \end{bmatrix} u, \qquad y = \begin{bmatrix} 1 & 0 \end{bmatrix} x. \tag{6.23}$$

The eigenvalues of this system are given by

$$\lambda = -\zeta\omega_0 \pm \sqrt{\omega_0^2(\zeta^2 - 1)},$$

and we see that the origin is a stable equilibrium point if $\omega_0 > 0$ and $\zeta > 0$. Note that the eigenvalues are complex if $\zeta < 1$ and real otherwise. Equations (6.22) and (6.23) can be used to describe many second-order systems, including damped oscillators, active filters and flexible structures, as shown in the examples below.

The form of the solution depends on the value of ζ, which is referred to as the *damping ratio* for the system. If $\zeta > 1$, we say that the system is *overdamped*, and the natural response ($u = 0$) of the system is given by

$$y(t) = \frac{\beta x_{10} + x_{20}}{\beta - \alpha} e^{-\alpha t} - \frac{\alpha x_{10} + x_{20}}{\beta - \alpha} e^{-\beta t},$$

where $\alpha = \omega_0(\zeta + \sqrt{\zeta^2 - 1})$ and $\beta = \omega_0(\zeta - \sqrt{\zeta^2 - 1})$. We see that the response consists of the sum of two exponentially decaying signals. If $\zeta = 1$, then the system is *critically* damped and solution becomes

$$y(t) = e^{-\zeta \omega_0 t}\left(x_{10} + (x_{20} + \zeta \omega_0 x_{10})t\right).$$

Note that this is still asymptotically stable as long as $\omega_0 > 0$, although the second term in the solution is increasing with time (but more slowly than the decaying exponential that is multiplying it).

Finally, if $0 < \zeta < 1$, then the solution is oscillatory and equation (6.22) is said to be *underdamped*. The parameter ω_0 is referred to as the *natural frequency* of the system, stemming from the fact that for small ζ, the eigenvalues of the system are approximately $\lambda = -\zeta \omega_0 \pm j\omega_0$. The natural response of the system is given by

$$y(t) = e^{-\zeta \omega_0 t}\left(x_{10} \cos \omega_d t + \left(\frac{\zeta \omega_0}{\omega_d} x_{10} + \frac{1}{\omega_d} x_{20}\right) \sin \omega_d t\right),$$

where $\omega_d = \omega_0 \sqrt{1 - \zeta^2}$ is called the *damped frequency*. For $\zeta \ll 1$, $\omega_d \approx \omega_0$ defines the oscillation frequency of the solution and ζ gives the damping rate relative to ω_0.

Because of the simple form of a second-order system, it is possible to solve for the step and frequency responses in analytical form. The solution for the step response depends on the magnitude of ζ:

$$
\begin{aligned}
y(t) &= k\left(1 - e^{-\zeta \omega_0 t} \cos \omega_d t - \frac{\zeta}{\sqrt{1 - \zeta^2}} e^{-\zeta \omega_0 t} \sin \omega_d t\right), \quad \zeta < 1; \\
y(t) &= k\left(1 - e^{-\omega_0 t}(1 + \omega_0 t)\right), \quad \zeta = 1; \\
y(t) &= k\left(1 - \frac{1}{2}\left(\frac{\zeta}{\sqrt{\zeta^2 - 1}} + 1\right)e^{-\omega_0 t(\zeta - \sqrt{\zeta^2 - 1})}\right. \\
&\quad \left. + \frac{1}{2}\left(\frac{\zeta}{\sqrt{\zeta^2 - 1}} - 1\right)e^{-\omega_0 t(\zeta + \sqrt{\zeta^2 - 1})}\right), \quad \zeta > 1,
\end{aligned}
\tag{6.24}
$$

where we have taken $x(0) = 0$. Note that for the lightly damped case ($\zeta < 1$) we have an oscillatory solution at frequency ω_d.

Step responses of systems with $k = 1$ and different values of ζ are shown in Figure 6.8. The shape of the response is determined by ζ, and the speed of the response is determined by ω_0 (included in the time axis scaling): the response is faster if ω_0 is larger.

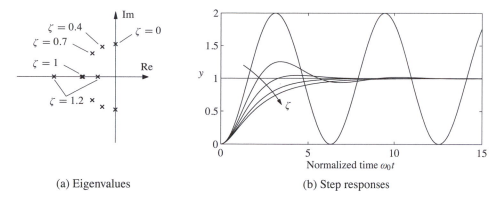

(a) Eigenvalues   (b) Step responses

Figure 6.8: Step response for a second-order system. Normalized step responses h for the system (6.23) for $\zeta = 0, 0.4, 0.7, 1$ and 1.2. As the damping ratio is increased, the rise time of the system gets longer, but there is less overshoot. The horizontal axis is in scaled units $\omega_0 t$; higher values of ω_0 result in a faster response (rise time and settling time).

In addition to the explicit form of the solution, we can also compute the properties of the step response that were defined in Section 5.3. For example, to compute the maximum overshoot for an underdamped system, we rewrite the output as

$$y(t) = k\left(1 - \frac{1}{\sqrt{1-\zeta^2}}e^{-\zeta\omega_0 t}\sin(\omega_d t + \varphi)\right), \qquad (6.25)$$

where $\varphi = \arccos\zeta$. The maximum overshoot will occur at the first time in which the derivative of y is zero, which can be shown to be

$$M_p = e^{-\pi\zeta/\sqrt{1-\zeta^2}}.$$

Similar computations can be done for the other characteristics of a step response. Table 6.1 summarizes the calculations.

The frequency response for a second-order system can also be computed ex-

Table 6.1: Properties of the step response for a second-order system with $0 < \zeta < 1$.

Property	Value	$\zeta = 0.5$	$\zeta = 1/\sqrt{2}$	$\zeta = 1$
Steady-state value	k	k	k	k
Rise time	$T_r = 1/\omega_0 \cdot e^{\varphi/\tan\varphi}$	$1.8/\omega_0$	$2.2/\omega_0$	$2.7/\omega_0$
Overshoot	$M_p = e^{-\pi\zeta/\sqrt{1-\zeta^2}}$	16%	4%	0%
Settling time (2%)	$T_s \approx 4/\zeta\omega_0$	$8.0/\omega_0$	$5.9/\omega_0$	$5.8/\omega_0$

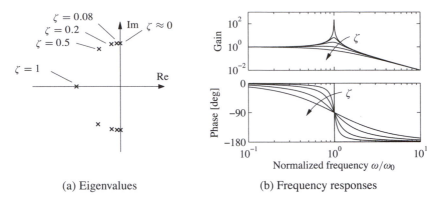

(a) Eigenvalues   (b) Frequency responses

Figure 6.9: Frequency response of a second-order system (6.23). (a) Eigenvalues as a function of ζ. (b) Frequency response as a function of ζ. The upper curve shows the gain ratio M, and the lower curve shows the phase shift θ. For small ζ there is a large peak in the magnitude of the frequency response and a rapid change in phase centered at $\omega = \omega_0$. As ζ is increased, the magnitude of the peak drops and the phase changes more smoothly between $0°$ and $-180°$.

plicitly and is given by

$$Me^{j\theta} = \frac{k\omega_0^2}{(i\omega)^2 + 2\zeta\omega_0(i\omega) + \omega_0^2} = \frac{k\omega_0^2}{\omega_0^2 - \omega^2 + 2i\zeta\omega_0\omega}.$$

A graphical illustration of the frequency response is given in Figure 6.9. Notice the resonant peak that increases with decreasing ζ. The peak is often characterized by is Q-value, defined as $Q = 1/2\zeta$. The properties of the frequency response for a second-order system are summarized in Table 6.2.

Example 6.6 Drug administration

To illustrate the use of these formulas, consider the two-compartment model for drug administration, described in Section 3.6. The dynamics of the system are

$$\frac{dc}{dt} = \begin{bmatrix} -k_0 - k_1 & k_1 \\ k_2 & -k_2 \end{bmatrix} c + \begin{bmatrix} b_0 \\ 0 \end{bmatrix} u, \qquad y = \begin{bmatrix} 0 & 1 \end{bmatrix} x,$$

where c_1 and c_2 are the concentrations of the drug in each compartment, $k_i, i = 0, \ldots, 2$ and b_0 are parameters of the system, u is the flow rate of the drug into

Table 6.2: Properties of the frequency response for a second-order system with $0 < \zeta < 1$.

Property	Value	$\zeta = 0.1$	$\zeta = 0.5$	$\zeta = 1/\sqrt{2}$
Zero frequency gain	M_0	k	k	k
Bandwidth	ω_b	$1.54\,\omega_0$	$1.27\,\omega_0$	ω_0
Resonant peak gain	M_r	$1.54\,k$	$1.27\,k$	k
Resonant frequency	ω_{mr}	ω_0	$0.707\omega_0$	0

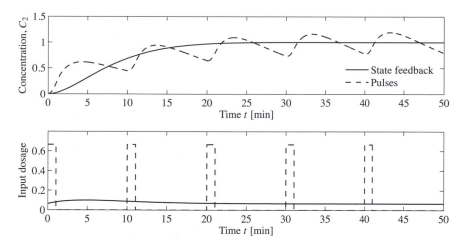

Figure 6.10: Open loop versus closed loop drug administration. Comparison between drug administration using a sequence of doses versus continuously monitoring the concentrations and adjusting the dosage continuously. In each case, the concentration is (approximately) maintained at the desired level, but the closed loop system has substantially less variability in drug concentration.

compartment 1 and y is the concentration of the drug in compartment 2. We assume that we can measure the concentrations of the drug in each compartment, and we would like to design a feedback law to maintain the output at a given reference value r.

We choose $\zeta = 0.9$ to minimize the overshoot and choose the rise time to be $T_r = 10$ min. Using the formulas in Table 6.1, this gives a value for $\omega_0 = 0.22$. We can now compute the gain to place the eigenvalues at this location. Setting $u = -Kx + k_r r$, the closed loop eigenvalues for the system satisfy

$$\lambda(s) = -0.198 \pm 0.0959i.$$

Choosing $k_1 = -0.2027$ and $k_2 = 0.2005$ gives the desired closed loop behavior. Equation (6.13) gives the reference gain $k_r = 0.0645$. The response of the controller is shown in Figure 6.10 and compared with an open loop strategy involving administering periodic doses of the drug. $\qquad \nabla$

Higher-Order Systems

Our emphasis so far has considered only second-order systems. For higher-order systems, eigenvalue assignment is considerably more difficult, especially when trying to account for the many trade-offs that are present in a feedback design.

One of the other reasons why second-order systems play such an important role in feedback systems is that even for more complicated systems the response is often characterized by the *dominant eigenvalues*. To define these more precisely, consider a system with eigenvalues $\lambda_j, j = 1, \ldots, n$. We define the *damping ratio* for a complex eigenvalue λ to be

$$\zeta = \frac{-\mathrm{Re}\,\lambda}{|\lambda|}.$$

We say that a complex conjugate pair of eigenvalues λ, λ^* is a *dominant pair* if it has the lowest damping ratio compared with all other eigenvalues of the system.

Assuming that a system is stable, the dominant pair of eigenvalues tends to be the most important element of the response. To see this, assume that we have a system in Jordan form with a simple Jordan block corresponding to the dominant pair of eigenvalues:

$$\frac{dz}{dt} = \begin{bmatrix} \lambda & & & & \\ & \lambda^* & & & \\ & & J_2 & & \\ & & & \ddots & \\ & & & & J_k \end{bmatrix} z + Bu, \qquad y = Cz.$$

(Note that the state z may be complex because of the Jordan transformation.) The response of the system will be a linear combination of the responses from each of the individual Jordan subsystems. As we see from Figure 6.8, for $\zeta < 1$ the subsystem with the slowest response is precisely the one with the smallest damping ratio. Hence, when we add the responses from each of the individual subsystems, it is the dominant pair of eigenvalues that will be the primary factor after the initial transients due to the other terms in the solution die out. While this simple analysis does not always hold (e.g., if some nondominant terms have larger coefficients because of the particular form of the system), it is often the case that the dominant eigenvalues determine the (step) response of the system.

The only formal requirement for eigenvalue assignment is that the system be reachable. In practice there are many other constraints because the selection of eigenvalues has a strong effect on the magnitude and rate of change of the control signal. Large eigenvalues will in general require large control signals as well as fast changes of the signals. The capability of the actuators will therefore impose constraints on the possible location of closed loop eigenvalues. These issues will be discussed in depth in Chapters 11 and 12.

We illustrate some of the main ideas using the balance system as an example.

Example 6.7 Balance system
Consider the problem of stabilizing a balance system, whose dynamics were given in Example 6.2. The dynamics are given by

$$A = \begin{bmatrix} 0 & 0 & 1 & 0 \\ 0 & 0 & 0 & 1 \\ 0 & m^2 l^2 g/\mu & -c J_t/\mu & -\gamma J_t lm/\mu \\ 0 & M_t mgl/\mu & -clm/\mu & -\gamma M_t/\mu \end{bmatrix}, \qquad B = \begin{bmatrix} 0 \\ 0 \\ J_t/\mu \\ lm/\mu \end{bmatrix},$$

where $M_t = M + m$, $J_t = J + ml^2$, $\mu = M_t J_t - m^2 l^2$ and we have left c and γ

nonzero. We use the following parameters for the system (corresponding roughly to a human being balanced on a stabilizing cart):

$$M = 10 \text{ kg}, \qquad m = 80 \text{ kg}, \qquad c = 0.1 \text{ N s/m},$$
$$J = 100 \text{ kg m}^2/\text{s}^2, \qquad l = 1 \text{ m}, \qquad \gamma = 0.01 \text{ N m s}, \qquad g = 9.8 \text{ m/s}^2.$$

The eigenvalues of the open loop dynamics are given by $\lambda \approx 0, 4.7, -1.9 \pm 2.7i$. We have verified already in Example 6.2 that the system is reachable, and hence we can use state feedback to stabilize the system and provide a desired level of performance.

To decide where to place the closed loop eigenvalues, we note that the closed loop dynamics will roughly consist of two components: a set of fast dynamics that stabilize the pendulum in the inverted position and a set of slower dynamics that control the position of the cart. For the fast dynamics, we look to the natural period of the pendulum (in the hanging-down position), which is given by $\omega_0 = \sqrt{mgl/(J + ml^2)} \approx 2.1$ rad/s. To provide a fast response we choose a damping ratio of $\zeta = 0.5$ and try to place the first pair of eigenvalues at $\lambda_{1,2} \approx -\zeta\omega_0 \pm \omega_0 \approx -1 \pm 2i$, where we have used the approximation that $\sqrt{1 - \zeta^2} \approx 1$. For the slow dynamics, we choose the damping ratio to be 0.7 to provide a small overshoot and choose the natural frequency to be 0.5 to give a rise time of approximately 5 s. This gives eigenvalues $\lambda_{3,4} = -0.35 \pm 0.35i$.

The controller consists of a feedback on the state and a feedforward gain for the reference input. The feedback gain is given by

$$K = \begin{bmatrix} -15.6 & 1730 & -50.1 & 443 \end{bmatrix},$$

which can be computed using Theorem 6.3 or using the MATLAB `place` command. The feedforward gain is $k_r = -1/(C(A - BK)^{-1}B) = -15.5$. The step response for the resulting controller (applied to the linearized system) is given in Figure 6.11a. While the step response gives the desired characteristics, the input required (bottom left) is excessively large, almost three times the force of gravity at its peak.

To provide a more realistic response, we can redesign the controller to have slower dynamics. We see that the peak of the input force occurs on the fast time scale, and hence we choose to slow this down by a factor of 3, leaving the damping ratio unchanged. We also slow down the second set of eigenvalues, with the intuition that we should move the position of the cart more slowly than we stabilize the pendulum dynamics. Leaving the damping ratio for the slow dynamics unchanged at 0.7 and changing the frequency to 1 (corresponding to a rise time of approximately 10 s), the desired eigenvalues become

$$\lambda = \{-0.33 \pm 0.66i, -0.18 \pm 0.18i\}.$$

The performance of the resulting controller is shown in Figure 6.11b. ∇

As we see from this example, it can be difficult to determine where to place the eigenvalues using state feedback. This is one of the principal limitations of this

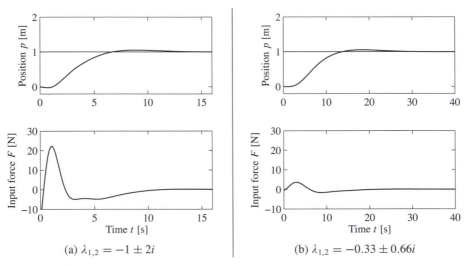

(a) $\lambda_{1,2} = -1 \pm 2i$ (b) $\lambda_{1,2} = -0.33 \pm 0.66i$

Figure 6.11: State feedback control of a balance system. The step response of a controller designed to give fast performance is shown in (a). Although the response characteristics (top left) look very good, the input magnitude (bottom left) is very large. A less aggressive controller is shown in (b). Here the response time is slowed down, but the input magnitude is much more reasonable. Both step responses are applied to the linearized dynamics.

approach, especially for systems of higher dimension. Optimal control techniques, such as the linear quadratic regulator problem discussed next, are one approach that is available. One can also focus on the frequency response for performing the design, which is the subject of Chapters 8–12.

 Linear Quadratic Regulators

As an alternative to selecting the closed loop eigenvalue locations to accomplish a certain objective, the gains for a state feedback controller can instead be chosen is by attempting to optimize a cost function. This can be particularly useful in helping balance the performance of the system with the magnitude of the inputs required to achieve that level of performance.

The infinite horizon, linear quadratic regulator (LQR) problem is one of the most common optimal control problems. Given a multi-input linear system

$$\frac{dx}{dt} = Ax + Bu, \qquad x \in \mathbb{R}^n, \, u \in \mathbb{R}^p,$$

we attempt to minimize the quadratic cost function

$$\tilde{J} = \int_0^\infty \left(x^T Q_x x + u^T Q_u u \right) dt, \tag{6.26}$$

where $Q_x \geq 0$ and $Q_u > 0$ are symmetric, positive (semi-) definite matrices of the appropriate dimensions. This cost function represents a trade-off between the distance of the state from the origin and the cost of the control input. By choosing the matrices Q_x and Q_u, we can balance the rate of convergence of the solutions with the cost of the control.

The solution to the LQR problem is given by a linear control law of the form

$$u = -Q_u^{-1} B^T P x,$$

where $P \in \mathbb{R}^{n \times n}$ is a positive definite, symmetric matrix that satisfies the equation

$$PA + A^T P - PBQ_u^{-1}B^T P + Q_x = 0. \tag{6.27}$$

Equation (6.27) is called the *algebraic Riccati equation* and can be solved numerically (e.g., using the `lqr` command in MATLAB).

One of the key questions in LQR design is how to choose the weights Q_x and Q_u. To guarantee that a solution exists, we must have $Q_x \geq 0$ and $Q_u > 0$. In addition, there are certain "observability" conditions on Q_x that limit its choice. Here we assume $Q_x > 0$ to ensure that solutions to the algebraic Riccati equation always exist.

To choose specific values for the cost function weights Q_x and Q_u, we must use our knowledge of the system we are trying to control. A particularly simple choice is to use diagonal weights

$$Q_x = \begin{bmatrix} q_1 & & 0 \\ & \ddots & \\ 0 & & q_n \end{bmatrix}, \qquad Q_u = \begin{bmatrix} \rho_1 & & 0 \\ & \ddots & \\ 0 & & \rho_n \end{bmatrix}.$$

For this choice of Q_x and Q_u, the individual diagonal elements describe how much each state and input (squared) should contribute to the overall cost. Hence, we can take states that should remain small and attach higher weight values to them. Similarly, we can penalize an input versus the states and other inputs through choice of the corresponding input weight ρ.

Example 6.8 Vectored thrust aircraft

Consider the original dynamics of the system (2.26), written in state space form as

$$\frac{dz}{dt} = \begin{bmatrix} z_4 \\ z_5 \\ z_6 \\ -g \sin\theta - \frac{c}{m} z_4 \\ -g \cos\theta - \frac{c}{m} z_5 \\ 0 \end{bmatrix} + \begin{bmatrix} 0 \\ 0 \\ 0 \\ \frac{1}{m} \cos\theta \, F_1 - \frac{1}{m} \sin\theta \, F_2 \\ \frac{1}{m} \sin\theta \, F_1 + \frac{1}{m} \cos\theta \, F_2 \\ \frac{r}{J} F_1 \end{bmatrix}$$

(see Example 5.4). The system parameters are $m = 4$ kg, $J = 0.0475$ kg m^2, $r = 0.25$ m, $g = 9.8$ m/s^2, $c = 0.05$ N s/m, which corresponds to a scaled model of the system. The equilibrium point for the system is given by $F_1 = 0$, $F_2 = mg$ and $z_e = (x_e, y_e, 0, 0, 0, 0)$. To derive the linearized model near an equilibrium point, we compute the linearization according to equation (5.34):

$$
A = \begin{bmatrix} 0 & 0 & 0 & 1 & 0 & 0 \\ 0 & 0 & 0 & 0 & 1 & 0 \\ 0 & 0 & 0 & 0 & 0 & 1 \\ 0 & 0 & -g & -c/m & 0 & 0 \\ 0 & 0 & 0 & 0 & -c/m & 0 \\ 0 & 0 & 0 & 0 & 0 & 0 \end{bmatrix}, \qquad B = \begin{bmatrix} 0 & 0 \\ 0 & 0 \\ 0 & 0 \\ 1/m & 0 \\ 0 & 1/m \\ r/J & 0 \end{bmatrix},
$$

$$
C = \begin{bmatrix} 1 & 0 & 0 & 0 & 0 & 0 \\ 0 & 1 & 0 & 0 & 0 & 0 \end{bmatrix}, \qquad D = \begin{bmatrix} 0 & 0 \\ 0 & 0 \end{bmatrix}.
$$

Letting $z = z - z_e$ and $v = u - u_e$, the linearized system is given by

$$
\frac{dz}{dt} = Az + Bv, \qquad y = Cx.
$$

It can be verified that the system is reachable.

To compute a linear quadratic regulator for the system, we write the cost function as

$$
J = \int_0^\infty (z^T Q_z z + v^T Q_v v)\,dt,
$$

where $z = z - z_e$ and $v = u - u_e$ represent the local coordinates around the desired equilibrium point (z_e, u_e). We begin with diagonal matrices for the state and input costs:

$$
Q_z = \begin{bmatrix} 1 & 0 & 0 & 0 & 0 & 0 \\ 0 & 1 & 0 & 0 & 0 & 0 \\ 0 & 0 & 1 & 0 & 0 & 0 \\ 0 & 0 & 0 & 1 & 0 & 0 \\ 0 & 0 & 0 & 0 & 1 & 0 \\ 0 & 0 & 0 & 0 & 0 & 1 \end{bmatrix}, \qquad Q_v = \begin{bmatrix} 1 & 0 \\ 0 & 1 \end{bmatrix}.
$$

This gives a control law of the form $v = -Kz$, which can then be used to derive the control law in terms of the original variables:

$$
u = v + u_e = -K(z - z_e) + u_e.
$$

As computed in Example 5.4, the equilibrium points have $u_e = (0, mg)$ and $z_e = (x_e, y_e, 0, 0, 0, 0)$. The response of the controller to a step change in the desired position is shown in Figure 6.12a. The response can be tuned by adjusting the weights in the LQR cost. Figure 6.12b shows the response in the x direction for different choices of the weight ρ. ∇

Linear quadratic regulators can also be designed for discrete-time systems, as illustrated by the following example.

Example 6.9 Web server control

Consider the web server example given in Section 3.4, where a discrete-time model for the system was given. We wish to design a control law that sets the server parameters so that the average server processor load is maintained at a desired

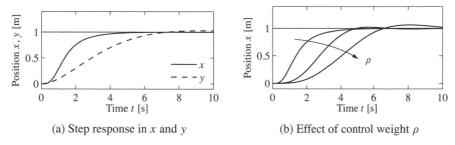

(a) Step response in x and y (b) Effect of control weight ρ

Figure 6.12: Step response for a vectored thrust aircraft. The plot in (a) shows the x and y positions of the aircraft when it is commanded to move 1 m in each direction. In (b) the x motion is shown for control weights $\rho = 1, 10^2, 10^4$. A higher weight of the input term in the cost function causes a more sluggish response.

level. Since other processes may be running on the server, the web server must adjust its parameters in response to changes in the load.

A block diagram for the control system is shown in Figure 6.13. We focus on the special case where we wish to control only the processor load using both the `KeepAlive` and `MaxClients` parameters. We also include a "disturbance" on the measured load that represents the use of the processing cycles by other processes running on the server. The system has the same basic structure as the generic control system in Figure 6.5, with the variation that the disturbance enters after the process dynamics.

The dynamics of the system are given by a set of difference equations of the form

$$x[k+1] = Ax[k] + Bu[k], \qquad y_{\text{cpu}}[k] = C_{\text{cpu}}x[k] + d_{\text{cpu}}[k],$$

where $x = (x_{\text{cpu}}, x_{\text{mem}})$ is the state, $u = (u_{\text{ka}}, u_{\text{mc}})$ is the input, d_{cpu} is the processing load from other processes on the computer and y_{cpu} is the total processor load.

We choose our controller to be a state feedback controller of the form

$$u = -K \begin{bmatrix} y_{\text{cpu}} \\ x_{\text{mem}} \end{bmatrix} + k_r r_{\text{cpu}},$$

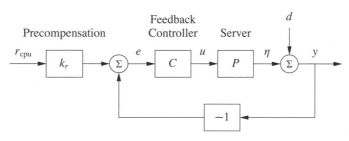

Figure 6.13: Feedback control of a web server. The controller sets the values of the web server parameters based on the difference between the nominal parameters (determined by $k_r r$) and the current load y_{cpu}. The disturbance d represents the load due to other processes running on the server. Note that the measurement is taken after the disturbance so that we measure the total load on the server.

where r_{cpu} is the desired processor load. Note that we have used the measured processor load y_{cpu} instead of the state to ensure that we adjust the system operation based on the actual load. (This modification is necessary because of the nonstandard way in which the disturbance enters the process dynamics.)

The feedback gain matrix K can be chosen by any of the methods described in this chapter. Here we use a linear quadratic regulator, with the cost function given by

$$Q_x = \begin{bmatrix} 5 & 0 \\ 0 & 1 \end{bmatrix}, \qquad Q_u = \begin{bmatrix} 1/50^2 & 0 \\ 0 & 1/1000^2 \end{bmatrix}.$$

The cost function for the state Q_x is chosen so that we place more emphasis on the processor load versus the memory use. The cost function for the inputs Q_u is chosen so as to normalize the two inputs, with a KeepAlive timeout of 50 s having the same weight as a MaxClients value of 1000. These values are squared since the cost associated with the inputs is given by $u^T Q_u u$. Using the dynamics in Section 3.4 and the dlqr command in MATLAB, the resulting gains become

$$K = \begin{bmatrix} -22.3 & 10.1 \\ 382.7 & 77.7 \end{bmatrix}.$$

As in the case of a continuous-time control system, the reference gain k_r is chosen to yield the desired equilibrium point for the system. Setting $x[k+1] = x[k] = x_e$, the steady-state equilibrium point and output for a given reference input r are given by

$$x_e = (A - BK)x_e + Bk_r r, \qquad y_e = Cx_e.$$

This is a matrix differential equation in which k_r is a column vector that sets the two inputs values based on the desired reference. If we take the desired output to be of the form $y_e = (r, 0)$, then we must solve

$$\begin{bmatrix} 1 \\ 0 \end{bmatrix} = C(A - BK - I)^{-1} Bk_r.$$

Solving this equation for k_r, we obtain

$$k_r = \left(\left(C(A - BK - I)^{-1} B \right) \right)^{-1} \begin{bmatrix} 1 \\ 0 \end{bmatrix} = \begin{bmatrix} 49.3 \\ 539.5 \end{bmatrix}.$$

The dynamics of the closed loop system are illustrated in Figure 6.14. We apply a change in load of $d_{cpu} = 0.3$ at time $t = 10$ s, forcing the controller to adjust the operation of the server to attempt to maintain the desired load at 0.57. Note that both the KeepAlive and MaxClients parameters are adjusted. Although the load is decreased, it remains approximately 0.2 above the desired steady state. (Better results can be obtained using the techniques of the next section.)   ∇

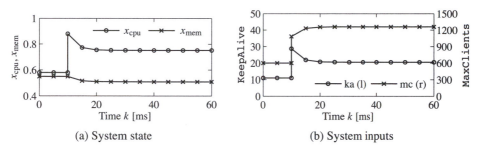

Figure 6.14: Web server with LQR control. The plot in (a) shows the state of the system under a change in external load applied at $k = 10$ ms. The corresponding web server parameters (system inputs) are shown in (b). The controller is able to reduce the effect of the disturbance by approximately 40%.

6.4 Integral Action

Controllers based on state feedback achieve the correct steady-state response to command signals by careful calibration of the gain k_r. However, one of the primary uses of feedback is to allow good performance in the presence of uncertainty, and hence requiring that we have an *exact* model of the process is undesirable. An alternative to calibration is to make use of integral feedback, in which the controller uses an integrator to provide zero steady-state error. The basic concept of integral feedback was given in Section 1.5 and in Section 3.1; here we provide a more complete description and analysis.

The basic approach in integral feedback is to create a state within the controller that computes the integral of the error signal, which is then used as a feedback term. We do this by augmenting the description of the system with a new state z:

$$\frac{d}{dt} \begin{bmatrix} x \\ z \end{bmatrix} = \begin{bmatrix} Ax + Bu \\ y - r \end{bmatrix} = \begin{bmatrix} Ax + Bu \\ Cx - r \end{bmatrix}. \tag{6.28}$$

The state z is seen to be the integral of the difference between the the actual output y and desired output r. Note that if we find a compensator that stabilizes the system, then we will necessarily have $\dot{z} = 0$ in steady state and hence $y = r$ in steady state.

Given the augmented system, we design a state space controller in the usual fashion, with a control law of the form

$$u = -Kx - k_i z + k_r r, \tag{6.29}$$

where K is the usual state feedback term, k_i is the integral term and k_r is used to set the nominal input for the desired steady state. The resulting equilibrium point for the system is given as

$$x_e = -(A - BK)^{-1} B(k_r r - k_i z_e).$$

Note that the value of z_e is not specified but rather will automatically settle to the value that makes $\dot{z} = y - r = 0$, which implies that at equilibrium the output will equal the reference value. This holds independently of the specific values of A,

B and K as long as the system is stable (which can be done through appropriate choice of K and k_i).

The final compensator is given by

$$u = -Kx - k_i z + k_r r, \qquad \frac{dz}{dt} = y - r,$$

where we have now included the dynamics of the integrator as part of the specification of the controller. This type of compensator is known as a *dynamic compensator* since it has its own internal dynamics. The following example illustrates the basic approach.

Example 6.10 Cruise control

Consider the cruise control example introduced in Section 3.1 and considered further in Example 5.11. The linearized dynamics of the process around an equilibrium point v_e, u_e are given by

$$\frac{dx}{dt} = ax - b_g\theta + bw, \qquad y = v = x + v_e,$$

where $x = v - v_e, w = u - u_e, m$ is the mass of the car and θ is the angle of the road. The constant a depends on the throttle characteristic and is given in Example 5.11.

If we augment the system with an integrator, the process dynamics become

$$\frac{dx}{dt} = ax - b_g\theta + bw, \qquad \frac{dz}{dt} = y - v_r = v_e + x - v_r,$$

or, in state space form,

$$\frac{d}{dt}\begin{bmatrix} x \\ z \end{bmatrix} = \begin{bmatrix} a & 0 \\ 1 & 0 \end{bmatrix}\begin{bmatrix} x \\ z \end{bmatrix} + \begin{bmatrix} b \\ 0 \end{bmatrix}u + \begin{bmatrix} -b_g \\ 0 \end{bmatrix}\theta + \begin{bmatrix} 0 \\ v_e - v_r \end{bmatrix}.$$

Note that when the system is at equilibrium, we have that $\dot{z} = 0$, which implies that the vehicle speed $v = v_e + x$ should be equal to the desired reference speed v_r. Our controller will be of the form

$$\frac{dz}{dt} = y - v_r, \qquad u = -k_p x - k_i z + k_r v_r,$$

and the gains k_p, k_i and k_r will be chosen to stabilize the system and provide the correct input for the reference speed.

Assume that we wish to design the closed loop system to have the characteristic polynomial

$$\lambda(s) = s^2 + a_1 s + a_2.$$

Setting the disturbance $\theta = 0$, the characteristic polynomial of the closed loop system is given by

$$\det(sI - (A - BK)) = s^2 + (bk_p - a)s + bk_i,$$

and hence we set

$$k_p = \frac{a_1 + a}{b}, \qquad k_i = \frac{a_2}{b}, \qquad k_r = -1/(C(A - BK)^{-1}B) = \frac{a}{b}.$$

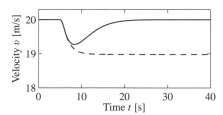
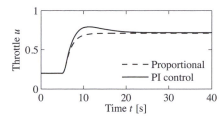

Figure 6.15: Velocity and throttle for a car with cruise control based on proportional (dashed) and PI control (solid). The PI controller is able to adjust the throttle to compensate for the effect of the hill and maintain the speed at the reference value of $v_r = 20$ m/s.

The resulting controller stabilizes the system and hence brings $\dot{z} = y - v_r$ to zero, resulting in perfect tracking. Notice that even if we have a small error in the values of the parameters defining the system, as long as the closed loop eigenvalues are still stable, then the tracking error will approach zero. Thus the exact calibration required in our previous approach (using k_r) is not needed here. Indeed, we can even choose $k_r = 0$ and let the feedback controller do all of the work.

Integral feedback can also be used to compensate for constant disturbances. Figure 6.15 shows the results of a simulation in which the car encounters a hill with angle $\theta = 4°$ at $t = 8$ s. The stability of the system is not affected by this external disturbance, and so we once again see that the car's velocity converges to the reference speed. This ability to handle constant disturbances is a general property of controllers with integral feedback (see Exercise 6.4).   ∇

6.5 Further Reading

The importance of state models and state feedback was discussed in the seminal paper by Kalman [113], where the state feedback gain was obtained by solving an optimization problem that minimized a quadratic loss function. The notions of reachability and observability (Chapter 7) are also due to Kalman [115] (see also [82, 118]). Kalman defines controllability and reachability as the ability to reach the origin and an arbitrary state, respectively [117]. We note that in most textbooks the term "controllability" is used instead of "reachability," but we prefer the latter term because it is more descriptive of the fundamental property of being able to reach arbitrary states. Most undergraduate textbooks on control contain material on state space systems, including, for example, Franklin, Powell and Emami-Naeini [79] and Ogata [162]. Friedland's textbook [80] covers the material in the previous, current and next chapter in considerable detail, including the topic of optimal control.

Exercises

6.1 (Double integrator) Consider the double integrator. Find a piecewise constant control strategy that drives the system from the origin to the state $x = (1, 1)$.

6.2 (Reachability from nonzero initial state) Extend the argument in Section 6.1 to show that if a system is reachable from an initial state of zero, it is reachable from a nonzero initial state.

6.3 (Unreachable systems) Consider the system shown in Figure 6.3. Write the dynamics of the two systems as

$$\frac{dx}{dt} = Ax + Bu, \qquad \frac{dz}{dt} = Az + Bu.$$

If x and z have the same initial condition, they will always have the same state regardless of the input that is applied. Show that this violates the definition of reachability and further show that the reachability matrix W_r is not full rank.

6.4 (Integral feedback for rejecting constant disturbances) Consider a linear system of the form

$$\frac{dx}{dt} = Ax + Bu + Fd, \qquad y = Cx$$

where u is a scalar and d is a disturbance that enters the system through a disturbance vector $F \in \mathbb{R}^n$. Assume that the matrix A is invertible and the zero frequency gain $CA^{-1}B$ is nonzero. Show that integral feedback can be used to compensate for a constant disturbance by giving zero steady-state output error even when $d \neq 0$.

6.5 (Rear-steered bicycle) A simple model for a bicycle was given by equation (3.5) in Section 3.2. A model for a bicycle with rear-wheel steering is obtained by reversing the sign of the velocity in the model. Determine the conditions under which this systems is reachable and explain any situations in which the system is not reachable.

6.6 (Characteristic polynomial for reachable canonical form) Show that the characteristic polynomial for a system in reachable canonical form is given by equation (6.7) and that

$$\frac{d^n z_k}{dt^n} + a_1 \frac{d^{n-1} z_k}{dt^{n-1}} + \cdots + a_{n-1} \frac{dz_k}{dt} + a_n z_k = \frac{d^{n-k} u}{dt^{n-k}},$$

where z_k is the kth state.

6.7 (Reachability matrix for reachable canonical form) Consider a system in reachable canonical form. Show that the inverse of the reachability matrix is given by

$$\tilde{W}_r^{-1} = \begin{bmatrix} 1 & a_1 & a_2 & \cdots & a_n \\ 0 & 1 & a_1 & \cdots & a_{n-1} \\ 0 & 0 & 1 & \ddots & \vdots \\ \vdots & & & \ddots & a_1 \\ 0 & 0 & 0 & \cdots & 1 \end{bmatrix}.$$

6.8 (Non-maintainable equilibria) Consider the normalized model of a pendulum on a cart

$$\frac{d^2x}{dt^2} = u, \qquad \frac{d^2\theta}{dt^2} = -\theta + u,$$

where x is cart position and θ is pendulum angle. Can the angle $\theta = \theta_0$ for $\theta_0 \neq 0$ be maintained?

6.9 (Eigenvalue assignment for unreachable system) Consider the system

$$\frac{dx}{dt} = \begin{bmatrix} 0 & 1 \\ 0 & 0 \end{bmatrix} x + \begin{bmatrix} 1 \\ 0 \end{bmatrix} u, \qquad y = \begin{bmatrix} 1 & 0 \end{bmatrix} x,$$

with the control law

$$u = -k_1 x_1 - k_2 x_2 + k_r r.$$

Show that eigenvalues of the system cannot be assigned to arbitrary values.

6.10 (Cayley–Hamilton theorem) Let $A \in \mathbb{R}^{n \times n}$ be a matrix with characteristic polynomial $\lambda(s) = \det(sI - A) = s^n + a_1 s^{n-1} + \cdots + a_{n-1}s + a_n$. Assume that the matrix A can be diagonalized and show that it satisfies

$$\lambda(A) = A^n + a_1 A^{n-1} + \cdots + a_{n-1}A + a_n I = 0,$$

Use the result to show that $A^k, k \geq n$, can be rewritten in terms of powers of A of order less than n.

6.11 (Motor drive) Consider the normalized model of the motor drive in Exercise 2.10. Using the following normalized parameters,

$$J_1 = 10/9, \qquad J_2 = 10, \qquad c = 0.1, \qquad k = 1, \qquad k_I = 1,$$

verify that the eigenvalues of the open loop system are $0, 0, -0.05 \pm i$. Design a state feedback that gives a closed loop system with eigenvalues $-2, -1$ and $-1 \pm i$. This choice implies that the oscillatory eigenvalues will be well damped and that the eigenvalues at the origin are replaced by eigenvalues on the negative real axis. Simulate the responses of the closed loop system to step changes in the command signal for θ_2 and a step change in a disturbance torque on the second rotor.

6.12 (Whipple bicycle model) Consider the Whipple bicycle model given by equation (3.7) in Section 3.2. Using the parameters from the companion web site, the model is unstable at the velocity $v = 5$ m/s and the open loop eigenvalues are -1.84, -14.29 and $1.30 \pm 4.60i$. Find the gains of a controller that stabilizes the bicycle and gives closed loop eigenvalues at -2, -10 and $-1 \pm i$. Simulate the response of the system to a step change in the steering reference of 0.002 rad.

6.13 (Atomic force microscope) Consider the model of an AFM in contact mode given in Example 5.9:

$$\frac{dx}{dt} = \begin{bmatrix} 0 & 1 & 0 & 0 \\ -k_2/(m_1+m_2) & -c_2/(m_1+m_2) & 1/m_2 & 0 \\ 0 & 0 & 0 & \omega_3 \\ 0 & 0 & -\omega_3 & -2\zeta_3\omega_3 \end{bmatrix} x + \begin{bmatrix} 0 \\ 0 \\ 0 \\ \omega_3 \end{bmatrix} u,$$

$$y = \frac{m_2}{m_1+m_2} \begin{bmatrix} \frac{m_1 k_2}{m_1+m_2} & \frac{m_1 c_2}{m_1+m_2} & 1 & 0 \end{bmatrix} x.$$

Use the MATLAB script afm_data.m from the companion web site to generate the system matrices.

(a) Compute the reachability matrix of the system and numerically determine its rank. Scale the model by using milliseconds instead of seconds as time units. Repeat the calculation of the reachability matrix and its rank.

(b) Find a state feedback controller that gives a closed loop system with complex poles having damping ratio 0.707. Use the scaled model for the computations.

(c) Compute state feedback gains using linear quadratic control. Experiment by using different weights. Compute the gains for $q_1 = q_2 = 0, q_3 = q_4 = 1$ and $\rho_1 = 0.1$ and explain the result. Choose $q_1 = q_2 = q_3 = q_4 = 1$ and explore what happens to the feedback gains and closed loop eigenvalues when you change ρ_1. Use the scaled system for this computation.

6.14 Consider the second-order system

$$\frac{d^2 y}{dt^2} + 0.5\frac{dy}{dt} + y = a\frac{du}{dt} + u.$$

Let the initial conditions be zero.

(a) Show that the initial slope of the unit step response is a. Discuss what it means when $a < 0$.

(b) Show that there are points on the unit step response that are invariant with a. Discuss qualitatively the effect of the parameter a on the solution.

(c) Simulate the system and explore the effect of a on the rise time and overshoot.

6.15 (Bryson's rule) Bryson and Ho [47] have suggested the following method for choosing the matrices Q_x and Q_u in equation (6.26). Start by choosing Q_x and Q_u as diagonal matrices whose elements are the inverses of the squares of the maxima of the corresponding variables. Then modify the elements to obtain a compromise among response time, damping and control effort. Apply this method to the motor drive in Exercise 6.11. Assume that the largest values of the φ_1 and φ_2 are 1, the largest values of $\dot{\varphi}_1$ and $\dot{\varphi}_2$ are 2 and the largest control signal is 10. Simulate the closed loop system for $\varphi_2(0) = 1$ and all other states are initialized to 0. Explore the effects of different values of the diagonal elements for Q_x and Q_u.

Chapter Seven
Output Feedback

One may separate the problem of physical realization into two stages: computation of the "best approximation" $\hat{x}(t_1)$ of the state from knowledge of $y(t)$ for $t \leq t_1$ and computation of $u(t_1)$ given $\hat{x}(t_1)$.

R. E. Kalman, "Contributions to the Theory of Optimal Control," 1960 [113].

In this chapter we show how to use output feedback to modify the dynamics of the system through the use of observers. We introduce the concept of observability and show that if a system is observable, it is possible to recover the state from measurements of the inputs and outputs to the system. We then show how to design a controller with feedback from the observer state. An important concept is the separation principle quoted above, which is also proved. The structure of the controllers derived in this chapter is quite general and is obtained by many other design methods.

7.1 Observability

In Section 6.2 of the previous chapter it was shown that it is possible to find a state feedback law that gives desired closed loop eigenvalues provided that the system is reachable and that all the states are measured. For many situations, it is highly unrealistic to assume that all the states are measured. In this section we investigate how the state can be estimated by using a mathematical model and a few measurements. It will be shown that computation of the states can be carried out by a dynamical system called an *observer*.

Definition of Observability

Consider a system described by a set of differential equations

$$\frac{dx}{dt} = Ax + Bu, \qquad y = Cx + Du, \tag{7.1}$$

where $x \in \mathbb{R}^n$ is the state, $u \in \mathbb{R}^p$ the input and $y \in \mathbb{R}^q$ the measured output. We wish to estimate the state of the system from its inputs and outputs, as illustrated in Figure 7.1. In some situations we will assume that there is only one measured signal, i.e., that the signal y is a scalar and that C is a (row) vector. This signal may be corrupted by noise n, although we shall start by considering the noise-free case. We write $\hat{x}$ for the state estimate given by the observer.

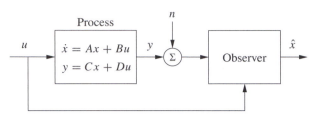

Figure 7.1: Block diagram for an observer. The observer uses the process measurement y (possibly corrupted by noise n) and the input u to estimate the current state of the process, denoted $\hat{x}$.

Definition 7.1 (Observability). A linear system is *observable* if for any $T > 0$ it is possible to determine the state of the system $x(T)$ through measurements of $y(t)$ and $u(t)$ on the interval $[0, T]$.

The definition above holds for nonlinear systems as well, and the results discussed here have extensions to the nonlinear case.

The problem of observability is one that has many important applications, even outside feedback systems. If a system is observable, then there are no "hidden" dynamics inside it; we can understand everything that is going on through observation (over time) of the inputs and outputs. As we shall see, the problem of observability is of significant practical interest because it will determine if a set of sensors is sufficient for controlling a system. Sensors combined with a mathematical model can also be viewed as a "virtual sensor" that gives information about variables that are not measured directly. The process of reconciling signals from many sensors with mathematical models is also called *sensor fusion*.

Testing for Observability

When discussing reachability in the last chapter, we neglected the output and focused on the state. Similarly, it is convenient here to initially neglect the input and focus on the autonomous system

$$\frac{dx}{dt} = Ax, \qquad y = Cx. \tag{7.2}$$

We wish to understand when it is possible to determine the state from observations of the output.

The output itself gives the projection of the state on vectors that are rows of the matrix C. The observability problem can immediately be solved if the matrix C is invertible. If the matrix is not invertible, we can take derivatives of the output to obtain

$$\frac{dy}{dt} = C\frac{dx}{dt} = CAx.$$

From the derivative of the output we thus get the projection of the state on vectors that are rows of the matrix CA. Proceeding in this way, we get

$$\begin{bmatrix} y \\ \dot{y} \\ \ddot{y} \\ \vdots \\ y^{(n-1)} \end{bmatrix} = \begin{bmatrix} C \\ CA \\ CA^2 \\ \vdots \\ CA^{n-1} \end{bmatrix} x. \tag{7.3}$$

We thus find that the state can be determined if the *observability matrix*

$$W_o = \begin{bmatrix} C \\ CA \\ CA^2 \\ \vdots \\ CA^{n-1} \end{bmatrix} \tag{7.4}$$

has n independent rows. It turns out that we need not consider any derivatives higher than $n - 1$ (this is an application of the Cayley–Hamilton theorem [Exercise 6.10]).

The calculation can easily be extended to systems with inputs. The state is then given by a linear combination of inputs and outputs and their higher derivatives. The observability criterion is unchanged. We leave this case as an exercise for the reader.

In practice, differentiation of the output can give large errors when there is measurement noise, and therefore the method sketched above is not particularly practical. We will address this issue in more detail in the next section, but for now we have the following basic result.

Theorem 7.1 (Observability rank condition). *A linear system of the form* (7.1) *is observable if and only if the observability matrix W_o is full rank.*

Proof. The sufficiency of the observability rank condition follows from the analysis above. To prove necessity, suppose that the system is observable but W_o is not full rank. Let $v \in \mathbb{R}^n$, $v \neq 0$, be a vector in the null space of W_o, so that $W_o v = 0$. If we let $x(0) = v$ be the initial condition for the system and choose $u = 0$, then the output is given by $y(t) = Ce^{At}v$. Since e^{At} can be written as a power series in A and since A^n and higher powers can be rewritten in terms of lower powers of A (by the Cayley–Hamilton theorem), it follows that the output will be identically zero (the reader should fill in the missing steps if this is not clear). However, if both the input and output of the system are 0, then a valid estimate of the state is $\hat{x} = 0$ for all time, which is clearly incorrect since $x(0) = v \neq 0$. Hence by contradiction we must have that W_o is full rank if the system is observable. $\square$

Example 7.1 Compartment model
Consider the two-compartment model in Figure 3.18a, but assume that the concentration in the first compartment can be measured. The system is described by the

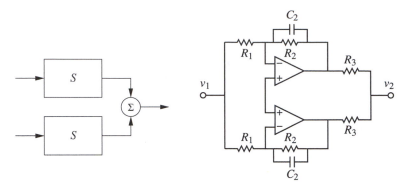

Figure 7.2: An unobservable system. Two identical subsystems have outputs that add together to form the overall system output. The individual states of the subsystem cannot be determined since the contributions of each to the output are not distinguishable. The circuit diagram on the right is an example of such a system.

linear system

$$\frac{dc}{dt} = \begin{bmatrix} -k_0 - k_1 & k_1 \\ k_2 & -k_2 \end{bmatrix} c + \begin{bmatrix} b_0 \\ 0 \end{bmatrix} u, \qquad y = \begin{bmatrix} 1 & 0 \end{bmatrix} x.$$

The first compartment represents the drug concentration in the blood plasma, and the second compartment the drug concentration in the tissue where it is active. To determine if it is possible to find the concentration in the tissue compartment from a measurement of blood plasma, we investigate the observability of the system by forming the observability matrix

$$W_o = \begin{bmatrix} C \\ CA \end{bmatrix} = \begin{bmatrix} 1 & 0 \\ -k_0 - k_1 & k_1 \end{bmatrix}.$$

The rows are linearly independent if $k_1 \neq 0$, and under this condition it is thus possible to determine the concentration of the drug in the active compartment from measurements of the drug concentration in the blood.   ∇

It is useful to have an understanding of the mechanisms that make a system unobservable. Such a system is shown in Figure 7.2. The system is composed of two identical systems whose outputs are added. It seems intuitively clear that it is not possible to deduce the states from the output since we cannot deduce the individual output contributions from the sum. This can also be seen formally (Exercise 7.2).

Observable Canonical Form

As in the case of reachability, certain canonical forms will be useful in studying observability. A linear single-input, single-output state space system is in *observable*

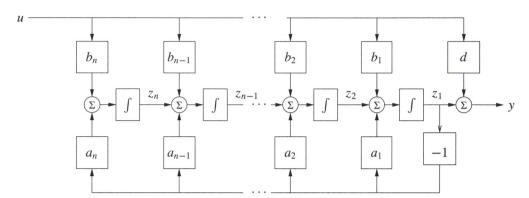

Figure 7.3: Block diagram of a system in observable canonical form. The states of the system are represented by individual integrators whose inputs are a weighted combination of the next integrator in the chain, the first state (rightmost integrator) and the system input. The output is a combination of the first state and the input.

canonical form if its dynamics are given by

$$\frac{dz}{dt} = \begin{bmatrix} -a_1 & 1 & 0 & \cdots & 0 \\ -a_2 & 0 & 1 & & 0 \\ \vdots & & & \ddots & \\ -a_{n-1} & 0 & 0 & & 1 \\ -a_n & 0 & 0 & \cdots & 0 \end{bmatrix} z + \begin{bmatrix} b_1 \\ b_2 \\ \vdots \\ b_{n-1} \\ b_n \end{bmatrix} u,$$

$$y = \begin{bmatrix} 1 & 0 & 0 & \cdots & 0 \end{bmatrix} z + Du.$$

The definition can be extended to systems with many inputs; the only difference is that the vector multiplying u is replaced by a matrix.

Figure 7.3 is a block diagram for a system in observable canonical form. As in the case of reachable canonical form, we see that the coefficients in the system description appear directly in the block diagram. The characteristic polynomial for a system in observable canonical form is

$$\lambda(s) = s^n + a_1 s^{n-1} + \cdots + a_{n-1}s + a_n. \tag{7.5}$$

It is possible to reason about the observability of a system in observable canonical form by studying the block diagram. If the input u and the output y are available, the state z_1 can clearly be computed. Differentiating z_1, we obtain the input to the integrator that generates z_1, and we can now obtain $z_2 = \dot{z}_1 + a_1 z_1 - b_1 u$. Proceeding in this way, we can compute all states. The computation will, however, require that the signals be differentiated.

To check observability more formally, we compute the observability matrix for a system in observable canonical form, which is given by

$$
W_o = \begin{bmatrix} 1 & 0 & 0 & \cdots & 0 \\ -a_1 & 1 & 0 & \cdots & 0 \\ -a_1^2 - a_1 a_2 & -a_1 & 1 & & 0 \\ \vdots & \vdots & & \ddots & \vdots \\ * & * & & \cdots & 1 \end{bmatrix},
$$

where $*$ represents an entry whose exact value is not important. The rows of this matrix are linearly independent (since it is lower triangular), and hence W_o is full rank. A straightforward but tedious calculation shows that the inverse of the observability matrix has a simple form given by

$$
W_o^{-1} = \begin{bmatrix} 1 & 0 & 0 & \cdots & 0 \\ a_1 & 1 & 0 & \cdots & 0 \\ a_2 & a_1 & 1 & \cdots & 0 \\ \vdots & \vdots & & \ddots & \vdots \\ a_{n-1} & a_{n-2} & a_{n-3} & \cdots & 1 \end{bmatrix}.
$$

As in the case of reachability, it turns out that if a system is observable then there always exists a transformation T that converts the system into observable canonical form. This is useful for proofs since it lets us assume that a system is in observable canonical form without any loss of generality. The observable canonical form may be poorly conditioned numerically.

7.2 State Estimation

Having defined the concept of observability, we now return to the question of how to construct an observer for a system. We will look for observers that can be represented as a linear dynamical system that takes the inputs and outputs of the system we are observing and produces an estimate of the system's state. That is, we wish to construct a dynamical system of the form

$$
\frac{d\hat{x}}{dt} = F\hat{x} + Gu + Hy,
$$

where u and y are the input and output of the original system and $\hat{x} \in \mathbb{R}^n$ is an estimate of the state with the property that $\hat{x}(t) \to x(t)$ as $t \to \infty$.

The Observer

We consider the system in equation (7.1) with D set to zero to simplify the exposition:

$$
\frac{dx}{dt} = Ax + Bu, \qquad y = Cx. \tag{7.6}
$$

We can attempt to determine the state simply by simulating the equations with the correct input. An estimate of the state is then given by

$$\frac{d\hat{x}}{dt} = A\hat{x} + Bu. \tag{7.7}$$

To find the properties of this estimate, introduce the estimation error $\tilde{x} = x - \hat{x}$. It follows from equations (7.6) and (7.7) that

$$\frac{d\tilde{x}}{dt} = A\tilde{x}.$$

If matrix A has all its eigenvalues in the left half-plane, the error $\tilde{x}$ will go to zero, and hence equation (7.7) is a dynamical system whose output converges to the state of the system (7.6).

The observer given by equation (7.7) uses only the process input u; the measured signal does not appear in the equation. We must also require that the system be stable, and essentially our estimator converges because the state of both the observer and the estimator are going zero. This is not very useful in a control design context since we want to have our estimate converge quickly to a nonzero state so that we can make use of it in our controller. We will therefore attempt to modify the observer so that the output is used and its convergence properties can be designed to be fast relative to the system's dynamics. This version will also work for unstable systems.

Consider the observer

$$\frac{d\hat{x}}{dt} = A\hat{x} + Bu + L(y - C\hat{x}). \tag{7.8}$$

This can be considered as a generalization of equation (7.7). Feedback from the measured output is provided by adding the term $L(y - C\hat{x})$, which is proportional to the difference between the observed output and the output predicted by the observer. It follows from equations (7.6) and (7.8) that

$$\frac{d\tilde{x}}{dt} = (A - LC)\tilde{x}.$$

If the matrix L can be chosen in such a way that the matrix $A - LC$ has eigenvalues with negative real parts, the error $\tilde{x}$ will go to zero. The convergence rate is determined by an appropriate selection of the eigenvalues.

Notice the similarity between the problems of finding a state feedback and finding the observer. State feedback design by eigenvalue assignment is equivalent to finding a matrix K so that $A - BK$ has given eigenvalues. Designing an observer with prescribed eigenvalues is equivalent to finding a matrix L so that $A - LC$ has given eigenvalues. Since the eigenvalues of a matrix and its transpose are the same we can establish the following equivalences:

$$A \leftrightarrow A^T, \qquad B \leftrightarrow C^T, \qquad K \leftrightarrow L^T, \qquad W_r \leftrightarrow W_o^T.$$

The observer design problem is the *dual* of the state feedback design problem. Using the results of Theorem 6.3, we get the following theorem on observer design.

Theorem 7.2 (Observer design by eigenvalue assignment). *Consider the system given by*

$$\frac{dx}{dt} = Ax + Bu, \qquad y = Cx, \tag{7.9}$$

with one input and one output. Let $\lambda(s) = s^n + a_1 s^{n-1} + \cdots + a_{n-1} s + a_n$ *be the characteristic polynomial for A. If the system is observable, then the dynamical system*

$$\frac{d\hat{x}}{dt} = A\hat{x} + Bu + L(y - C\hat{x}) \tag{7.10}$$

is an observer for the system, with L chosen as

$$L = W_o^{-1} \widetilde{W}_o \begin{bmatrix} p_1 - a_1 \\ p_2 - a_2 \\ \vdots \\ p_n - a_n \end{bmatrix} \tag{7.11}$$

and the matrices W_o *and* $\widetilde{W}_o$ *given by*

$$W_o = \begin{bmatrix} C \\ CA \\ \vdots \\ CA^{n-1} \end{bmatrix}, \qquad \widetilde{W}_o = \begin{bmatrix} 1 & 0 & 0 & \cdots & 0 & 0 \\ a_1 & 1 & 0 & \cdots & 0 & 0 \\ a_2 & a_1 & 1 & & 0 & 0 \\ \vdots & \vdots & & \ddots & & \vdots \\ a_{n-2} & a_{n-3} & a_{n-4} & & 1 & 0 \\ a_{n-1} & a_{n-2} & a_{n-3} & \cdots & a_1 & 1 \end{bmatrix}^{-1}.$$

The resulting observer error $\tilde{x} = x - \hat{x}$ *is governed by a differential equation having the characteristic polynomial*

$$p(s) = s^n + p_1 s^{n-1} + \cdots + p_n.$$

The dynamical system (7.10) is called an *observer* for (the states of) the system (7.9) because it will generate an approximation of the states of the system from its inputs and outputs. This form of an observer is a much more useful form than the one given by pure differentiation in equation (7.3).

Example 7.2 Compartment model
Consider the compartment model in Example 7.1, which is characterized by the matrices

$$A = \begin{bmatrix} -k_0 - k_1 & k_1 \\ k_2 & -k_2 \end{bmatrix}, \qquad B = \begin{bmatrix} b_0 \\ 0 \end{bmatrix}, \qquad C = \begin{bmatrix} 1 & 0 \end{bmatrix}.$$

The observability matrix was computed in Example 7.1, where we concluded that the system was observable if $k_1 \neq 0$. The dynamics matrix has the characteristic polynomial

$$\lambda(s) = \det \begin{bmatrix} s + k_0 + k_1 & -k_1 \\ -k_2 & s + k_2 \end{bmatrix} = s^2 + (k_0 + k_1 + k_2)s + k_0 k_2.$$

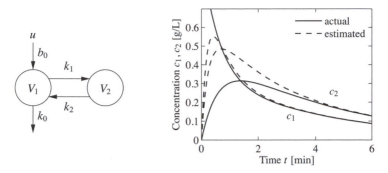

Figure 7.4: Observer for a two compartment system. A two compartment model is shown on the left. The observer measures the input concentration u and output concentration $y = c_1$ to determine the compartment concentrations, shown on the right. The true concentrations are shown by solid lines and the estimates generated by the observer by dashed lines.

Let the desired characteristic polynomial of the observer be $s^2 + p_1 s + p_2$, and equation (7.11) gives the observer gain

$$
L = \begin{bmatrix} 1 & 0 \\ -k_0 - k_1 & k_1 \end{bmatrix}^{-1} \begin{bmatrix} 1 & 0 \\ k_0 + k_1 + k_2 & 1 \end{bmatrix}^{-1} \begin{bmatrix} p_1 - k_0 - k_1 - k_2 \\ p_2 - k_0 k_2 \end{bmatrix}
$$

$$
= \begin{bmatrix} p_1 - k_0 - k_1 - k_2 \\ (p_2 - p_1 k_2 + k_1 k_2 + k_2^2)/k_1 \end{bmatrix}.
$$

Notice that the observability condition $k_1 \neq 0$ is essential. The behavior of the observer is illustrated by the simulation in Figure 7.4b. Notice how the observed concentrations approach the true concentrations. ∇

The observer is a dynamical system whose inputs are the process input u and the process output y. The rate of change of the estimate is composed of two terms. One term, $A\hat{x} + Bu$, is the rate of change computed from the model with $\hat{x}$ substituted for x. The other term, $L(y - \hat{y})$, is proportional to the difference $e = y - \hat{y}$ between measured output y and its estimate $\hat{y} = C\hat{x}$. The observer gain L is a matrix that tells how the error e is weighted and distributed among the states. The observer thus combines measurements with a dynamical model of the system. A block diagram of the observer is shown in Figure 7.5.

Computing the Observer Gain

For simple low-order problems it is convenient to introduce the elements of the observer gain L as unknown parameters and solve for the values required to give the desired characteristic polynomial, as illustrated in the following example.

Example 7.3 Vehicle steering
The normalized linear model for vehicle steering derived in Examples 5.12 and 6.4 gives the following state space model dynamics relating lateral path deviation y to

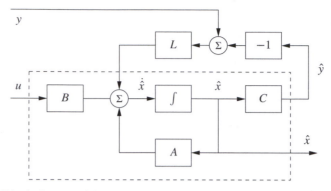

Figure 7.5: Block diagram of the observer. The observer takes the signals y and u as inputs and produces an estimate x. Notice that the observer contains a copy of the process model that is driven by $y - \hat{y}$ through the observer gain L.

steering angle u:

$$\frac{dx}{dt} = \begin{bmatrix} 0 & 1 \\ 0 & 0 \end{bmatrix} x + \begin{bmatrix} \gamma \\ 1 \end{bmatrix} u, \qquad y = \begin{bmatrix} 1 & 0 \end{bmatrix} x. \qquad (7.12)$$

Recall that the state x_1 represents the lateral path deviation and that x_2 represents the turning rate. We will now derive an observer that uses the system model to determine the turning rate from the measured path deviation.

The observability matrix is

$$W_o = \begin{bmatrix} 1 & 0 \\ 0 & 1 \end{bmatrix},$$

i.e., the identity matrix. The system is thus observable, and the eigenvalue assignment problem can be solved. We have

$$A - LC = \begin{bmatrix} -l_1 & 1 \\ -l_2 & 0 \end{bmatrix},$$

which has the characteristic polynomial

$$\det(sI - A + LC) = \det \begin{bmatrix} s + l_1 & -1 \\ l_2 & s \end{bmatrix} = s^2 + l_1 s + l_2.$$

Assuming that we want to have an observer with the characteristic polynomial

$$s^2 + p_1 s + p_2 = s^2 + 2\zeta_o \omega_o s + \omega_o^2,$$

the observer gains should be chosen as

$$l_1 = p_1 = 2\zeta_o \omega_o, \qquad l_2 = p_2 = \omega_o^2.$$

The observer is then

$$\frac{d\hat{x}}{dt} = A\hat{x} + Bu + L(y - C\hat{x}) = \begin{bmatrix} 0 & 1 \\ 0 & 0 \end{bmatrix} \hat{x} + \begin{bmatrix} \gamma \\ 1 \end{bmatrix} u + \begin{bmatrix} l_1 \\ l_2 \end{bmatrix} (y - \hat{x}_1).$$

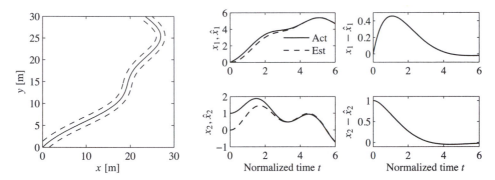

Figure 7.6: Simulation of an observer for a vehicle driving on a curvy road (left). The observer has an initial velocity error. The plots on the middle show the lateral deviation x_1, the lateral velocity x_2 by solid lines and their estimates $\hat{x}_1$ and $\hat{x}_2$ by dashed lines. The plots on the right show the estimation errors.

A simulation of the observer for a vehicle driving on a curvy road is simulated in Figure 7.6. The vehicle length is the time unit in the normalized model. The figure shows that the observer error settles in about 3 vehicle lengths.   ∇

For systems of high order we have to use numerical calculations. The duality between the design of a state feedback and the design of an observer means that the computer algorithms for state feedback can also be used for the observer design; we simply use the transpose of the dynamics matrix and the output matrix. The MATLAB command `acker`, which essentially is a direct implementation of the calculations given in Theorem 7.2, can be used for systems with one output. The MATLAB command `place` can be used for systems with many outputs. It is also better conditioned numerically.

7.3 Control Using Estimated State

In this section we will consider a state space system of the form

$$\frac{dx}{dt} = Ax + Bu, \qquad y = Cx. \tag{7.13}$$

Notice that we have assumed that there is no direct term in the system ($D = 0$). This is often a realistic assumption. The presence of a direct term in combination with a controller having proportional action creates an algebraic loop, which will be discussed in Section 8.3. The problem can be solved even if there is a direct term, but the calculations are more complicated.

We wish to design a feedback controller for the system where only the output is measured. As before, we will assume that u and y are scalars. We also assume that the system is reachable and observable. In Chapter 6 we found a feedback of the form

$$u = -Kx + k_r r$$

for the case that all states could be measured, and in Section 7.2 we developed an observer that can generate estimates of the state $\hat{x}$ based on inputs and outputs. In this section we will combine the ideas of these sections to find a feedback that gives desired closed loop eigenvalues for systems where only outputs are available for feedback.

If all states are not measurable, it seems reasonable to try the feedback

$$u = -K\hat{x} + k_r r, \tag{7.14}$$

where $\hat{x}$ is the output of an observer of the state, i.e.,

$$\frac{d\hat{x}}{dt} = A\hat{x} + Bu + L(y - C\hat{x}). \tag{7.15}$$

Since the system (7.13) and the observer (7.15) are both of state dimension n, the closed loop system has state dimension $2n$ with state $(x, \hat{x})$. The evolution of the states is described by equations (7.13)–(7.15). To analyze the closed loop system, the state variable $\hat{x}$ is replaced by

$$\tilde{x} = x - \hat{x}. \tag{7.16}$$

Subtraction of equation (7.15) from equation (7.13) gives

$$\frac{d\tilde{x}}{dt} = Ax - A\hat{x} - L(Cx - C\hat{x}) = A\tilde{x} - LC\tilde{x} = (A - LC)\tilde{x}.$$

Returning to the process dynamics, introducing u from equation (7.14) into equation (7.13) and using equation (7.16) to eliminate $\hat{x}$ gives

$$\frac{dx}{dt} = Ax + Bu = Ax - BK\hat{x} + Bk_r r = Ax - BK(x - \tilde{x}) + Bk_r r$$

$$= (A - BK)x + BK\tilde{x} + Bk_r r.$$

The closed loop system is thus governed by

$$\frac{d}{dt}\begin{bmatrix} x \\ \tilde{x} \end{bmatrix} = \begin{bmatrix} A - BK & BK \\ 0 & A - LC \end{bmatrix}\begin{bmatrix} x \\ \tilde{x} \end{bmatrix} + \begin{bmatrix} Bk_r \\ 0 \end{bmatrix} r. \tag{7.17}$$

Notice that the state $\tilde{x}$, representing the observer error, is not affected by the reference signal r. This is desirable since we do not want the reference signal to generate observer errors.

Since the dynamics matrix is block diagonal, we find that the characteristic polynomial of the closed loop system is

$$\lambda(s) = \det(sI - A + BK)\det(sI - A + LC).$$

This polynomial is a product of two terms: the characteristic polynomial of the closed loop system obtained with state feedback and the characteristic polynomial of the observer error. The feedback (7.14) that was motivated heuristically thus provides a neat solution to the eigenvalue assignment problem. The result is summarized as follows.

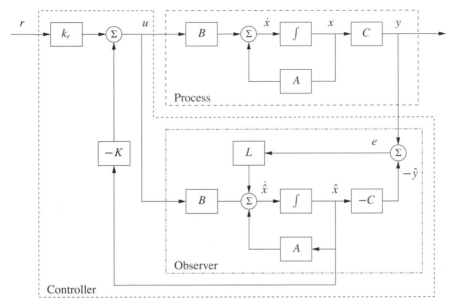

Figure 7.7: Block diagram of an observer-based control system. The observer uses the measured output y and the input u to construct an estimate of the state. This estimate is used by a state feedback controller to generate the corrective input. The controller consists of the observer and the state feedback; the observer is identical to that in Figure 7.5.

Theorem 7.3 (Eigenvalue assignment by output feedback). *Consider the system*

$$\frac{dx}{dt} = Ax + Bu, \qquad y = Cx.$$

The controller described by

$$\frac{d\hat{x}}{dt} = A\hat{x} + Bu + L(y - C\hat{x}) = (A - BK - LC)\hat{x} + Ly,$$

$$u = -K\hat{x} + k_r r$$

gives a closed loop system with the characteristic polynomial

$$\lambda(s) = \det{(sI - A + BK)} \det{(sI - A + LC)}.$$

This polynomial can be assigned arbitrary roots if the system is reachable and observable.

The controller has a strong intuitive appeal: it can be thought of as being composed of two parts, one state feedback and one observer. The dynamics of the controller are generated by the observer. The feedback gain K can be computed as if all state variables can be measured, and it depends on only A and B. The observer gain L depends on only A and C. The property that the eigenvalue assignment for output feedback can be separated into an eigenvalue assignment for a state feedback and an observer is called the *separation principle*.

A block diagram of the controller is shown in Figure 7.7. Notice that the con-

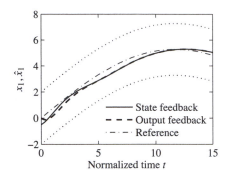 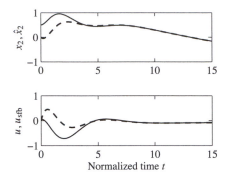

Figure 7.8: Simulation of a vehicle driving on a curvy road with a controller based on state feedback and an observer. The left plot shows the lane boundaries (dotted), the vehicle position (solid) and its estimate (dashed), the upper right plot shows the velocity (solid) and its estimate (dashed), and the lower right plot shows the control signal using state feedback (solid) and the control signal using the estimated state (dashed).

troller contains a dynamical model of the plant. This is called the *internal model principle*: the controller contains a model of the process being controlled.

Example 7.4 Vehicle steering

Consider again the normalized linear model for vehicle steering in Example 6.4. The dynamics relating the steering angle u to the lateral path deviation y is given by the state space model (7.12). Combining the state feedback derived in Example 6.4 with the observer determined in Example 7.3, we find that the controller is given by

$$\frac{d\hat{x}}{dt} = A\hat{x} + Bu + L(y - C\hat{x}) = \begin{bmatrix} 0 & 1 \\ 0 & 0 \end{bmatrix}\hat{x} + \begin{bmatrix} \gamma \\ 1 \end{bmatrix}u + \begin{bmatrix} l_1 \\ l_2 \end{bmatrix}(y - \hat{x}_1),$$

$$u = -K\hat{x} + k_r r = k_1(r - \hat{x}_1) - k_2\hat{x}_2.$$

Elimination of the variable u gives

$$\frac{d\hat{x}}{dt} = (A - BK - LC)\hat{x} + Ly + Bk_r r$$

$$= \begin{bmatrix} -l_1 - \gamma k_1 & 1 - \gamma k_2 \\ -k_1 - l_2 & -k_2 \end{bmatrix}\hat{x} + \begin{bmatrix} l_1 \\ l_2 \end{bmatrix}y + \begin{bmatrix} \gamma \\ 1 \end{bmatrix}k_1 r.$$

The controller is a dynamical system of second order, with two inputs y and r and one output u. Figure 7.8 shows a simulation of the system when the vehicle is driven along a curvy road. Since we are using a normalized model, the length unit is the vehicle length and the time unit is the time it takes to travel one vehicle length. The estimator is initialized with all states equal to zero but the real system has an initial velocity of 0.5. The figures show that the estimates converge quickly to their true values. The vehicle tracks the desired path, which is in the middle of the road, but there are errors because the road is irregular. The tracking error can be improved by introducing feedforward (Section 7.5).   ∇

7.4 Kalman Filtering

One of the principal uses of observers in practice is to estimate the state of a system in the presence of *noisy* measurements. We have not yet treated noise in our analysis, and a full treatment of stochastic dynamical systems is beyond the scope of this text. In this section, we present a brief introduction to the use of stochastic systems analysis for constructing observers. We work primarily in discrete time to avoid some of the complications associated with continuous-time random processes and to keep the mathematical prerequisites to a minimum. This section assumes basic knowledge of random variables and stochastic processes; see Kumar and Varaiya [131] or Åström [15] for the required material.

Consider a discrete-time linear system with dynamics

$$x[k+1] = Ax[k] + Bu[k] + Fv[k], \qquad y[k] = Cx[k] + w[k], \qquad (7.18)$$

where $v[k]$ and $w[k]$ are Gaussian white noise processes satisfying

$$E\{v[k]\} = 0, \qquad\qquad E\{w[k]\} = 0,$$

$$E\{v[k]v^T[j]\} = \begin{cases} 0 & k \neq j \\ R_v & k = j, \end{cases} \qquad E\{w[k]w^T[j]\} = \begin{cases} 0 & k \neq j \\ R_w & k = j, \end{cases} \qquad (7.19)$$

$$E\{v[k]w^T[j]\} = 0.$$

$E\{v[k]\}$ represents the expected value of $v[k]$ and $E\{v[k]v^T[j]\}$ the correlation matrix. The matrices R_v and R_w are the covariance matrices for the process disturbance v and measurement noise w. We assume that the initial condition is also modeled as a Gaussian random variable with

$$E\{x[0]\} = x_0, \qquad E\{x[0]x^T[0]\} = P_0. \qquad (7.20)$$

We would like to find an estimate $\hat{x}[k]$ that minimizes the mean square error $E\{(x[k] - \hat{x}[k])(x[k] - \hat{x}[k])^T\}$ given the measurements $\{y(\tau) : 0 \leq \tau \leq t\}$. We consider an observer in the same basic form as derived previously:

$$\hat{x}[k+1] = A\hat{x}[k] + Bu[k] + L[k](y[k] - C\hat{x}[k]). \qquad (7.21)$$

The following theorem summarizes the main result.

Theorem 7.4 (Kalman, 1961). *Consider a random process $x[k]$ with dynamics given by equation (7.18) and noise processes and initial conditions described by equations (7.19) and (7.20). The observer gain L that minimizes the mean square error is given by*

$$L[k] = AP[k]C^T(R_w + CP[k]C^T)^{-1},$$

where

$$P[k+1] = (A - LC)P[k](A - LC)^T + FR_vF^T + LR_wL^T$$
$$P_0 = E\{x[0]x^T[0]\}. \qquad (7.22)$$

Before we prove this result, we reflect on its form and function. First, note that the Kalman filter has the form of a *recursive* filter: given mean square error

$P[k] = E\{(x[k] - \hat{x}[k])(x[k] - \hat{x}[k])^T\}$ at time k, we can compute how the estimate and error *change*. Thus we do not need to keep track of old values of the output. Furthermore, the Kalman filter gives the estimate $\hat{x}[k]$ *and* the error covariance $P[k]$, so we can see how reliable the estimate is. It can also be shown that the Kalman filter extracts the maximum possible information about output data. If we form the residual between the measured output and the estimated output,

$$e[k] = y[k] - C\hat{x}[k],$$

we can show that for the Kalman filter the correlation matrix is

$$R_e(j, k) = E\{e[j]e^T[k]\} = W[k]\delta_{jk}, \qquad \delta_{jk} = \begin{cases} 1 & j = k \\ 0 & j \neq k. \end{cases}$$

In other words, the error is a white noise process, so there is no remaining dynamic information content in the error.

The Kalman filter is extremely versatile and can be used even if the process, noise or disturbances are nonstationary. When the system is stationary and *if $P[k]$* converges, then the observer gain is constant:

$$L = APC^T(R_w + CPC^T),$$

where P satisfies

$$P = APA^T + FR_vF^T - APC^T(R_w + CPC^T)^{-1}CPA^T.$$

We see that the optimal gain depends on both the process noise and the measurement noise, but in a nontrivial way. Like the use of LQR to choose state feedback gains, the Kalman filter permits a systematic derivation of the observer gains given a description of the noise processes. The solution for the constant gain case is solved by the `dlqe` command in MATLAB.

Proof of theorem. We wish to minimize the mean square of the error $E\{(x[k] - \hat{x}[k])(x[k] - \hat{x}[k])^T\}$. We will define this quantity as $P[k]$ and then show that it satisfies the recursion given in equation (7.22). By definition,

$$\begin{aligned} P[k+1] &= E\{(x[k+1] - \hat{x}[k+1])(x[k+1] - \hat{x}[k+1])^T\} \\ &= (A - LC)P[k](A - LC)^T + FR_vF^T + LR_wL^T \\ &= AP[k]A^T + FR_vF^T - AP[k]C^TL^T - LCP[k]A^T \\ &\quad + L(R_w + CP[k]C^T)L^T. \end{aligned}$$

Letting $R_\epsilon = (R_w + CP[k]C^T)$, we have

$$\begin{aligned} P[k+1] &= AP[k]A^T + FR_vF^T - AP[k]C^TL^T - LCP[k]A^T + LR_\epsilon L^T \\ &= AP[k]A^T + FR_vF^T + (L - AP[k]C^TR_\epsilon^{-1})R_\epsilon(L - AP[k]C^TR_\epsilon^{-1})^T \\ &\quad - AP[k]C^TR_\epsilon^{-1}CP^T[k]A^T. \end{aligned}$$

To minimize this expression, we choose $L = AP[k]C^TR_\epsilon^{-1}$, and the theorem is proved.   □

The Kalman filter can also be applied to continuous-time stochastic processes. The mathematical derivation of this result requires more sophisticated tools, but the final form of the estimator is relatively straightforward.

Consider a continuous stochastic system

$$\frac{dx}{dt} = Ax + Bu + Fv, \qquad E\{v(s)v^T(t)\} = R_v(t)\delta(t-s),$$

$$y = Cx + w, \qquad E\{w(s)w^T(t)\} = R_w(t)\delta(t-s),$$

where $\delta(\tau)$ is the unit impulse function. Assume that the disturbance v and noise w are zero mean and Gaussian (but not necessarily stationary):

$$\text{pdf}(v) = \frac{1}{\sqrt[n]{2\pi}\sqrt{\det R_v}}e^{-\frac{1}{2}v^T R_v^{-1}v}, \quad \text{pdf}(w) = \frac{1}{\sqrt[n]{2\pi}\sqrt{\det R_w}}e^{-\frac{1}{2}w^T R_w^{-1}w}.$$

We wish to find the estimate $\hat{x}(t)$ that minimizes the mean square error $E\{(x(t) - \hat{x}(t))(x(t) - \hat{x}(t))^T\}$ given $\{y(\tau) : 0 \le \tau \le t\}$.

Theorem 7.5 (Kalman–Bucy, 1961). *The optimal estimator has the form of a linear observer*

$$\frac{d\hat{x}}{dt} = A\hat{x} + Bu + L(y - C\hat{x}),$$

where $L(t) = P(t)C^T R_w^{-1}$ *and* $P(t) = E\{(x(t)-\hat{x}(t))(x(t)-\hat{x}(t))^T\}$ *and satisfies*

$$\frac{dP}{dt} = AP + PA^T - PC^T R_w^{-1}(t)CP + FR_v(t)F^T, \quad P[0] = E\{x[0]x^T[0]\}.$$

As in the discrete case, when the system is stationary and if $P(t)$ converges, the observer gain is constant:

$$L = PC^T R_w^{-1} \quad \text{where} \quad AP + PA^T - PC^T R_w^{-1}CP + FR_v F^T = 0.$$

The second equation is the *algebraic Riccati equation*.

Example 7.5 Vectored thrust aircraft
We consider the lateral dynamics of the system, consisting of the subsystems whose states are given by $z = (x, \theta, \dot{x}, \dot{\theta})$. To design a Kalman filter for the system, we must include a description of the process disturbances and the sensor noise. We thus augment the system to have the form

$$\frac{dz}{dt} = Az + Bu + Fv, \qquad y = Cz + w,$$

where F represents the structure of the disturbances (including the effects of nonlinearities that we have ignored in the linearization), w represents the disturbance source (modeled as zero mean, Gaussian white noise) and v represents that measurement noise (also zero mean, Gaussian and white).

For this example, we choose F as the identity matrix and choose disturbances v_i, $i = 1, \ldots, n$, to be independent disturbances with covariance given by $R_{ii} = 0.1$, $R_{ij} = 0, i \ne j$. The sensor noise is a single random variable which we model as

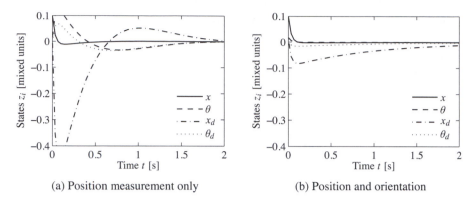

(a) Position measurement only   (b) Position and orientation

Figure 7.9: Kalman filter design for a vectored thrust aircraft. In the first design (a) only
the lateral position of the aircraft is measured. Adding a direct measurement of the roll
angle produces a much better observer (b). The initial condition for both simulations is
$(0.1, 0.0175, 0.01, 0)$.

having covariance $R_w = 10^{-4}$. Using the same parameters as before, the resulting
Kalman gain is given by

$$L = \begin{bmatrix} 37.0 \\ -46.9 \\ 185 \\ -31.6 \end{bmatrix}.$$

The performance of the estimator is shown in Figure 7.9a. We see that while the
estimator converges to the system state, it contains significant overshoot in the state
estimate, which can lead to poor performance in a closed loop setting.

To improve the performance of the estimator, we explore the impact of adding a
new output measurement. Suppose that instead of measuring just the output position
x, we also measure the orientation of the aircraft θ. The output becomes

$$y = \begin{bmatrix} 1 & 0 & 0 & 0 \\ 0 & 1 & 0 & 0 \end{bmatrix} z + \begin{bmatrix} w_1 \\ w_2 \end{bmatrix},$$

and if we assume that w_1 and w_2 are independent noise sources each with covariance
$R_{w_i} = 10^{-4}$, then the optimal estimator gain matrix becomes

$$L = \begin{bmatrix} 32.6 & -0.150 \\ -0.150 & 32.6 \\ 32.7 & -9.79 \\ -0.0033 & 31.6 \end{bmatrix}.$$

These gains provide good immunity to noise and high performance, as illustrated
in Figure 7.9b.   ∇

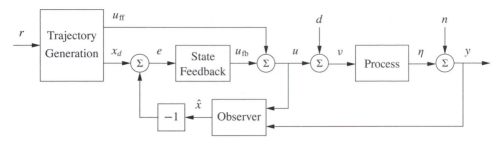

Figure 7.10: Block diagram of a controller based on a structure with two degrees of freedom which combines feedback and feedforward. The controller consists of a trajectory generator, state feedback and an observer. The trajectory generation subsystem computes a feedforward command u_{ff} along with the desired state x_d. The state feedback controller uses the estimated state and desired state to compute a corrective input u_{fb}.

7.5 A General Controller Structure

State estimators and state feedback are important components of a controller. In this section, we will add feedforward to arrive at a general controller structure that appears in many places in control theory and is the heart of most modern control systems. We will also briefly sketch how computers can be used to implement a controller based on output feedback.

Feedforward

In this chapter and the previous one we have emphasized feedback as a mechanism for minimizing tracking error; reference values were introduced simply by adding them to the state feedback through a gain k_r. A more sophisticated way of doing this is shown by the block diagram in Figure 7.10, where the controller consists of three parts: an observer that computes estimates of the states based on a model and measured process inputs and outputs, a state feedback, and a trajectory generator that generates the desired behavior of all states x_d and a feedforward signal u_{ff}. Under the ideal conditions of no disturbances and no modeling errors the signal u_{ff} generates the desired behavior x_d when applied to the process. The signal x_d can be generated by a system that gives the desired response of the state. To generate the the signal u_{ff}, we must also have a model of the inverse of the process dynamics.

To get some insight into the behavior of the system, we assume that there are no disturbances and that the system is in equilibrium with a constant reference signal and with the observer state $\hat{x}$ equal to the process state x. When the reference signal is changed, the signals u_{ff} and x_d will change. The observer tracks the state perfectly because the initial state was correct. The estimated state $\hat{x}$ is thus equal to the desired state x_d, and the feedback signal $u_{\mathrm{fb}} = L(x_d - \hat{x})$ will also be zero. All action is thus created by the signals from the trajectory generator. If there are some disturbances or some modeling errors, the feedback signal will attempt to correct the situation.

This controller is said to have *two degrees of freedom* because the responses to command signals and disturbances are decoupled. Disturbance responses are governed by the observer and the state feedback, while the response to command signals is governed by the trajectory generator (feedforward).

For an analytic description we start with the full nonlinear dynamics of the process

$$\frac{dx}{dt} = f(x, u), \qquad y = h(x, u). \tag{7.23}$$

Assume that the trajectory generator is able to compute a desired trajectory (x_d, u_{ff}) that satisfies the dynamics (7.23) and satisfies $r = h(x_d, u_{ff})$. To design the controller, we construct the error system. Let $z = x - x_d$ and $v = u - u_{ff}$ and compute the dynamics for the error:

$$\dot{z} = \dot{x} - \dot{x}_d = f(x, u) - f(x_d, u_{ff})$$
$$= f(z + x_d, v + u_{ff}) - f(x_d, u_{ff}) =: F(z, v, x_d(t), u_{ff}(t)).$$

In general, this system is time-varying. Note that $z = -e$ in Figure 7.10 due to the convention of using negative feedback in the block diagram.

For trajectory tracking, we can assume that e is small (if our controller is doing a good job), and so we can linearize around $z = 0$:

$$\frac{dz}{dt} \approx A(t)z + B(t)v, \quad A(t) = \left.\frac{\partial F}{\partial z}\right|_{(x_d(t), u_{ff}(t))}, \quad B(t) = \left.\frac{\partial F}{\partial v}\right|_{(x_d(t), u_{ff}(t))}.$$

It is often the case that $A(t)$ and $B(t)$ depend only on x_d, in which case it is convenient to write $A(t) = A(x_d)$ and $B(t) = B(x_d)$.

Assume now that x_d and u_{ff} are either constant or slowly varying (with respect to the performance criterion). This allows us to consider just the (constant) linear system given by $(A(x_d), B(x_d))$. If we design a state feedback controller $K(x_d)$ for each x_d, then we can regulate the system using the feedback

$$v = -K(x_d)z.$$

Substituting back the definitions of e and v, our controller becomes

$$u = -K(x_d)(x - x_d) + u_{ff}.$$

This form of controller is called a *gain scheduled* linear controller with *feedforward* u_{ff}.

Finally, we consider the observer. The full nonlinear dynamics can be used for the prediction portion of the observer and the linearized system for the correction term:

$$\frac{d\hat{x}}{dt} = f(\hat{x}, u) + L(\hat{x})(y - h(\hat{x}, u)),$$

where $L(\hat{x})$ is the observer gain obtained by linearizing the system around the currently estimated state. This form of the observer is known as an *extended Kalman filter* and has proved to be a very effective means of estimating the state of a nonlinear system.

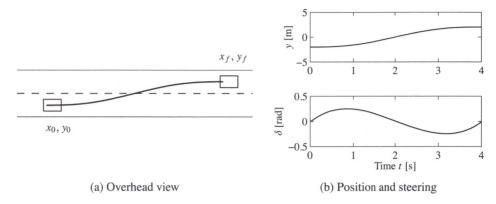

(a) Overhead view (b) Position and steering

Figure 7.11: Trajectory generation for changing lanes. We wish to change from the left lane to the right lane over a distance of 30 m in 4 s. The planned trajectory in the xy plane is shown in (a) and the lateral position y and the steering angle δ over the maneuver time interval are shown in (b).

There are many ways to generate the feedforward signal, and there are also many different ways to compute the feedback gain K and the observer gain L. Note that once again the internal model principle applies: the controller contains a model of the system to be controlled through the observer.

Example 7.6 Vehicle steering

To illustrate how we can use a two degree-of-freedom design to improve the performance of the system, consider the problem of steering a car to change lanes on a road, as illustrated in Figure 7.11a.

We use the non-normalized form of the dynamics, where were derived in Example 2.8. Using the center of the rear wheels as the reference ($\alpha = 0$), the dynamics can be written as

$$\frac{dx}{dt} = \cos\theta v, \qquad \frac{dy}{dt} = \sin\theta v, \qquad \frac{d\theta}{dt} = \frac{v}{b}\tan\delta,$$

where v is the forward velocity of the vehicle and δ is the steering angle. To generate a trajectory for the system, we note that we can solve for the states and inputs of the system given x, y by solving the following sets of equations:

$$\dot{x} = v\cos\theta, \qquad \ddot{x} = \dot{v}\cos\theta - v\dot{\theta}\sin\theta,$$
$$\dot{y} = v\sin\theta, \qquad \ddot{y} = \dot{v}\sin\theta + v\dot{\theta}\cos\theta, \qquad (7.24)$$
$$\dot{\theta} = (v/b)\tan\delta.$$

This set of five equations has five unknowns (θ, $\dot{\theta}$, v, $\dot{v}$ and δ) that can be solved using trigonometry and linear algebra. It follows that we can compute a feasible trajectory for the system given any path $x(t)$, $y(t)$. (This special property of a system is known as *differential flatness* [73, 74].)

To find a trajectory from an initial state (x_0, y_0, θ_0) to a final state (x_f, y_f, θ_f)

at a time T, we look for a path $x(t)$, $y(t)$ that satisfies

$$
\begin{aligned}
x(0) &= x_0, & x(T) &= x_f, \\
y(0) &= y_0, & y(T) &= y_f, \\
\dot{x}(0)\sin\theta_0 - \dot{y}(0)\cos\theta_0 &= 0, & \dot{x}(T)\sin\theta_f - \dot{y}(T)\cos\theta_f &= 0, \\
\dot{y}(0)\sin\theta_0 + \dot{x}(0)\cos\theta_0 &= v_0, & \dot{y}(T)\sin\theta_f + \dot{x}(T)\cos\theta_f &= v_f.
\end{aligned}
\tag{7.25}
$$

One such trajectory can be found by choosing $x(t)$ and $y(t)$ to have the form

$$
x_d(t) = \alpha_0 + \alpha_1 t + \alpha_2 t^2 + \alpha_3 t^3, \qquad y_d(t) = \beta_0 + \beta_1 t + \beta_2 t^2 + \beta_3 t^3.
$$

Substituting these equations into equation (7.25), we are left with a set of linear equations that can be solved for $\alpha_i, \beta_i, i = 0, 1, 2, 3$. This gives a feasible trajectory for the system by using equation (7.24) to solve for θ_d, v_d and δ_d.

Figure 7.11b shows a sample trajectory generated by a set of higher-order equations that also set the initial and final steering angle to zero. Notice that the feedforward input is quite different from 0, allowing the controller to command a steering angle that executes the turn in the absence of errors. ∇

Kalman's Decomposition of a Linear System

In this chapter and the previous one we have seen that two fundamental properties of a linear input/output system are reachability and observability. It turns out that these two properties can be used to classify the dynamics of a system. The key result is Kalman's decomposition theorem, which says that a linear system can be divided into four subsystems: Σ_{ro} which is reachable and observable, $\Sigma_{r\bar{o}}$ which is reachable but not observable, $\Sigma_{\bar{r}o}$ which is not reachable but is observable and $\Sigma_{\bar{r}\bar{o}}$ which is neither reachable nor observable.

We will first consider this in the special case of systems where the matrix A has distinct eigenvalues. In this case we can find a set of coordinates such that the A matrix is diagonal and, with some additional reordering of the states, the system can be written as

$$
\frac{dx}{dt} = \begin{bmatrix} A_{ro} & 0 & 0 & 0 \\ 0 & A_{r\bar{o}} & 0 & 0 \\ 0 & 0 & A_{\bar{r}o} & 0 \\ 0 & 0 & 0 & A_{\bar{r}\bar{o}} \end{bmatrix} x + \begin{bmatrix} B_{ro} \\ B_{r\bar{o}} \\ 0 \\ 0 \end{bmatrix} u,
\tag{7.26}
$$

$$
y = \begin{bmatrix} C_{ro} & 0 & C_{\bar{r}o} & 0 \end{bmatrix} x + Du.
$$

All states x_k such that $B_k \neq 0$ are reachable, and all states such that $C_k \neq 0$ are observable. If we set the initial state to zero (or equivalently look at the steady-state response if A is stable), the states given by $x_{\bar{r}o}$ and $x_{\bar{r}\bar{o}}$ will be zero and $x_{r\bar{o}}$ does not affect the output. Hence the output y can be determined from the system

$$
\frac{dx_{ro}}{dt} = A_{ro}x_{ro} + B_{ro}u, \qquad y = C_{ro}x_{ro} + Du.
$$

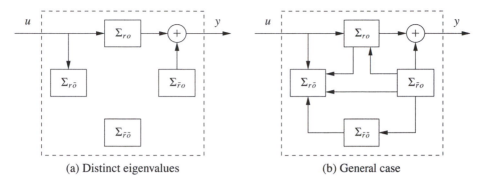

(a) Distinct eigenvalues   (b) General case

Figure 7.12: Kalman's decomposition of a linear system. The decomposition in (a) is for a system with distinct eigenvalues and the one in (b) is the general case. The system is broken into four subsystems, representing the various combinations of reachable and observable states. The input/output relationship only depends on the subset of states that are both reachable and observable.

Thus from the input/output point of view, it is only the reachable and observable dynamics that matter. A block diagram of the system illustrating this property is given in Figure 7.12a.

The general case of the Kalman decomposition is more complicated and requires some additional linear algebra; see the original paper by Kalman, Ho and Narendra [118]. The key result is that the state space can still be decomposed into four parts, but there will be additional coupling so that the equations have the form

$$
\frac{dx}{dt} = \begin{bmatrix} A_{ro} & 0 & * & 0 \\ * & A_{r\bar{o}} & * & * \\ 0 & 0 & A_{\bar{r}o} & 0 \\ 0 & 0 & * & A_{\bar{r}\bar{o}} \end{bmatrix} x + \begin{bmatrix} B_{ro} \\ B_{r\bar{o}} \\ 0 \\ 0 \end{bmatrix} u,
$$
$$
y = \begin{bmatrix} C_{ro} & 0 & C_{\bar{r}o} & 0 \end{bmatrix} x,
$$
(7.27)

where $*$ denotes block matrices of appropriate dimensions. The input/output response of the system is given by

$$
\frac{dx_{ro}}{dt} = A_{ro}x_{ro} + B_{ro}u, \qquad y = C_{ro}x_{ro} + Du, \tag{7.28}
$$

which are the dynamics of the reachable and observable subsystem Σ_{ro}. A block diagram of the system is shown in Figure 7.12b.

The following example illustrates Kalman's decomposition.

Example 7.7 System and controller with feedback from observer states
Consider the system

$$
\frac{dx}{dt} = Ax + Bu, \qquad y = Cx.
$$

The following controller, based on feedback from the observer state, was given in

Theorem 7.3:

$$\frac{d\hat{x}}{dt} = A\hat{x} + Bu + L(y - C\hat{x}), \qquad u = -K\hat{x} + k_r r.$$

Introducing the states x and $\tilde{x} = x - \hat{x}$, the closed loop system can be written as

$$\frac{d}{dt}\begin{bmatrix} x \\ \tilde{x} \end{bmatrix} = \begin{bmatrix} A - BK & BK \\ 0 & A - LC \end{bmatrix}\begin{bmatrix} x \\ \tilde{x} \end{bmatrix} + \begin{bmatrix} Bk_r \\ 0 \end{bmatrix}r, \qquad y = \begin{bmatrix} C & 0 \end{bmatrix}x,$$

which is a Kalman decomposition like the one shown in Figure 7.12b with only two subsystems Σ_{ro} and $\Sigma_{\bar{r}o}$. The subsystem Σ_{ro}, with state x, is reachable and observable, and the subsystem $\Sigma_{\bar{r}o}$, with state $\tilde{x}$, is not reachable but observable. It is natural that the state $\tilde{x}$ is not reachable from the reference signal r because it would not make sense to design a system where changes in the command signal could generate observer errors. The relationship between the reference r and the output y is given by

$$\frac{dx}{dt} = (A - BK)x + Bk_r r, \qquad y = Cx,$$

which is the same relationship as for a system with full state feedback. ∇

Computer Implementation

The controllers obtained so far have been described by ordinary differential equations. They can be implemented directly using analog components, whether electronic circuits, hydraulic valves or other physical devices. Since in modern engineering applications most controllers are implemented using computers, we will briefly discuss how this can be done.

A computer-controlled system typically operates periodically: every cycle, signals from the sensors are sampled and converted to digital form by the A/D converter, the control signal is computed and the resulting output is converted to analog form for the actuators, as shown in Figure 7.13. To illustrate the main principles of how to implement feedback in this environment, we consider the controller described by equations (7.14) and (7.15), i.e.,

$$\frac{d\hat{x}}{dt} = A\hat{x} + Bu + L(y - C\hat{x}), \qquad u = -K\hat{x} + k_r r.$$

The second equation consists only of additions and multiplications and can thus be implemented directly on a computer. The first equation can be implemented by approximating the derivative by a difference

$$\frac{d\hat{x}}{dt} \approx \frac{\hat{x}(t_{k+1}) - \hat{x}(t_k)}{h} = A\hat{x}(t_k) + Bu(t_k) + L(y(t_k) - C\hat{x}(t_k)),$$

where t_k are the sampling instants and $h = t_{k+1} - t_k$ is the sampling period. Rewriting the equation to isolate $\hat{x}(t_{k+1})$, we get the difference equation

$$\hat{x}(t_{k+1}) = \hat{x}(t_k) + h\big(A\hat{x}(t_k) + Bu(t_k) + L(y(t_k) - C\hat{x}(t_k))\big). \qquad (7.29)$$

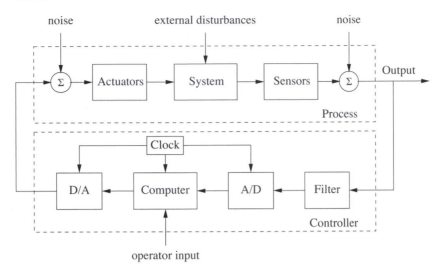

Figure 7.13: Components of a computer-controlled system. The controller consists of analog-to-digital (A/D) and digital-to-analog (D/A) converters, as well as a computer that implements the control algorithm. A system clock controls the operation of the controller, synchronizing the A/D, D/A and computing processes. The operator input is also fed to the computer as an external input.

The calculation of the estimated state at time t_{k+1} requires only addition and multiplication and can easily be done by a computer. A section of pseudocode for the program that performs this calculation is

```
% Control algorithm - main loop
r = adin(ch1)                    % read reference
y = adin(ch2)                    % get process output
u = -K*xhat + kr*r               % compute control variable
daout(ch1, u)                    % set analog output
xhat = xhat + h*(A*x+B*u+L*(y-C*x))  % update state estimate
```

The program runs periodically at a fixed rate h. Notice that the number of computations between reading the analog input and setting the analog output has been minimized by updating the state after the analog output has been set. The program has an array of states `xhat` that represents the state estimate. The choice of sampling period requires some care.

There are more sophisticated ways of approximating a differential equation by a difference equation. If the control signal is constant between the sampling instants, it is possible to obtain exact equations; see [18].

There are several practical issues that also must be dealt with. For example, it is necessary to filter measured signals before they are sampled so that the filtered signal has little frequency content above $f_s/2$, where f_s is the sampling frequency. This avoids a phenomena knows as *aliasing*. If controllers with integral action are used, it is also necessary to provide protection so that the integral does not become too large when the actuator saturates. This issue, called *integrator windup*, is studied in more detail in Chapter 10. Care must also be taken so that parameter changes do not cause disturbances.

7.6 Further Reading

The notion of observability is due to Kalman [115] and, combined with the dual notion of reachability, it was a major stepping stone toward establishing state space control theory beginning in the 1960s. The observer first appeared as the Kalman filter, in the paper by Kalman [114] on the discrete-time case and Kalman and Bucy [116] on the continuous-time case. Kalman also conjectured that the controller for output feedback could be obtained by combining a state feedback with an observer; see the quote in the beginning of this chapter. This result was formally proved by Josep and Tou [111] and Gunckel and Franklin [93]. The combined result is known as the linear quadratic Gaussian control theory; a compact treatment is given in the books by Anderson and Moore [7] and Åström [15]. Much later it was shown that solutions to robust control problems also had a similar structure but with different ways of computing observer and state feedback gains [65]. The general controller structure discussed in Section 7.5, which combines feedback and feedforward, was described by Horowitz in 1963 [102]. The particular form in Figure 7.10 appeared in [18], which also treats digital implementation of the controller. The hypothesis that motion control in humans is based on a combination of feedback and feedforward was proposed by Ito in 1970 [107].

Exercises

7.1 (Coordinate transformations) Consider a system under a coordinate transformation $z = Tx$, where $T \in \mathbb{R}^{n \times n}$ is an invertible matrix. Show that the observability matrix for the transformed system is given by $\widetilde{W}_o = W_o T^{-1}$ and hence observability is independent of the choice of coordinates.

7.2 Show that the system depicted in Figure 7.2 is not observable.

7.3 (Observable canonical form) Show that if a system is observable, then there exists a change of coordinates $z = Tx$ that puts the transformed system into observable canonical form.

7.4 (Bicycle dynamics) The linearized model for a bicycle is given in equation (3.5), which has the form

$$J\frac{d^2\varphi}{dt^2} - \frac{Dv_0}{b}\frac{d\delta}{dt} = mgh\varphi + \frac{mv_0^2 h}{b}\delta,$$

where φ is the tilt of the bicycle and δ is the steering angle. Give conditions under which the system is observable and explain any special situations where it loses observability.

7.5 (Integral action) The model (7.1) assumes that the input $u = 0$ corresponds to $x = 0$. In practice, it is very difficult to know the value of the control signal that gives a precise value of the state or the output because this would require a perfectly calibrated system. One way to avoid this assumption is to assume that the model is given by

$$\frac{dx}{dt} = Ax + B(u + u_0), \qquad y = Cx + Du,$$

where u_0 is an unknown constant that can be modeled as $du_0/dt = 0$. Consider u_0 as an additional state variable and derive a controller based on feedback from the observed state. Show that the controller has integral action and that it does not require a perfectly calibrated system.

7.6 (Vectored thrust aircraft) The lateral dynamics of the vectored thrust aircraft example described in Example 6.8 can be obtained by considering the motion described by the states $z = (x, \theta, \dot{x}, \dot{\theta})$. Construct an estimator for these dynamics by setting the eigenvalues of the observer into a *Butterworth pattern* with $\lambda_{bw} = -3.83 \pm 9.24i$, $-9.24 \pm 3.83i$. Using this estimator combined with the state space controller computed in Example 6.8, plot the step response of the closed loop system.

7.7 (Uniqueness of observers) Show that the design of an observer by eigenvalue assignment is unique for single-output systems. Construct examples that show that the problem is not necessarily unique for systems with many outputs.

7.8 (Observers using differentiation) Consider the linear system (7.2), and assume that the observability matrix W_o is invertible. Show that

$$\hat{x} = W_o^{-1} \begin{bmatrix} y & \dot{y} & \ddot{y} & \cdots & y^{(n-1)} \end{bmatrix}^T$$

is an observer. Show that it has the advantage of giving the state instantaneously but that it also has some severe practical drawbacks.

7.9 (Observer for Teorell's compartment model) Teorell's compartment model, shown in Figure 3.17, has the following state space representation:

$$\frac{dx}{dt} = \begin{bmatrix} -k_1 & 0 & 0 & 0 & 0 \\ k_1 & -k_2 - k_4 & 0 & k_3 & 0 \\ 0 & k_4 & 0 & 0 & 0 \\ 0 & k_2 & 0 & -k_3 - k_5 & 0 \\ 0 & 0 & 0 & k_5 & 0 \end{bmatrix} x + \begin{bmatrix} 1 \\ 0 \\ 0 \\ 0 \\ 0 \end{bmatrix} u,$$

where representative parameters are $k_1 = 0.02$, $k_2 = 0.1$, $k_3 = 0.05$, $k_4 = k_5 = 0.005$. The concentration of a drug that is active in compartment 5 is measured in the bloodstream (compartment 2). Determine the compartments that are observable from measurement of concentration in the bloodstream and design an estimator for these concentrations base on eigenvalue assignment. Choose the closed loop eigenvalues -0.03, -0.05 and -0.1. Simulate the system when the input is a pulse injection.

7.10 (Observer design for motor drive) Consider the normalized model of the motor drive in Exercise 2.10 where the open loop system has the eigenvalues $0, 0, -0.05 \pm i$. A state feedback that gave a closed loop system with eigenvalues in $-2, -1$ and $-1 \pm i$ was designed in Exercise 6.11. Design an observer for the system that has eigenvalues $-4, -2$ and $-2 \pm 2i$. Combine the observer with the state feedback from Exercise 6.11 to obtain an output feedback and simulate the complete system.

7.11 (Feedforward design for motor drive) Consider the normalized model of the motor drive in Exercise 2.10. Design the dynamics of the block labeled "trajectory generation" in Figure 7.10 so that the dynamics relating the output η to the reference signal r has the dynamics

$$\frac{d^3 y_m}{dt^3} + a_{m1}\frac{d^2 y_m}{dt^2} + a_{m2}\frac{dy_m}{dt} + a_{m3}y_m = a_{m3}r, \qquad (7.30)$$

with parameters $a_{m1} = 2.5\omega_m$, $a_{m2} = 2.5\omega_m^2$ and $a_{m3} = \omega_m^3$. Discuss how the largest value of the feedforward signal for a unit step in the command signal depends on ω_m.

7.12 (Whipple bicycle model) Consider the Whipple bicycle model given by equation (3.7) in Section 3.2. A state feedback for the system was designed in Exercise 6.12. Design an observer and an output feedback for the system.

 7.13 (Discrete-time random walk) Suppose that we wish to estimate the position of a particle that is undergoing a random walk in one dimension (i.e., along a line). We model the position of the particle as

$$x[k+1] = x[k] + u[k],$$

where x is the position of the particle and u is a white noise processes with $E\{u[i]\} = 0$ and $E\{u[i]\,u[j]\} = R_u\delta(i-j)$. We assume that we can measure x subject to additive, zero-mean, Gaussian white noise with covariance 1.

(a) Compute the expected value and covariance of the particle as a function of k.

(b) Construct a Kalman filter to estimate the position of the particle given the noisy measurements of its position. Compute the steady-state expected value and covariance of the error of your estimate.

(c) Suppose that $E\{u[0]\} = \mu \neq 0$ but is otherwise unchanged. How would your answers to parts (a) and (b) change?

7.14 (Kalman decomposition) Consider a linear system characterized by the matrices

$$A = \begin{bmatrix} -2 & 1 & -1 & 2 \\ 1 & -3 & 0 & 2 \\ 1 & 1 & -4 & 2 \\ 0 & 1 & -1 & -1 \end{bmatrix}, \quad B = \begin{bmatrix} 2 \\ 2 \\ 2 \\ 1 \end{bmatrix}, \quad C = \begin{bmatrix} 0 & 1 & -1 & 0 \end{bmatrix}, \quad D = 0.$$

Construct a Kalman decomposition for the system. (Hint: Try to diagonalize.)

Chapter Eight
Transfer Functions

The typical regulator system can frequently be described, in essentials, by differential equations of no more than perhaps the second, third or fourth order. ...In contrast, the order of the set of differential equations describing the typical negative feedback amplifier used in telephony is likely to be very much greater. As a matter of idle curiosity, I once counted to find out what the order of the set of equations in an amplifier I had just designed would have been, if I had worked with the differential equations directly. It turned out to be 55.

Henrik Bode, 1960 [41].

This chapter introduces the concept of the *transfer function*, which is a compact description of the input/output relation for a linear system. Combining transfer functions with block diagrams gives a powerful method for dealing with complex linear systems. The relationship between transfer functions and other descriptions of system dynamics is also discussed.

8.1 Frequency Domain Modeling

Figure 8.1 is a block diagram for a typical control system, consisting of a process to be controlled and a controller that combines feedback and feedforward. We saw in the previous two chapters how to analyze and design such systems using state space descriptions of the blocks. As mentioned in Chapter 2, an alternative approach is to focus on the input/output characteristics of the system. Since it is the inputs and outputs that are used to connect the systems, one could expect that this point of

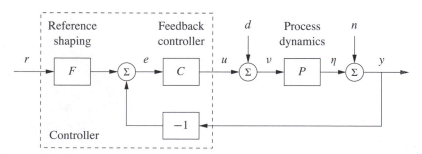

Figure 8.1: A block diagram for a feedback control system. The reference signal r is fed through a reference shaping block, which produces the signal that will be tracked. The error between this signal and the output is fed to a controller, which produces the input to the process. Disturbances and noise are included as external signals at the input and output of the process dynamics.

view would allow an understanding of the overall behavior of the system. Transfer functions are the main tool in implementing this point of view for linear systems.

The basic idea of the transfer function comes from looking at the frequency response of a system. Suppose that we have an input signal that is periodic. Then we can decompose this signal into the sum of a set of sines and cosines,

$$u(t) = \sum_{k=0}^{\infty} a_k \sin(k\omega t) + b_k \cos(k\omega t),$$

where ω is the fundamental frequency of the periodic input. Each of the terms in this input generates a corresponding sinusoidal output (in steady state), with possibly shifted magnitude and phase. The gain and phase at each frequency are determined by the frequency response given in equation (5.24):

$$G(s) = C(sI - A)^{-1}B + D, \tag{8.1}$$

where we set $s = i(k\omega)$ for each $k = 1, \ldots, \infty$ and $i = \sqrt{-1}$. If we know the steady-state frequency response $G(s)$, we can thus compute the response to any (periodic) signal using superposition.

The transfer function generalizes this notion to allow a broader class of input signals besides periodic ones. As we shall see in the next section, the transfer function represents the response of the system to an *exponential input*, $u = e^{st}$. It turns out that the form of the transfer function is precisely the same as that of equation (8.1). This should not be surprising since we derived equation (8.1) by writing sinusoids as sums of complex exponentials. Formally, the transfer function is the ratio of the Laplace transforms of output and input, although one does not have to understand the details of Laplace transforms in order to make use of transfer functions.

Modeling a system through its response to sinusoidal and exponential signals is known as *frequency domain modeling*. This terminology stems from the fact that we represent the dynamics of the system in terms of the generalized frequency s rather than the time domain variable t. The transfer function provides a complete representation of a linear system in the frequency domain.

The power of transfer functions is that they provide a particularly convenient representation in manipulating and analyzing complex linear feedback systems. As we shall see, there are many graphical representations of transfer functions that capture interesting properties of the underlying dynamics. Transfer functions also make it possible to express the changes in a system because of modeling error, which is essential when considering sensitivity to process variations of the sort discussed in Chapter 12. More specifically, using transfer functions, it is possible to analyze what happens when dynamic models are approximated by static models or when high-order models are approximated by low-order models. One consequence is that we can introduce concepts that express the degree of stability of a system.

While many of the concepts for state space modeling and analysis apply directly to nonlinear systems, frequency domain analysis applies primarily to linear systems. The notions of gain and phase can be generalized to nonlinear systems and, in particular, propagation of sinusoidal signals through a nonlinear system can approximately be captured by an analog of the frequency response called the describing function. These extensions of frequency response will be discussed in Section 9.5.

8.2 Derivation of the Transfer Function

As we have seen in previous chapters, the input/output dynamics of a linear system have two components: the initial condition response and the forced response. In addition, we can speak of the transient properties of the system and its steady-state response to an input. The transfer function focuses on the steady-state forced response to a given input and provides a mapping between inputs and their corresponding outputs. In this section, we will derive the transfer function in terms of the exponential response of a linear system.

Transmission of Exponential Signals

To formally compute the transfer function of a system, we will make use of a special type of signal, called an *exponential signal*, of the form e^{st}, where $s = \sigma + i\omega$ is a complex number. Exponential signals play an important role in linear systems. They appear in the solution of differential equations and in the impulse response of linear systems, and many signals can be represented as exponentials or sums of exponentials. For example, a constant signal is simply $e^{\alpha t}$ with $\alpha = 0$. Damped sine and cosine signals can be represented by

$$e^{(\sigma+i\omega)t} = e^{\sigma t}e^{i\omega t} = e^{\sigma t}(\cos \omega t + i \sin \omega t),$$

where $\sigma < 0$ determines the decay rate. Figure 8.2 gives examples of signals that can be represented by complex exponentials; many other signals can be represented by linear combinations of these signals. As in the case of sinusoidal signals, we will allow complex-valued signals in the derivation that follows, although in practice we always add together combinations of signals that result in real-valued functions.

To investigate how a linear system responds to an exponential input $u(t) = e^{st}$ we consider the state space system

$$\frac{dx}{dt} = Ax + Bu, \qquad y = Cx + Du. \tag{8.2}$$

Let the input signal be $u(t) = e^{st}$ and assume that $s \neq \lambda_j(A), j = 1, \ldots, n$, where $\lambda_j(A)$ is the jth eigenvalue of A. The state is then given by

$$x(t) = e^{At}x(0) + \int_0^t e^{A(t-\tau)}Be^{s\tau}\,d\tau = e^{At}x(0) + e^{At}\int_0^t e^{(sI-A)\tau}B\,d\tau.$$

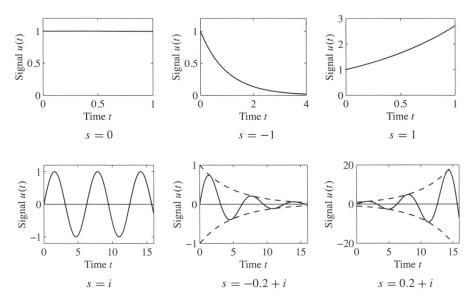

Figure 8.2: Examples of exponential signals. The top row corresponds to exponential signals with a real exponent, and the bottom row corresponds to those with complex exponents. The dashed line in the last two cases denotes the bounding envelope for the oscillatory signals. In each case, if the real part of the exponent is negative then the signal decays, while if the real part is positive then it grows.

As we saw in Section 5.3, if $s \neq \lambda(A)$, the integral can be evaluated and we get

$$x(t) = e^{At}x(0) + e^{At}(sI - A)^{-1}\left(e^{(sI-A)t} - I\right)B$$
$$= e^{At}\left(x(0) - (sI - A)^{-1}B\right) + (sI - A)^{-1}Be^{st}.$$

The output of equation (8.2) is thus

$$y(t) = Cx(t) + Du(t)$$
$$= Ce^{At}\left(x(0) - (sI - A)^{-1}B\right) + \left(C(sI - A)^{-1}B + D\right)e^{st}, \qquad (8.3)$$

a linear combination of the exponential functions e^{st} and e^{At}. The first term in equation (8.3) is the transient response of the system. Recall that e^{At} can be written in terms of the eigenvalues of A (using the Jordan form in the case of repeated eigenvalues), and hence the transient response is a linear combination of terms of the form $e^{\lambda_j t}$, where λ_j are eigenvalues of A. If the system is stable, then $e^{At} \to 0$ as $t \to \infty$ and this term dies away.

The second term of the output (8.3) is proportional to the input $u(t) = e^{st}$. This term is called the *pure exponential response*. If the initial state is chosen as

$$x(0) = (sI - A)^{-1}B,$$

then the output consists of only the pure exponential response and both the state and the output are proportional to the input:

$$x(t) = (sI - A)^{-1}Be^{st} = (sI - A)^{-1}Bu(t),$$
$$y(t) = (C(sI - A)^{-1}B + D)e^{st} = (C(sI - A)^{-1}B + D)u(t).$$

This is also the output we see in steady state, when the transients represented by the first term in equation (8.3) have died out. The map from the input to the output,

$$G_{yu}(s) = C(sI - A)^{-1}B + D, \tag{8.4}$$

is the *transfer function* from u to y for the system (8.2), and we can write $y(t) = G_{yu}(s)u(t)$ for the case that $u(t) = e^{st}$. Compare with the definition of frequency response given by equation (5.24).

An important point in the derivation of the transfer function is the fact that we have restricted s so that $s \neq \lambda_j(A)$, the eigenvalues of A. At those values of s, we see that the response of the system is singular (since $sI - A$ will fail to be invertible). If $s = \lambda_j(A)$, the response of the system to the exponential input $u = e^{\lambda_j t}$ is $y = p(t)e^{\lambda_j t}$, where $p(t)$ is a polynomial of degree less than or equal to the multiplicity of the eigenvalue λ_j (see Exercise 8.2).

Example 8.1 Damped oscillator
Consider the response of a damped linear oscillator, whose state space dynamics were studied in Section 6.3:

$$\frac{dx}{dt} = \begin{bmatrix} 0 & \omega_0 \\ -\omega_0 & -2\zeta\omega_0 \end{bmatrix} x + \begin{bmatrix} 0 \\ k\omega_0 \end{bmatrix} u, \qquad y = \begin{bmatrix} 1 & 0 \end{bmatrix} x. \tag{8.5}$$

This system is stable if $\zeta > 0$, and so we can look at the steady-state response to an input $u = e^{st}$,

$$
\begin{aligned}
G_{yu}(s) = C(sI - A)^{-1}B &= \begin{bmatrix} 1 & 0 \end{bmatrix} \begin{bmatrix} s & -\omega_0 \\ \omega_0 & s + 2\zeta\omega_0 \end{bmatrix}^{-1} \begin{bmatrix} 0 \\ k\omega_0 \end{bmatrix} \\
&= \begin{bmatrix} 1 & 0 \end{bmatrix} \left(\frac{1}{s^2 + 2\zeta\omega_0 s + \omega_0^2} \begin{bmatrix} s & \omega_0 \\ -\omega_0 & s + 2\zeta\omega_0 \end{bmatrix} \right) \begin{bmatrix} 0 \\ k\omega_0 \end{bmatrix} \\
&= \frac{k\omega_0^2}{s^2 + 2\zeta\omega_0 s + \omega_0^2}.
\end{aligned}
\tag{8.6}
$$

To compute the steady-state response to a step function, we set $s = 0$ and we see that

$$u = 1 \qquad \Longrightarrow \qquad y = G_{yu}(0)u = k.$$

If we wish to compute the steady-state response to a sinusoid, we write

$$u = \sin \omega t = \frac{1}{2}\left(ie^{-i\omega t} - ie^{i\omega t}\right),$$

$$y = \frac{1}{2}\left(iG_{yu}(-i\omega)e^{-i\omega t} - iG_{yu}(i\omega)e^{i\omega t}\right).$$

We can now write $G(i\omega)$ in terms of its magnitude and phase,

$$G(i\omega) = \frac{k\omega_0^2}{s^2 + 2\zeta\omega_0 s + \omega_0^2} = Me^{i\theta},$$

where the magnitude (or gain) M and phase θ are given by

$$M = \frac{k\omega_0^2}{\sqrt{(\omega_0^2 - \omega^2)^2 + (2\zeta\omega_0\omega)^2}}, \qquad \frac{\sin\theta}{\cos\theta} = \frac{-2\zeta\omega_0\omega}{\omega_0^2 - \omega^2}.$$

We can also make use of the fact that $G(-i\omega)$ is given by its complex conjugate $G^*(i\omega)$, and it follows that $G(-i\omega) = Me^{-i\theta}$. Substituting these expressions into our output equation, we obtain

$$y = \frac{1}{2}\left(i(Me^{-i\theta})e^{-i\omega t} - i(Me^{i\theta})e^{i\omega t}\right)$$

$$= M \cdot \frac{1}{2}\left(ie^{-i(\omega t + \theta)} - ie^{i(\omega t + \theta)}\right) = M\sin(\omega t + \theta).$$

The responses to other signals can be computed by writing the input as an appropriate combination of exponential responses and using linearity.   ∇

Coordinate Changes

The matrices A, B and C in equation (8.2) depend on the choice of coordinate system for the states. Since the transfer function relates input to outputs, it should be invariant to coordinate changes in the state space. To show this, consider the model (8.2) and introduce new coordinates z by the transformation $z = Tx$, where T is a nonsingular matrix. The system is then described by

$$\frac{dz}{dt} = T(Ax + Bu) = TAT^{-1}z + TBu =: \tilde{A}z + \tilde{B}u,$$

$$y = Cx + Du = CT^{-1}z + Du =: \tilde{C}z + Du.$$

This system has the same form as equation (8.2), but the matrices A, B and C are different:

$$\tilde{A} = TAT^{-1}, \qquad \tilde{B} = TB, \qquad \tilde{C} = CT^{-1}. \tag{8.7}$$

Computing the transfer function of the transformed model, we get

$$\tilde{G}(s) = \tilde{C}(sI - \tilde{A})^{-1}\tilde{B} + \tilde{D} = CT^{-1}(sI - TAT^{-1})^{-1}TB + D$$

$$= C\left(T^{-1}(sI - TAT^{-1})T\right)^{-1}B + D = C(sI - A)^{-1}B + D = G(s),$$

which is identical to the transfer function (8.4) computed from the system description (8.2). The transfer function is thus invariant to changes of the coordinates in the state space.

 Another property of the transfer function is that it corresponds to the portion of the state space dynamics that is both reachable and observable. In particular, if we make use of the Kalman decomposition (Section 7.5), then the transfer function depends only on the dynamics in the reachable and observable subspace Σ_{ro} (Exercise 8.7).

Transfer Functions for Linear Systems

Consider a linear input/output system described by the controlled differential equation

$$\frac{d^n y}{dt^n} + a_1 \frac{d^{n-1} y}{dt^{n-1}} + \cdots + a_n y = b_0 \frac{d^m u}{dt^m} + b_1 \frac{d^{m-1} u}{dt^{m-1}} + \cdots + b_m u, \qquad (8.8)$$

where u is the input and y is the output. This type of description arises in many applications, as described briefly in Section 2.2; bicycle dynamics and AFM modeling are two specific examples. Note that here we have generalized our previous system description to allow both the input and its derivatives to appear.

To determine the transfer function of the system (8.8), let the input be $u(t) = e^{st}$. Since the system is linear, there is an output of the system that is also an exponential function $y(t) = y_0 e^{st}$. Inserting the signals into equation (8.8), we find

$$(s^n + a_1 s^{n-1} + \cdots + a_n) y_0 e^{st} = (b_0 s^m + b_1 s^{m-1} \cdots + b_m) e^{-st},$$

and the response of the system can be completely described by two polynomials

$$a(s) = s^n + a_1 s^{n-1} + \cdots + a_n, \qquad b(s) = b_0 s^m + b_1 s^{m-1} + \cdots + b_m. \qquad (8.9)$$

The polynomial $a(s)$ is the characteristic polynomial of the ordinary differential equation. If $a(s) \neq 0$, it follows that

$$y(t) = y_0 e^{st} = \frac{b(s)}{a(s)} e^{st}. \qquad (8.10)$$

The transfer function of the system (8.8) is thus the rational function

$$G(s) = \frac{b(s)}{a(s)}, \qquad (8.11)$$

where the polynomials $a(s)$ and $b(s)$ are given by equation (8.9). Notice that the transfer function for the system (8.8) can be obtained by inspection since the coefficients of $a(s)$ and $b(s)$ are precisely the coefficients of the derivatives of u and y. The *order* of the transfer function is defined as the order of the denominator polynomial.

Equations (8.8)–(8.11) can be used to compute the transfer functions of many simple ordinary differential equations. Table 8.1 gives some of the more common forms. The first five of these follow directly from the analysis above. For the proportional-integral-derivative (PID) controller, we make use of the fact that the integral of an exponential input is given by $(1/s)e^{st}$.

The last entry in Table 8.1 is for a pure time delay, in which the output is identical to the input at an earlier time. Time delays appear in many systems: typical examples are delays in nerve propagation, communication and mass transport. A system with

Table 8.1: Transfer functions for some common ordinary differential equations.

Type	ODE	Transfer Function
Integrator	$\dot{y} = u$	$\dfrac{1}{s}$
Differentiator	$y = \dot{u}$	s
First-order system	$\dot{y} + ay = u$	$\dfrac{1}{s+a}$
Double integrator	$\ddot{y} = u$	$\dfrac{1}{s^2}$
Damped oscillator	$\ddot{y} + 2\zeta\omega_0\dot{y} + \omega_0^2 y = u$	$\dfrac{1}{s^2 + 2\zeta\omega_0 s + \omega_0^2}$
PID controller	$y = k_p u + k_d \dot{u} + k_i \int u$	$k_p + k_d s + \dfrac{k_i}{s}$
Time delay	$y(t) = u(t - \tau)$	$e^{-\tau s}$

a time delay has the input/output relation

$$y(t) = u(t - \tau). \tag{8.12}$$

As before, let the input be $u(t) = e^{st}$. Assuming that there is an output of the form $y(t) = y_0 e^{st}$ and inserting into equation (8.12), we get

$$y(t) = y_0 e^{st} = e^{s(t-\tau)} = e^{-s\tau} e^{st} = e^{-s\tau} u(t).$$

The transfer function of a time delay is thus $G(s) = e^{-s\tau}$, which is not a rational function but is analytic except at infinity. (A complex function is *analytic* in a region if it has no singularities in the region.)

Example 8.2 Electrical circuit elements

Modeling of electrical circuits is a common use of transfer functions. Consider, for example, a resistor modeled by Ohm's law $V = IR$, where V is the voltage across the resister, I is the current through the resistor and R is the resistance value. If we consider current to be the input and voltage to be the output, the resistor has the transfer function $Z(s) = R$. $Z(s)$ is also called the *impedance* of the circuit element.

Next we consider an inductor whose input/output characteristic is given by

$$L\frac{dI}{dt} = V.$$

Letting the current be $I(t) = e^{st}$, we find that the voltage is $V(t) = Lse^{st}$ and the transfer function of an inductor is thus $Z(s) = Ls$. A capacitor is characterized by

$$C\frac{dV}{dt} = I,$$

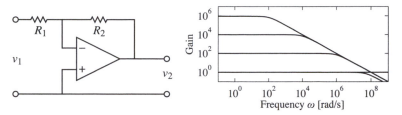

Figure 8.3: Stable amplifier based on negative feedback around an operational amplifier. The block diagram on the left shows a typical amplifier with low-frequency gain R_2/R_1. If we model the dynamic response of the op amp as $G(s) = ak/(s + a)$, then the gain falls off at frequency $\omega = aR_1k/R_2$, as shown in the gain curves on the right. The frequency response is computed for $k = 10^7$, $a = 10$ rad/s, $R_2 = 10^6$ Ω, and $R_1 = 1, 10^2, 10^4$ and 10^6 Ω.

and a similar analysis gives a transfer function from current to voltage of $Z(s) = 1/(Cs)$. Using transfer functions, complex electrical circuits can be analyzed algebraically by using the complex impedance $Z(s)$ just as one would use the resistance value in a resistor network. ∇

Example 8.3 Operational amplifier circuit

To further illustrate the use of exponential signals, we consider the operational amplifier circuit introduced in Section 3.3 and reproduced in Figure 8.3a. The model introduced in Section 3.3 is a simplification because the linear behavior of the amplifier was modeled as a constant gain. In reality there are significant dynamics in the amplifier, and the static model $v_{\text{out}} = -kv$ (equation (3.10)) should therefore be replaced by a dynamic model. In the linear range of the amplifier, we can model the operational amplifier as having a steady-state frequency response

$$\frac{v_{\text{out}}}{v} = -\frac{ak}{s + a} =: -G(s). \tag{8.13}$$

This response corresponds to a first-order system with time constant $1/a$. The parameter k is called the *open loop gain*, and the product ak is called the *gain-bandwidth product*; typical values for these parameters are $k = 10^7$ and $ak = 10^7$–10^9 rad/s.

Since all of the elements of the circuit are modeled as being linear, if we drive the input v_1 with an exponential signal e^{st}, then in steady state all signals will be exponentials of the same form. This allows us to manipulate the equations describing the system in an algebraic fashion. Hence we can write

$$\frac{v_1 - v}{R_1} = \frac{v - v_2}{R_2} \quad \text{and} \quad v_2 = -G(s)v, \tag{8.14}$$

using the fact that the current into the amplifier is very small, as we did in Section 3.3. Eliminating v between these equations gives the following transfer function of the system

$$\frac{v_2}{v_1} = \frac{-R_2 G(s)}{R_1 + R_2 + R_1 G(s)} = \frac{-R_2 ak}{R_1 ak + (R_1 + R_2)(s + a)}.$$

The low-frequency gain is obtained by setting $s = 0$, hence

$$G_{v_2 v_1}(0) = \frac{-kR_2}{(k+1)R_1 + R_2} \approx -\frac{R_2}{R_1},$$

which is the result given by (3.11) in Section 3.3. The bandwidth of the amplifier circuit is

$$\omega_b = a\frac{R_1(k+1) + R_2}{R_1 + R_2} \approx a\frac{R_1 k}{R_2},$$

where the approximation holds for $R_2/R_1 \gg 1$. The gain of the closed loop system drops off at high frequencies as $R_2 k/(\omega(R_1 + R_2))$. The frequency response of the transfer function is shown in Figure 8.3b for $k = 10^7$, $a = 10$ rad/s, $R_2 = 10^6$ Ω and $R_1 = 1, 10^2, 10^4$ and 10^6 Ω.

Note that in solving this example, we bypassed explicitly writing the signals as $v = v_0 e^{st}$ and instead worked directly with v, assuming it was an exponential. This shortcut is handy in solving problems of this sort and when manipulating block diagrams. A comparison with Section 3.3, where we made the same calculation when $G(s)$ was a constant, shows analysis of systems using transfer functions is as easy as using static systems. The calculations are the same if the resistances R_1 and R_2 are replaced by impedances, as discussed in Example 8.2.   ∇

 Although we have focused thus far on ordinary differential equations, transfer functions can also be used for other types of linear systems. We illustrate this via an example of a transfer function for a partial differential equation.

Example 8.4 Heat propagation
Consider the problem of one-dimensional heat propagation in a semi-infinite metal rod. Assume that the input is the temperature at one end and that the output is the temperature at a point along the rod. Let $\theta(x, t)$ be the temperature at position x and time t. With a proper choice of length scales and units, heat propagation is described by the partial differential equation

$$\frac{\partial \theta}{\partial t} = \frac{\partial^2 \theta}{\partial^2 x}, \tag{8.15}$$

and the point of interest can be assumed to have $x = 1$. The boundary condition for the partial differential equation is

$$\theta(0, t) = u(t).$$

To determine the transfer function we choose the input as $u(t) = e^{st}$. Assume that there is a solution to the partial differential equation of the form $\theta(x, t) = \psi(x)e^{st}$ and insert this into equation (8.15) to obtain

$$s\psi(x) = \frac{d^2\psi}{dx^2},$$

with boundary condition $\psi(0) = e^{st}$. This ordinary differential equation (with independent variable x) has the solution

$$\psi(x) = Ae^{x\sqrt{s}} + Be^{-x\sqrt{s}}.$$

Matching the boundary conditions gives $A = 0$ and $B = e^{st}$, so the solution is

$$y(t) = \theta(1, t) = \psi(1)e^{st} = e^{-\sqrt{s}}e^{st} = e^{-\sqrt{s}}u(t).$$

The system thus has the transfer function $G(s) = e^{-\sqrt{s}}$. As in the case of a time delay, the transfer function is not a rational function but is an analytic function. ∇

Gains, Poles and Zeros

The transfer function has many useful interpretations and the features of a transfer function are often associated with important system properties. Three of the most important features are the gain and the locations of the poles and zeros.

The *zero frequency gain* of a system is given by the magnitude of the transfer function at $s = 0$. It represents the ratio of the steady-state value of the output with respect to a step input (which can be represented as $u = e^{st}$ with $s = 0$). For a state space system, we computed the zero frequency gain in equation (5.22):

$$G(0) = D - CA^{-1}B.$$

For a system written as a linear differential equation

$$\frac{d^n y}{dt^n} + a_1 \frac{d^{n-1} y}{dt^{n-1}} + \cdots + a_n y = b_0 \frac{d^m u}{dt^m} + b_1 \frac{d^{m-1} u}{dt^{m-1}} + \cdots + b_m u,$$

if we assume that the input and output of the system are constants y_0 and u_0, then we find that $a_n y_0 = b_m u_0$. Hence the zero frequency gain is

$$G(0) = \frac{y_0}{u_0} = \frac{b_m}{a_n}. \tag{8.16}$$

Next consider a linear system with the rational transfer function

$$G(s) = \frac{b(s)}{a(s)}.$$

The roots of the polynomial $a(s)$ are called the *poles* of the system, and the roots of $b(s)$ are called the *zeros* of the system. If p is a pole, it follows that $y(t) = e^{pt}$ is a solution of equation (8.8) with $u = 0$ (the homogeneous solution). A pole p corresponds to a *mode* of the system with corresponding modal solution e^{pt}. The unforced motion of the system after an arbitrary excitation is a weighted sum of modes.

Zeros have a different interpretation. Since the pure exponential output corresponding to the input $u(t) = e^{st}$ with $a(s) \neq 0$ is $G(s)e^{st}$, it follows that the pure exponential output is zero if $b(s) = 0$. Zeros of the transfer function thus block transmission of the corresponding exponential signals.

For a state space system with transfer function $G(s) = C(sI - A)^{-1}B + D$, the poles of the transfer function are the eigenvalues of the matrix A in the state space

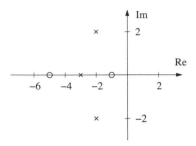

Figure 8.4: A pole zero diagram for a transfer function with zeros at -5 and -1 and poles at -3 and $-2 \pm 2j$. The circles represent the locations of the zeros, and the crosses the locations of the poles. A complete characterization requires we also specify the gain of the system.

model. One easy way to see this is to notice that the value of $G(s)$ is unbounded when s is an eigenvalue of a system since this is precisely the set of points where the characteristic polynomial $\lambda(s) = \det(sI - A) = 0$ (and hence $sI - A$ is noninvertible). It follows that the poles of a state space system depend only on the matrix A, which represents the intrinsic dynamics of the system. We say that a transfer function is stable if all of its poles have negative real part.

To find the zeros of a state space system, we observe that the zeros are complex numbers s such that the input $u(t) = u_0 e^{st}$ gives zero output. Inserting the pure exponential response $x(t) = x_0 e^{st}$ and $y(t) = 0$ in equation (8.2) gives

$$se^{st} x_0 = A x_0 e^{st} + B u_0 e^{st} \qquad 0 = C e^{st} x_0 + D e^{st} u_0,$$

which can be written as

$$\begin{bmatrix} A - sI & B \\ C & D \end{bmatrix} \begin{bmatrix} x_0 \\ u_0 \end{bmatrix} e^{st} = 0.$$

This equation has a solution with nonzero x_0, u_0 only if the matrix on the left does not have full rank. The zeros are thus the values s such that the matrix

$$\begin{bmatrix} A - sI & B \\ C & D \end{bmatrix} \tag{8.17}$$

loses rank.

Since the zeros depend on A, B, C and D, they therefore depend on how the inputs and outputs are coupled to the states. Notice in particular that if the matrix B has full row rank, then the matrix in equation (8.17) has n linearly independent rows for all values of s. Similarly there are n linearly independent columns if the matrix C has full column rank. This implies that systems where the matrix B or C is square and full rank do not have zeros. In particular it means that a system has no zeros if it is fully actuated (each state can be controlled independently) or if the full state is measured.

A convenient way to view the poles and zeros of a transfer function is through a *pole zero diagram*, as shown in Figure 8.4. In this diagram, each pole is marked with a cross, and each zero with a circle. If there are multiple poles or zeros at a

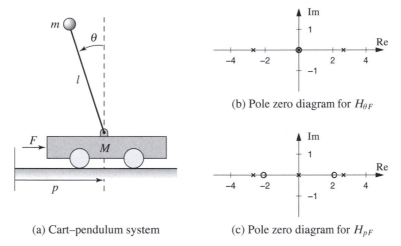

(a) Cart–pendulum system   (c) Pole zero diagram for H_{pF}

Figure 8.5: Poles and zeros for a balance system. The balance system (a) can be modeled around its vertical equilibrium point by a fourth order linear system. The poles and zeros for the transfer functions $H_{\theta F}$ and H_{pF} are shown in (b) and (c), respectively.

fixed location, these are often indicated with overlapping crosses or circles (or other annotations). Poles in the left half-plane correspond to stable modes of the system, and poles in the right half-plane correspond to unstable modes. We thus call a pole in the left-half plane a *stable pole* and a pole in the right-half plane an *unstable pole*. A similar terminology is used for zeros, even though the zeros do not directly related to stability or instability of the system. Notice that the gain must also be given to have a complete description of the transfer function.

Example 8.5 Balance system
Consider the dynamics for a balance system, shown in Figure 8.5. The transfer function for a balance system can be derived directly from the second-order equations, given in Example 2.1:

$$M_t \frac{d^2 p}{dt^2} - ml \frac{d^2 \theta}{dt^2} \cos\theta + c \frac{dp}{dt} + ml \sin\theta \Big(\frac{d\theta}{dt}\Big)^2 = F,$$

$$-ml \cos\theta \frac{d^2 p}{dt^2} + J_t \frac{d^2 \theta}{dt^2} - mgl \sin\theta + \gamma \dot\theta = 0.$$

If we assume that θ and $\dot\theta$ are small, we can approximate this nonlinear system by a set of linear second-order differential equations,

$$M_t \frac{d^2 p}{dt^2} - ml \frac{d^2 \theta}{dt^2} + c \frac{dp}{dt} = F,$$

$$-ml \frac{d^2 p}{dt^2} + J_t \frac{d^2 \theta}{dt^2} + \gamma \frac{d\theta}{dt} - mgl\theta = 0.$$

If we let F be an exponential signal, the resulting response satisfies

$$M_t s^2 p - m l s^2 \theta + c s p = F,$$
$$J_t s^2 \theta - m l s^2 p + \gamma s \theta - m g l \theta = 0,$$

where all signals are exponential signals. The resulting transfer functions for the position of the cart and the orientation of the pendulum are given by solving for p and θ in terms of F to obtain

$$H_{\theta F} = \frac{m l s}{(M_t J_t - m^2 l^2)s^4 + (\gamma M_t + c J_t)s^2 + (c\gamma - M_t m g l)s - m g l c},$$

$$H_{pF} = \frac{J_t s^2 + \gamma s - m g l}{(M_t J_t - m^2 l^2)s^4 + (\gamma M_t + c J_t)s^3 + (c\gamma - M_t m g l)s^2 - m g l c s},$$

where each of the coefficients is positive. The pole zero diagrams for these two transfer functions are shown in Figure 8.5 using the parameters from Example 6.7.

If we assume the damping is small and set $c = 0$ and $\gamma = 0$, we obtain

$$H_{\theta F} = \frac{m l}{(M_t J_t - m^2 l^2)s^2 - M_t m g l},$$

$$H_{pF} = \frac{J_t s^2 - m g l}{s^2 \big((M_t J_t - m^2 l^2)s^2 - M_t m g l\big)}.$$

This gives nonzero poles and zeros at

$$p = \pm \sqrt{\frac{m g l M_t}{M_t J_t - m^2 l^2}} \approx \pm 2.68, \qquad z = \pm \sqrt{\frac{m g l}{J_t}} \approx \pm 2.09.$$

We see that these are quite close to the pole and zero locations in Figure 8.5.   ∇

8.3 Block Diagrams and Transfer Functions

The combination of block diagrams and transfer functions is a powerful way to represent control systems. Transfer functions relating different signals in the system can be derived by purely algebraic manipulations of the transfer functions of the blocks using *block diagram algebra*. To show how this can be done, we will begin with simple combinations of systems.

Consider a system that is a cascade combination of systems with the transfer functions $G_1(s)$ and $G_2(s)$, as shown in Figure 8.6a. Let the input of the system be $u = e^{st}$. The pure exponential output of the first block is the exponential signal $G_1 u$, which is also the input to the second system. The pure exponential output of the second system is

$$y = G_2(G_1 u) = (G_2 G_1)u.$$

The transfer function of the series connection is thus $G = G_2 G_1$, i.e., the product of the transfer functions. The order of the individual transfer functions is due to the fact that we place the input signal on the right-hand side of this expression,

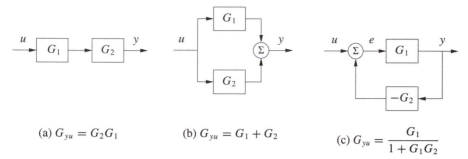

(a) $G_{yu} = G_2 G_1$ (b) $G_{yu} = G_1 + G_2$ (c) $G_{yu} = \dfrac{G_1}{1 + G_1 G_2}$

Figure 8.6: Interconnections of linear systems. Series (a), parallel (b) and feedback (c) connections are shown. The transfer functions for the composite systems can be derived by algebraic manipulations assuming exponential functions for all signals.

hence we first multiply by G_1 and then by G_2. Unfortunately, this has the opposite ordering from the diagrams that we use, where we typically have the signal flow from left to right, so one needs to be careful. The ordering is important if either G_1 or G_2 is a vector-valued transfer function, as we shall see in some examples.

Consider next a parallel connection of systems with the transfer functions G_1 and G_2, as shown in Figure 8.6b. Letting $u = e^{st}$ be the input to the system, the pure exponential output of the first system is then $y_1 = G_1 u$ and the output of the second system is $y_2 = G_2 u$. The pure exponential output of the parallel connection is thus

$$y = G_1 u + G_2 u = (G_1 + G_2)u,$$

and the transfer function for a parallel connection is $G = G_1 + G_2$.

Finally, consider a feedback connection of systems with the transfer functions G_1 and G_2, as shown in Figure 8.6c. Let $u = e^{st}$ be the input to the system, y be the pure exponential output, and e be the pure exponential part of the intermediate signal given by the sum of u and the output of the second block. Writing the relations for the different blocks and the summation unit, we find

$$y = G_1 e, \qquad e = u - G_2 y.$$

Elimination of e gives

$$y = G_1(u - G_2 y) \quad \Longrightarrow \quad (1 + G_1 G_2)y = G_1 u \quad \Longrightarrow \quad y = \frac{G_1}{1 + G_1 G_2}u.$$

The transfer function of the feedback connection is thus

$$G = \frac{G_1}{1 + G_1 G_2}.$$

These three basic interconnections can be used as the basis for computing transfer functions for more complicated systems.

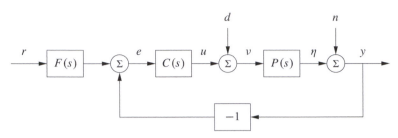

Figure 8.7: Block diagram of a feedback system. The inputs to the system are the reference signal r, the process disturbance d and the measurement noise n. The remaining signals in the system can all be chosen as possible outputs, and transfer functions can be used to relate the system inputs to the other labeled signals.

Control System Transfer Functions

Consider the system in Figure 8.7, which was given at the beginning of the chapter. The system has three blocks representing a process P, a feedback controller C and a feedforward controller F. Together, C and F define the *control law* for the system. There are three external signals: the reference (or command signal) r, the load disturbance d and the measurement noise n. A typical problem is to find out how the error e is related to the signals r, d and n.

To derive the relevant transfer functions we assume that all signals are exponential signals, drop the arguments of signals and transfer functions and trace the signals around the loop. We begin with the signal in which we are interested, in this case the control error e, given by

$$e = Fr - y.$$

The signal y is the sum of n and η, where η is the output of the process:

$$y = n + \eta, \qquad \eta = P(d + u), \qquad u = Ce.$$

Combining these equations gives

$$e = Fr - y = Fr - (n + \eta) = Fr - \left(n + P(d + u)\right)$$
$$= Fr - \left(n + P(d + Ce)\right),$$

and hence

$$e = Fr - n - Pd - PCe.$$

Finally, solving this equation for e gives

$$e = \frac{F}{1 + PC} r - \frac{1}{1 + PC} n - \frac{P}{1 + PC} d = G_{er}r + G_{en}n + G_{ed}d, \qquad (8.18)$$

and the error is thus the sum of three terms, depending on the reference r, the measurement noise n and the load disturbance d. The functions

$$G_{er} = \frac{F}{1 + PC}, \qquad G_{en} = \frac{-1}{1 + PC}, \qquad G_{ed} = \frac{-P}{1 + PC} \qquad (8.19)$$

are transfer functions from reference r, noise n and disturbance d to the error e.

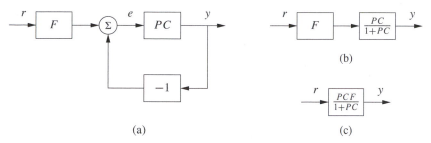

Figure 8.8: Example of block diagram algebra. The results from multiplying the process and controller transfer functions (from Figure 8.7) are shown in (a). Replacing the feedback loop with its transfer function equivalent yields (b), and finally multiplying the two remaining blocks gives the reference to output representation in (c).

We can also derive transfer functions by manipulating the block diagrams directly, as illustrated in Figure 8.8. Suppose we wish to compute the transfer function between the reference r and the output y. We begin by combining the process and controller blocks in Figure 8.7 to obtain the diagram in Figure 8.8a. We can now eliminate the feedback loop using the algebra for a feedback interconnection (Figure 8.8b) and then use the series interconnection rule to obtain

$$G_{yr} = \frac{PCF}{1 + PC}. \tag{8.20}$$

Similar manipulations can be used to obtain the other transfer functions (Exercise 8.8).

The derivation illustrates an effective way to manipulate the equations to obtain the relations between inputs and outputs in a feedback system. The general idea is to start with the signal of interest and to trace signals around the feedback loop until coming back to the signal we started with. With some practice, equations (8.18) and (8.19) can be written directly by inspection of the block diagram. Notice, for example, that all terms in equation (8.19) have the same denominators and that the numerators are the blocks that one passes through when going directly from input to output (ignoring the feedback). This type of rule can be used to compute transfer functions by inspection, although for systems with multiple feedback loops it can be tricky to compute them without writing down the algebra explicitly.

Example 8.6 Vehicle steering
Consider the linearized model for vehicle steering introduced in Example 5.12. In Examples 6.4 and 7.3 we designed a state feedback compensator and state estimator for the system. A block diagram for the resulting control system is given in Figure 8.9. Note that we have split the estimator into two components, $G_{\hat{x}u}(s)$ and $G_{\hat{x}y}(s)$, corresponding to its inputs u and y. The controller can be described as the sum of two (open loop) transfer functions

$$u = G_{uy}(s)y + G_{ur}(s)r.$$

The first transfer function, $G_{uy}(s)$, describes the feedback term and the second, $G_{ur}(s)$, describes the feedforward term. We call these *open loop* transfer functions

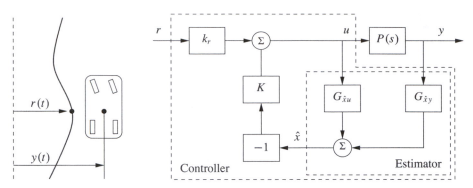

Figure 8.9: Block diagram for a steering control system. The control system is designed to maintain the lateral position of the vehicle along a reference curve (left). The structure of the control system is shown on the right as a block diagram of transfer functions. The estimator consists of two components that compute the estimated state $\hat{x}$ from the combination of the input u and output y of the process. The estimated state is fed through a state feedback controller and combined with a reference gain to obtain the commanded steering angle u.

because they represent the relationships between the signals without considering the dynamics of the process (e.g., removing $P(s)$ from the system description). To derive these functions, we compute the transfer functions for each block and then use block diagram algebra.

We begin with the estimator, which takes u and y as its inputs and produces an estimate $\hat{x}$. The dynamics for this process were derived in Example 7.3 and are given by

$$\frac{d\hat{x}}{dt} = (A - LC)\hat{x} + Ly + Bu,$$

$$\hat{x} = \underbrace{\left(sI - (A - LC)\right)^{-1} B}_{G_{\hat{x}u}} u + \underbrace{\left(sI - (A - LC)\right)^{-1} L}_{G_{\hat{x}y}} y.$$

Using the expressions for A, B, C and L from Example 7.3, we obtain

$$G_{\hat{x}u}(s) = \begin{bmatrix} \dfrac{\gamma s + 1}{s^2 + l_1 s + l_2} \\[2mm] \dfrac{s + l_1 - \gamma l_2}{s^2 + l_1 s + l_2} \end{bmatrix}, \qquad G_{\hat{x}y}(s) = \begin{bmatrix} \dfrac{l_1 s + l_2}{s^2 + l_1 s + l_2} \\[2mm] \dfrac{l_2 s}{s^2 + l_1 s + l_2} \end{bmatrix},$$

where l_1 and l_2 are the observer gains and γ is the scaled position of the center of mass from the rear wheels. The controller was a state feedback compensator, which can be viewed as a constant, multi-input, single-output transfer function of the form $u = -K\hat{x}$.

We can now proceed to compute the transfer function for the overall control system. Using block diagram algebra, we have

$$G_{uy}(s) = \frac{-K G_{\hat{x}y}(s)}{1 + K G_{\hat{x}u}(s)} = -\frac{s(k_1 l_1 + k_2 l_2) + k_1 l_2}{s^2 + s(\gamma k_1 + k_2 + l_1) + k_1 + l_2 + k_2 l_1 - \gamma k_2 l_2}$$

and

$$G_{ur}(s) = \frac{k_r}{1 + KG_{\hat{x}u}(s)} = \frac{k_1(s^2 + l_1 s + l_2)}{s^2 + s(\gamma k_1 + k_2 + l_1) + k_1 + l_2 + k_2 l_1 - \gamma k_2 l_2},$$

where k_1 and k_2 are the controller gains.

Finally, we compute the full closed loop dynamics. We begin by deriving the transfer function for the process $P(s)$. We can compute this directly from the state space description of the dynamics, which was given in Example 5.12. Using that description, we have

$$P(s) = G_{yu}(s) = C(sI - A)^{-1}B + D = \begin{bmatrix} 1 & 0 \end{bmatrix} \begin{bmatrix} s & -1 \\ 0 & s \end{bmatrix}^{-1} \begin{bmatrix} \gamma \\ 1 \end{bmatrix} = \frac{\gamma s + 1}{s^2}.$$

The transfer function for the full closed loop system between the input r and the output y is then given by

$$G_{yr} = \frac{k_r P(s)}{1 + P(s)G_{uy}(s)} = \frac{k_1(\gamma s + 1)}{s^2 + (k_1 \gamma + k_2)s + k_1}.$$

Note that the observer gains l_1 and l_2 do not appear in this equation. This is because we are considering steady-state analysis and, in steady state, the estimated state exactly tracks the state of the system assuming perfect models. We will return to this example in Chapter 12 to study the robustness of this particular approach.   ∇

Pole/Zero Cancellations

Because transfer functions are often polynomials in s, it can sometimes happen that the numerator and denominator have a common factor, which can be canceled. Sometimes these cancellations are simply algebraic simplifications, but in other situations they can mask potential fragilities in the model. In particular, if a pole/zero cancellation occurs because terms in separate blocks that just happen to coincide, the cancellation may not occur if one of the systems is slightly perturbed. In some situations this can result in severe differences between the expected behavior and the actual behavior.

To illustrate when we can have pole/zero cancellations, consider the block diagram in Figure 8.7 with $F = 1$ (no feedforward compensation) and C and P given by

$$C(s) = \frac{n_c(s)}{d_c(s)}, \qquad P(s) = \frac{n_p(s)}{d_p(s)}.$$

The transfer function from r to e is then given by

$$G_{er}(s) = \frac{1}{1 + PC} = \frac{d_c(s)d_p(s)}{d_c(s)d_p(s) + n_c(s)n_p(s)}.$$

If there are common factors in the numerator and denominator polynomials, then these terms can be factored out and eliminated from both the numerator and denominator. For example, if the controller has a zero at $s = -a$ and the process has a pole at $s = -a$, then we will have

$$G_{er}(s) = \frac{(s+a)d_c(s)d_p'(s)}{(s+a)d_c(s)d_p'(s) + (s+a)n_c'(s)n_p(s)} = \frac{d_c(s)d_p'(s)}{d_c(s)d_p'(s) + n_c'(s)n_p(s)},$$

where $n_c'(s)$ and $d_p'(s)$ represent the relevant polynomials with the term $s + a$ factored out. In the case when $a < 0$ (so that the zero or pole is in the right half-plane), we see that there is no impact on the transfer function G_{er}.

Suppose instead that we compute the transfer function from d to e, which represents the effect of a disturbance on the error between the reference and the output. This transfer function is given by

$$G_{ed}(s) = -\frac{d_c(s)n_p(s)}{(s+a)d_c(s)d_p'(s) + (s+a)n_c'(s)n_p(s)}.$$

Notice that if $a < 0$, then the pole is in the right half-plane and the transfer function G_{ed} is *unstable*. Hence, even though the transfer function from r to e appears to be okay (assuming a perfect pole/zero cancellation), the transfer function from d to e can exhibit unbounded behavior. This unwanted behavior is typical of an *unstable pole/zero cancellation*.

It turns out that the cancellation of a pole with a zero can also be understood in terms of the state space representation of the systems. Reachability or observability is lost when there are cancellations of poles and zeros (Exercise 8.11). A consequence is that the transfer function represents the dynamics only in the reachable and observable subspace of a system (see Section 7.5).

Example 8.7 Cruise control
The input/output response from throttle to velocity for the linearized model for a car has the transfer function $G(s) = b/(s-a), a < 0$. A simple (but not necessarily good) way to design a PI controller is to choose the parameters of the PI controller so that the controller zero at $s = -k_i/k_p$ cancels the process pole at $s = a$. The transfer function from reference to velocity is $G_{vr}(s) = bk_p/(s+bk_p)$, and control design is simply a matter of choosing the gain k_p. The closed loop system dynamics are of first order with the time constant $1/bk_p$.

Figure 8.10 shows the velocity error when the car encounters an increase in the road slope. A comparison with the controller used in Figure 3.3b (reproduced in dashed curves) shows that the controller based on pole/zero cancellation has very poor performance. The velocity error is larger, and it takes a long time to settle.

Notice that the control signal remains practically constant after $t = 15$ even if the error is large after that time. To understand what happens we will analyze the system. The parameters of the system are $a = -0.0101$ and $b = 1.32$, and the controller parameters are $k_p = 0.5$ and $k_i = 0.0051$. The closed loop time constant is $1/(bk_p) = 2.5$ s, and we would expect that the error would settle in about 10 s (4 time constants). The transfer functions from road slope to velocity and control

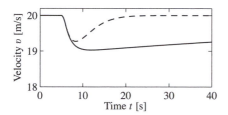
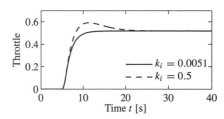

Figure 8.10: Car with PI cruise control encountering a sloping road. The velocity error is shown on the left and the throttle is shown on the right. Results with a PI controller with $k_p = 0.5$ and $k_i = 0.0051$, where the process pole $s = -0.0101$, is shown by solid lines, and a controller with $k_p = 0.5$ and $k_i = 0.5$ is shown by dashed lines. Compare with Figure 3.3b.

signals are

$$G_{v\theta}(s) = \frac{b_g k_p s}{(s-a)(s+bk_p)}, \qquad G_{u\theta}(s) = \frac{bk_p}{s+bk_p}.$$

Notice that the canceled mode $s = a = -0.0101$ appears in $G_{v\theta}$ but not in $G_{u\theta}$. The reason why the control signal remains constant is that the controller has a zero at $s = -0.0101$, which cancels the slowly decaying process mode. Notice that the error would diverge if the canceled pole was unstable. $\qquad \nabla$

The lesson we can learn from this example is that it is a bad idea to try to cancel unstable or slow process poles. A more detailed discussion of pole/zero cancellations is given in Section 12.4.

Algebraic Loops

When analyzing or simulating a system described by a block diagram, it is necessary to form the differential equations that describe the complete system. In many cases the equations can be obtained by combining the differential equations that describe each subsystem and substituting variables. This simple procedure cannot be used when there are closed loops of subsystems that all have a direct connection between inputs and outputs, known as an *algebraic loop*.

To see what can happen, consider a system with two blocks, a first-order nonlinear system,

$$\frac{dx}{dt} = f(x, u), \qquad y = h(x), \tag{8.21}$$

and a proportional controller described by $u = -ky$. There is no direct term since the function h does not depend on u. In that case we can obtain the equation for the closed loop system simply by replacing u by $-ky$ in (8.21) to give

$$\frac{dx}{dt} = f(x, -ky), \qquad y = h(x).$$

Such a procedure can easily be automated using simple formula manipulation.

The situation is more complicated if there is a direct term. If $y = h(x, u)$, then replacing u by $-ky$ gives

$$\frac{dx}{dt} = f(x, -ky), \qquad y = h(x, -ky).$$

To obtain a differential equation for x, the algebraic equation $y = h(x, -ky)$ must be solved to give $y = \alpha(x)$, which in general is a complicated task.

When algebraic loops are present, it is necessary to solve algebraic equations to obtain the differential equations for the complete system. Resolving algebraic loops is a nontrivial problem because it requires the symbolic solution of algebraic equations. Most block diagram-oriented modeling languages cannot handle algebraic loops, and they simply give a diagnosis that such loops are present. In the era of analog computing, algebraic loops were eliminated by introducing fast dynamics between the loops. This created differential equations with fast and slow modes that are difficult to solve numerically. Advanced modeling languages like Modelica use several sophisticated methods to resolve algebraic loops.

8.4 The Bode Plot

The frequency response of a linear system can be computed from its transfer function by setting $s = i\omega$, corresponding to a complex exponential

$$u(t) = e^{i\omega t} = \cos(\omega t) + i\sin(\omega t).$$

The resulting output has the form

$$y(t) = G(i\omega)e^{i\omega t} = Me^{i(\omega t + \varphi)} = M\cos(\omega t + \varphi) + iM\sin(\omega t + \varphi),$$

where M and φ are the gain and phase of G:

$$M = |G(i\omega)|, \qquad \varphi = \arctan\frac{\operatorname{Im} G(i\omega)}{\operatorname{Re} G(i\omega)}.$$

The phase of G is also called the *argument* of G, a term that comes from the theory of complex variables.

It follows from linearity that the response to a single sinusoid (sin or cos) is amplified by M and phase-shifted by φ. Note that $-\pi < \varphi \leq \pi$, so the arctangent must be taken respecting the signs of the numerator and denominator. It will often be convenient to represent the phase in degrees rather than radians. We will use the notation $\angle G(i\omega)$ for the phase in degrees and $\arg G(i\omega)$ for the phase in radians. In addition, while we always take $\arg G(i\omega)$ to be in the range $(-\pi, \pi]$, we will take $\angle G(i\omega)$ to be continuous, so that it can take on values outside the range of $-180°$ to $180°$.

The frequency response $G(i\omega)$ can thus be represented by two curves: the gain curve and the phase curve. The *gain curve* gives $|G(i\omega)|$ as a function of frequency ω, and the *phase curve* gives $\angle G(i\omega)$. One particularly useful way of drawing these

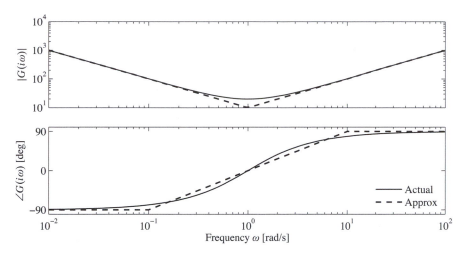

Figure 8.11: Bode plot of the transfer function $C(s) = 20 + 10/s + 10s$ corresponding to an ideal PID controller. The top plot is the gain curve and the bottom plot is the phase curve. The dashed lines show straight-line approximations of the gain curve and the corresponding phase curve.

curves is to use a log/log scale for the gain plot and a log/linear scale for the phase plot. This type of plot is called a *Bode plot* and is shown in Figure 8.11.

Sketching and Interpreting Bode Plots

Part of the popularity of Bode plots is that they are easy to sketch and interpret. Since the frequency scale is logarithmic, they cover the behavior of a linear system over a wide frequency range.

Consider a transfer function that is a rational functionof the form

$$G(s) = \frac{b_1(s)b_2(s)}{a_1(s)a_2(s)}.$$

We have

$$\log|G(s)| = \log|b_1(s)| + \log|b_2(s)| - \log|a_1(s)| - \log|a_2(s)|,$$

and hence we can compute the gain curve by simply adding and subtracting gains corresponding to terms in the numerator and denominator. Similarly,

$$\angle G(s) = \angle b_1(s) + \angle b_2(s) - \angle a_1(s) - \angle a_2(s),$$

and so the phase curve can be determined in an analogous fashion. Since a polynomial can be written as a product of terms of the type

$$k, \quad s, \quad s + a, \quad s^2 + 2\zeta\omega_0 s + \omega_0^2,$$

it suffices to be able to sketch Bode diagrams for these terms. The Bode plot of a complex system is then obtained by adding the gains and phases of the terms.

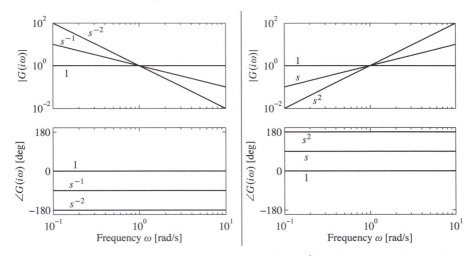

Figure 8.12: Bode plots of the transfer functions $G(s) = s^k$ for $k = -2, -1, 0, 1, 2$. On a log-log scale, the gain curve is a straight line with slope k. Using a log-linear scale, the phase curves for the transfer functions are constants, with phase equal to $90° \times k$

The simplest term in a transfer function is one of the form s^k, where $k > 0$ if the term appears in the numerator and $k < 0$ if the term is in the denominator. The gain and phase of the term are given by

$$\log |G(i\omega)| = k \log \omega, \quad \angle G(i\omega) = 90k.$$

The gain curve is thus a straight line with slope k, and the phase curve is a constant at $90° \times k$. The case when $k = 1$ corresponds to a differentiator and has slope 1 with phase $90°$. The case when $k = -1$ corresponds to an integrator and has slope -1 with phase $-90°$. Bode plots of the various powers of k are shown in Figure 8.12.

Consider next the transfer function of a first-order system, given by

$$G(s) = \frac{a}{s + a}.$$

We have

$$|G(s)| = \frac{|a|}{|s + a|}, \quad \angle G(s) = \angle(a) - \angle(s + a),$$

and hence

$$\log |G(i\omega)| = \log a - \frac{1}{2} \log (\omega^2 + a^2), \quad \angle G(i\omega) = -\frac{180}{\pi} \arctan \frac{\omega}{a}.$$

The Bode plot is shown in Figure 8.13a, with the magnitude normalized by the zero frequency gain. Both the gain curve and the phase curve can be approximated by

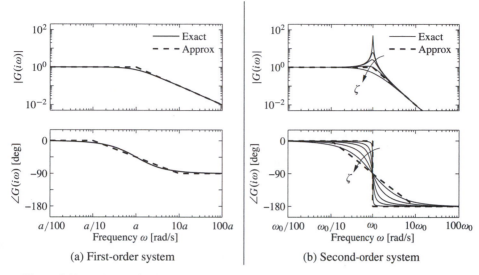

Figure 8.13: Bode plots for first- and second-order systems. (a) The first-order system $G(s) = a/(s+a)$ can be approximated by asymptotic curves (dashed) in both the gain and the frequency, with the breakpoint in the gain curve at $\omega = a$ and the phase decreasing by $90°$ over a factor of 100 in frequency. (b) The second-order system $G(s) = \omega_0^2/(s^2+2\zeta\omega_0 s + \omega_0^2)$ has a peak at frequency a and then a slope of -2 beyond the peak; the phase decreases from $0°$ to $-180°$. The height of the peak and the rate of change of phase depending on the damping ratio ζ ($\zeta = 0.02, 0.1, 0.2, 0.5$ and 1.0 shown).

the following straight lines

$$\log|G(i\omega)| \approx \begin{cases} 0 & \text{if } \omega < a \\ \log a - \log \omega & \text{if } \omega > a, \end{cases}$$

$$\angle G(i\omega) \approx \begin{cases} 0 & \text{if } \omega < a/10 \\ -45 - 45(\log\omega - \log a) & a/10 < \omega < 10a \\ -90 & \text{if } \omega > 10a. \end{cases}$$

The approximate gain curve consists of a horizontal line up to frequency $\omega = a$, called the *breakpoint* or *corner frequency*, after which the curve is a line of slope -1 (on a log-log scale). The phase curve is zero up to frequency $a/10$ and then decreases linearly by $45°$/decade up to frequency $10a$, at which point it remains constant at $90°$. Notice that a first-order system behaves like a constant for low frequencies and like an integrator for high frequencies; compare with the Bode plot in Figure 8.12.

Finally, consider the transfer function for a second-order system,

$$G(s) = \frac{\omega_0^2}{s^2 + 2\omega_0\zeta s + \omega_0^2},$$

for which we have

$$\log |G(i\omega)| = 2 \log \omega_0 - \frac{1}{2} \log \left(\omega^4 + 2\omega_0^2 \omega^2 (2\zeta^2 - 1) + \omega_0^4\right),$$

$$\angle G(i\omega) = -\frac{180}{\pi} \arctan \frac{2\zeta \omega_0 \omega}{\omega_0^2 - \omega^2}.$$

The gain curve has an asymptote with zero slope for $\omega \ll \omega_0$. For large values of ω the gain curve has an asymptote with slope -2. The largest gain $Q = \max_\omega |G(i\omega)| \approx 1/(2\zeta)$, called the *Q-value*, is obtained for $\omega \approx \omega_0$. The phase is zero for low frequencies and approaches $180°$ for large frequencies. The curves can be approximated with the following piecewise linear expressions

$$\log |G(i\omega)| \approx \begin{cases} 0 & \text{if } \omega \ll \omega_0 \\ 2 \log \omega_0 - 2 \log \omega & \text{if } \omega \gg \omega_0, \end{cases}$$

$$\angle G(i\omega) \approx \begin{cases} 0 & \text{if } \omega \ll \omega_0 \\ -180 & \text{if } \omega \gg \omega_0. \end{cases}$$

The Bode plot is shown in Figure 8.13b. Note that the asymptotic approximation is poor near $\omega = \omega_0$ and that the Bode plot depends strongly on ζ near this frequency.

Given the Bode plots of the basic functions, we can now sketch the frequency response for a more general system. The following example illustrates the basic idea.

Example 8.8 Asymptotic approximation for a transfer function
Consider the transfer function given by

$$G(s) = \frac{k(s+b)}{(s+a)(s^2 + 2\zeta \omega_0 s + \omega_0^2)}, \qquad a \ll b \ll \omega_0.$$

The Bode plot for this transfer function appears in Figure 8.14, with the complete transfer function shown as a solid line and the asymptotic approximation shown as a dashed line.

We begin with the gain curve. At low frequency, the magnitude is given by

$$G(0) = \frac{kb}{a\omega_0^2}.$$

When we reach $\omega = a$, the effect of the pole begins and the gain decreases with slope -1. At $\omega = b$, the zero comes into play and we increase the slope by 1, leaving the asymptote with net slope 0. This slope is used until the effect of the second-order pole is seen at $\omega = \omega_0$, at which point the asymptote changes to slope -2. We see that the gain curve is fairly accurate except in the region of the peak due to the second-order pole (since for this case ζ is reasonably small).

The phase curve is more complicated since the effect of the phase stretches out much further. The effect of the pole begins at $\omega = a/10$, at which point we change from phase 0 to a slope of $-45°$/decade. The zero begins to affect the phase at $\omega = b/10$, producing a flat section in the phase. At $\omega = 10a$ the phase

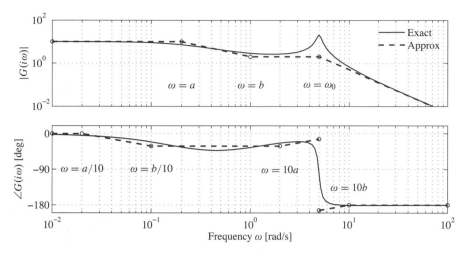

Figure 8.14: Asymptotic approximation to a Bode plot. The thin line is the Bode plot for the transfer function $G(s) = k(s + b)/(s + a)(s^2 + 2\zeta\omega_0 s + \omega_0^2)$, where $a \ll b \ll \omega_0$. Each segment in the gain and phase curves represents a separate portion of the approximation, where either a pole or a zero begins to have effect. Each segment of the approximation is a straight line between these points at a slope given by the rules for computing the effects of poles and zeros.

contributions from the pole end, and we are left with a slope of $+45°$/decade (from the zero). At the location of the second-order pole, $s \approx i\omega_0$, we get a jump in phase of $-180°$. Finally, at $\omega = 10b$ the phase contributions of the zero end, and we are left with a phase of -180 degrees. We see that the straight-line approximation for the phase is not as accurate as it was for the gain curve, but it does capture the basic features of the phase changes as a function of frequency.   ▽

The Bode plot gives a quick overview of a system. Since any signal can be decomposed into a sum of sinusoids, it is possible to visualize the behavior of a system for different frequency ranges. The system can be viewed as a filter that can change the amplitude (and phase) of the input signals according to the frequency response. For example, if there are frequency ranges where the gain curve has constant slope and the phase is close to zero, the action of the system for signals with these frequencies can be interpreted as a pure gain. Similarly, for frequencies where the slope is $+1$ and the phase close to $90°$, the action of the system can be interpreted as a differentiator, as shown in Figure 8.12.

Three common types of frequency responses are shown in Figure 8.15. The system in Figure 8.15a is called a *low-pass filter* because the gain is constant for low frequencies and drops for high frequencies. Notice that the phase is zero for low frequencies and $-180°$ for high frequencies. The systems in Figure 8.15b and c are called a *band-pass filter* and *high-pass filter* for similar reasons.

To illustrate how different system behaviors can be read from the Bode plots we consider the band-pass filter in Figure 8.15b. For frequencies around $\omega = \omega_0$, the signal is passed through with no change in gain. However, for frequencies well

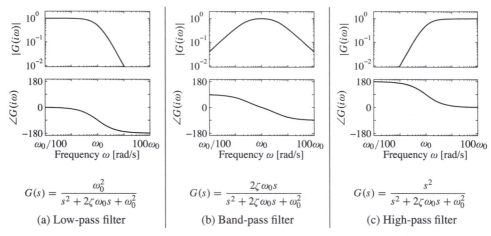

$$G(s) = \frac{\omega_0^2}{s^2 + 2\zeta\omega_0 s + \omega_0^2} \qquad G(s) = \frac{2\zeta\omega_0 s}{s^2 + 2\zeta\omega_0 s + \omega_0^2} \qquad G(s) = \frac{s^2}{s^2 + 2\zeta\omega_0 s + \omega_0^2}$$

(a) Low-pass filter   (b) Band-pass filter   (c) High-pass filter

Figure 8.15: Bode plots for low-pass, band-pass and high-pass filters. The top plots are the gain curves and the bottom plots are the phase curves. Each system passes frequencies in a different range and attenuates frequencies outside of that range.

below or well above ω_0, the signal is attenuated. The phase of the signal is also affected by the filter, as shown in the phase curve. For frequencies below $\omega_0/100$ there is a phase lead of $90°$, and for frequencies above $100\omega_0$ there is a phase lag of $90°$. These actions correspond to differentiation and integration of the signal in these frequency ranges.

Example 8.9 Transcriptional regulation
Consider a genetic circuit consisting of a single gene. We wish to study the response of the protein concentration to fluctuations in the mRNA dynamics. We consider two cases: a *constitutive promoter* (no regulation) and self-repression (negative feedback), illustrated in Figure 8.16. The dynamics of the system are given by

$$\frac{dm}{dt} = \alpha(p) - \gamma m - u, \qquad \frac{dp}{dt} = \beta m - \delta p,$$

where u is a disturbance term that affects mRNA transcription.

For the case of no feedback we have $\alpha(p) = \alpha_0$, and the system has an equilibrium point at $m_e = \alpha_0/\gamma$, $p_e = \beta\alpha_0/(\delta\gamma)$. The transfer function from v to p is given by

$$G_{pv}^{ol}(s) = \frac{-\beta}{(s+\gamma)(s+\delta)}.$$

For the case of negative regulation, we have

$$\alpha(p) = \frac{\alpha_1}{1 + kp^n} + \alpha_0,$$

and the equilibrium points satisfy

$$m_e = \frac{\delta}{\beta} p_e, \qquad \frac{\alpha}{1 + kp_e^n} + \alpha_0 = \gamma m_e = \frac{\gamma\delta}{\beta} p_e.$$

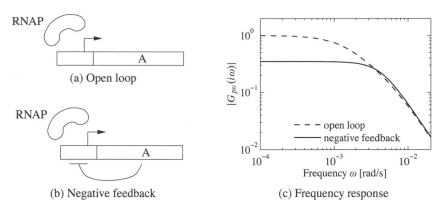

(a) Open loop

(b) Negative feedback

(c) Frequency response

Figure 8.16: Noise attenuation in a genetic circuit. The open loop system (a) consists of a constitutive promoter, while the closed loop circuit (b) is self-regulated with negative feedback (repressor). The frequency response for each circuit is shown in (c).

The resulting transfer function is given by

$$G_{pv}^{\mathrm{cl}}(s) = \frac{\beta}{(s + \gamma)(s + \delta) + \beta\sigma}, \qquad \sigma = \frac{n\alpha_1 k p_e^{n-1}}{(1 + k p_e^n)^2}.$$

Figure 8.16c shows the frequency response for the two circuits. We see that the feedback circuit attenuates the response of the system to disturbances with low-frequency content but slightly amplifies disturbances at high frequency (compared to the open loop system). Notice that these curves are very similar to the frequency response curves for the op amp shown in Figure 8.3b. ∇

Transfer Functions from Experiments

The transfer function of a system provides a summary of the input/output response and is very useful for analysis and design. However, modeling from first principles can be difficult and time-consuming. Fortunately, we can often build an input/output model for a given application by directly measuring the frequency response and fitting a transfer function to it. To do so, we perturb the input to the system using a sinusoidal signal at a fixed frequency. When steady state is reached, the amplitude ratio and the phase lag give the frequency response for the excitation frequency. The complete frequency response is obtained by sweeping over a range of frequencies.

By using correlation techniques it is possible to determine the frequency response very accurately, and an analytic transfer function can be obtained from the frequency response by curve fitting. The success of this approach has led to instruments and software that automate this process, called *spectrum analyzers*. We illustrate the basic concept through two examples.

Example 8.10 Atomic force microscope

To illustrate the utility of spectrum analysis, we consider the dynamics of the atomic force microscope, introduced in Section 3.5. Experimental determination of the

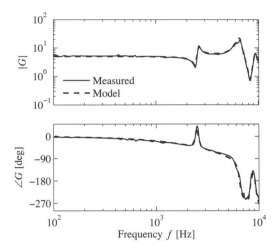

Figure 8.17: Frequency response of a preloaded piezoelectric drive for an atomic force microscope. The Bode plot shows the response of the measured transfer function (solid) and the fitted transfer function (dashed).

frequency response is particularly attractive for this system because its dynamics are very fast and hence experiments can be done quickly. A typical example is given in Figure 8.17, which shows an experimentally determined frequency response (solid line). In this case the frequency response was obtained in less than a second. The transfer function

$$G(s) = \frac{k\omega_2^2\omega_3^2\omega_5^2(s^2 + 2\zeta_1\omega_1 s + \omega_1^2)(s^2 + 2\zeta_4\omega_4 s + \omega_4^2)e^{-s\tau}}{\omega_1^2\omega_4^2(s^2 + 2\zeta_2\omega_2 s + \omega_2^2)(s^2 + 2\zeta_3\omega_3 s + \omega_3^2)(s^2 + 2\zeta_5\omega_5 s + \omega_5^2)},$$

with $\omega_k = 2\pi f_k$ and $f_1 = 2.42$ kHz, $\zeta_1 = 0.03$, $f_2 = 2.55$ kHz, $\zeta_2 = 0.03$, $f_3 = 6.45$ kHz, $\zeta_3 = 0.042$, $f_4 = 8.25$ kHz, $\zeta_4 = 0.025$, $f_5 = 9.3$ kHz, $\zeta_5 = 0.032$, $\tau = 10^{-4}$ s and $k = 5$, was fit to the data (dashed line). The frequencies associated with the zeros are located where the gain curve has minima, and the frequencies associated with the poles are located where the gain curve has local maxima. The relative damping ratios are adjusted to give a good fit to maxima and minima. When a good fit to the gain curve is obtained, the time delay is adjusted to give a good fit to the phase curve. The piezo drive is preloaded, and a simple model of its dynamics is derived in Exercise 3.7. The pole at 2.42 kHz corresponds to the trampoline mode derived in the exercise; the other resonances are higher modes.

$$\nabla$$

Example 8.11 Pupillary light reflex dynamics
The human eye is an organ that is easily accessible for experiments. It has a control system that adjusts the pupil opening to regulate the light intensity at the retina.

This control system was explored extensively by Stark in the 1960s [184]. To determine the dynamics, light intensity on the eye was varied sinusoidally and the pupil opening was measured. A fundamental difficulty is that the closed loop system is insensitive to internal system parameters, so analysis of a closed loop system thus

| (a) Closed loop | (b) Open loop | (c) High gain |

Figure 8.18: Light stimulation of the eye. In (a) the light beam is so large that it always covers the whole pupil, giving closed loop dynamics. In (b) the light is focused into a beam which is so narrow that it is not influenced by the pupil opening, giving open loop dynamics. In (c) the light beam is focused on the edge of the pupil opening, which has the effect of increasing the gain of the system since small changes in the pupil opening have a large effect on the amount of light entering the eye. From Stark [184].

gives little information about the internal properties of the system. Stark used a clever experimental technique that allowed him to investigate both open and closed loop dynamics. He excited the system by varying the intensity of a light beam focused on the eye and measured pupil area, as illustrated in Figure 8.18. By using a wide light beam that covers the whole pupil, the measurement gives the closed loop dynamics. The open loop dynamics were obtained by using a narrow beam, which is small enough that it is not influenced by the pupil opening. The result of one experiment for determining open loop dynamics is given in Figure 8.19. Fitting a transfer function to the gain curve gives a good fit for $G(s) = 0.17/(1 + 0.08s)^3$. This curve gives a poor fit to the phase curve as shown by the dashed curve in Figure 8.19. The fit to the phase curve is improved by adding a time delay, which leaves the gain curve unchanged while substantially modifying the phase curve. The final fit gives the model

$$G(s) = \frac{0.17}{(1 + 0.08s)^3} e^{-0.2s}.$$

The Bode plot of this is shown with solid curves in Figure 8.19. Modeling of the pupillary reflex from first principles is discussed in detail in [119].   ∇

Notice that for both the AFM drive and pupillary dynamics it is not easy to derive appropriate models from first principles. In practice, it is often fruitful to use a combination of analytical modeling and experimental identification of parameters. Experimental determination of frequency response is less attractive for systems with slow dynamics because the experiment takes a long time.

8.5 Laplace Transforms

Transfer functions are conventionally introduced using Laplace transforms, and in this section we derive the transfer function using this formalism. We assume basic familiarity with Laplace transforms; students who are not familiar with them can safely skip this section. A good reference for the mathematical material in this

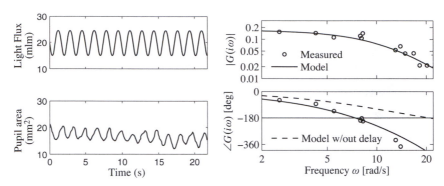

Figure 8.19: Sample curves from an open loop frequency response of the eye (left) and a Bode plot for the open loop dynamics (right). The solid curve shows a fit of the data using a third-order transfer function with time delay. The dashed curve in the Bode plot is the phase of the system without time delay, showing that the delay is needed to properly capture the phase. (Figure redrawn from the data of Stark [184].)

section is the classic book by Widder [198].

Traditionally, Laplace transforms were used to compute responses of linear systems to different stimuli. Today we can easily generate the responses using computers. Only a few elementary properties are needed for basic control applications. There is, however, a beautiful theory for Laplace transforms that makes it possible to use many powerful tools from the theory of functions of a complex variable to get deep insights into the behavior of systems.

Consider a function $f(t)$, $f : \mathbb{R}^+ \to \mathbb{R}$, that is integrable and grows no faster than $e^{s_0 t}$ for some finite $s_0 \in \mathbb{R}$ and large t. The Laplace transform maps f to a function $F = \mathcal{L}f : \mathbb{C} \to \mathbb{C}$ of a complex variable. It is defined by

$$F(s) = \int_0^\infty e^{-st} f(t)\, dt, \quad \operatorname{Re} s > s_0. \tag{8.22}$$

The transform has some properties that makes it well suited to deal with linear systems.

First we observe that the transform is linear because

$$\mathcal{L}(af + bg) = \int_0^\infty e^{-st}(af(t) + bg(t))\, dt$$
$$= a\int_0^\infty e^{-st} f(t)\, dt + b\int_0^\infty e^{-st} g(t)\, dt = a\mathcal{L}f + b\mathcal{L}g. \tag{8.23}$$

Next we calculate the Laplace transform of the derivative of a function. We have

$$\mathcal{L}\frac{df}{dt} = \int_0^\infty e^{-st} f'(t)\, dt = e^{-st} f(t)\Big|_0^\infty + s\int_0^\infty e^{-st} f(t)\, dt = -f(0) + s\mathcal{L}f,$$

where the second equality is obtained using integration by parts. We thus obtain

$$\mathcal{L}\frac{df}{dt} = s\mathcal{L}f - f(0) = sF(s) - f(0). \tag{8.24}$$

This formula is particularly simple if the initial conditions are zero because it follows that differentiation of a function corresponds to multiplication of the transform by s.

Since differentiation corresponds to multiplication by s, we can expect that integration corresponds to division by s. This is true, as can be seen by calculating the Laplace transform of an integral. Using integration by parts, we get

$$\mathcal{L} \int_0^t f(\tau)\, d\tau = \int_0^\infty \left(e^{-st} \int_0^t f(\tau)\, d\tau \right) dt$$

$$= -\frac{e^{-st}}{s} \int_0^t f(\tau)\, d\tau \Big|_0^\infty + \int_0^\infty \frac{e^{-s\tau}}{s} f(\tau)\, d\tau = \frac{1}{s} \int_0^\infty e^{-s\tau} f(\tau)\, d\tau,$$

hence

$$\mathcal{L} \int_0^t f(\tau)\, d\tau = \frac{1}{s} \mathcal{L} f = \frac{1}{s} F(s). \tag{8.25}$$

Next consider a linear time-invariant system with zero initial state. We saw in Section 5.3 that the relation between the input u and the output y is given by the convolution integral

$$y(t) = \int_0^\infty h(t - \tau) u(\tau)\, d\tau,$$

where $h(t)$ is the impulse response for the system. Taking the Laplace transform of this expression, we have

$$Y(s) = \int_0^\infty e^{-st} y(t)\, dt = \int_0^\infty e^{-st} \int_0^\infty h(t - \tau) u(\tau)\, d\tau\, dt$$

$$= \int_0^\infty \int_0^t e^{-s(t-\tau)} e^{-s\tau} h(t - \tau) u(\tau)\, d\tau\, dt$$

$$= \int_0^\infty e^{-s\tau} u(\tau)\, d\tau \int_0^\infty e^{-st} h(t)\, dt = H(s) U(s).$$

Thus, the input/output response is given by $Y(s) = H(s)U(s)$, where H, U and Y are the Laplace transforms of h, u and y. The system theoretic interpretation is that the Laplace transform of the output of a linear system is a product of two terms, the Laplace transform of the input $U(s)$ and the Laplace transform of the impulse response of the system $H(s)$. A mathematical interpretation is that the Laplace transform of a convolution is the product of the transforms of the functions that are convolved. The fact that the formula $Y(s) = H(s)U(s)$ is much simpler than a convolution is one reason why Laplace transforms have become popular in engineering.

We can also use the Laplace transform to derive the transfer function for a state space system. Consider, for example, a linear state space system described by

$$\frac{dx}{dt} = Ax + Bu, \qquad y = Cx + Du.$$

Taking Laplace transforms *under the assumption that all initial values are zero*

gives
$$sX(s) = AX(s) + BU(s) \qquad Y(s) = CX(s) + DU(s).$$

Elimination of $X(s)$ gives

$$Y(s) = \left(C(sI - A)^{-1}B + D\right)U(s). \tag{8.26}$$

The transfer function is $G(s) = C(sI - A)^{-1}B + D$ (compare with equation (8.4)).

8.6 Further Reading

The idea of characterizing a linear system by its steady-state response to sinusoids was introduced by Fourier in his investigation of heat conduction in solids [76]. Much later, it was used by the electrical engineer Steinmetz who introduced the $i\omega$ method for analyzing electrical circuits. Transfer functions were introduced via the Laplace transform by Gardner Barnes [81], who also used them to calculate the response of linear systems. The Laplace transform was very important in the early phase of control because it made it possible to find transients via tables (see, e.g., [110]). Combined with block diagrams, transfer functions and Laplace transforms provided powerful techniques for dealing with complex systems. Calculation of responses based on Laplace transforms is less important today, when responses of linear systems can easily be generated using computers. There are many excellent books on the use of Laplace transforms and transfer functions for modeling and analysis of linear input/output systems. Traditional texts on control such as [61], [79] and [162] are representative examples. Pole/zero cancellation was one of the mysteries of early control theory. It is clear that common factors can be canceled in a rational function, but cancellations have system theoretical consequences that were not clearly understood until Kalman's decomposition of a linear system was introduced [118]. In the following chapters, we will use transfer functions extensively to analyze stability and to describe model uncertainty.

Exercises

8.1 Let $G(s)$ be the transfer function for a linear system. Show that if we apply an input $u(t) = A\sin(\omega t)$, then the steady-state output is given by $y(t) = |G(i\omega)|A\sin(\omega t + \arg G(i\omega))$. (Hint: Start by showing that the real part of a complex number is a linear operation and then use this fact.)

8.2 Consider the system
$$\frac{dx}{dt} = ax + u.$$

Compute the exponential response of the system and use this to derive the transfer function from u to y. Show that when $s = a$, a pole of the transfer function, the response to the exponential input $u(t) = e^{st}$ is $x(t) = e^{at}x(0) + te^{at}$.

8.3 (Inverted pendulum) A model for an inverted pendulum was introduced in Example 2.2. Neglecting damping and linearizing the pendulum around the upright position gives a linear system characterized by the matrices

$$A = \begin{bmatrix} 0 & 1 \\ mgl/J_t & 0 \end{bmatrix}, \quad B = \begin{bmatrix} 0 \\ 1/J_t \end{bmatrix}, \quad C = \begin{bmatrix} 1 & 0 \end{bmatrix}, \quad D = 0.$$

Determine the transfer function of the system.

8.4 (Solutions corresponding to poles and zeros) Consider the differential equation

$$\frac{d^n y}{dt^n} + a_1 \frac{d^{n-1} y}{dt^{n-1}} + \cdots + a_n y = b_1 \frac{d^{n-1} u}{dt^{n-1}} + b_2 \frac{d^{n-2} u}{dt^{n-2}} + \cdots + b_n u.$$

(a) Let λ be a root of the characteristic polynomial

$$s^n + a_1 s^{n-1} + \cdots + a_n = 0.$$

Show that if $u(t) = 0$, the differential equation has the solution $y(t) = e^{\lambda t}$.

(b) Let κ be a zero of the polynomial

$$b(s) = b_1 s^{n-1} + b_2 s^{n-2} + \cdots + b_n.$$

Show that if the input is $u(t) = e^{\kappa t}$, then there is a solution to the differential equation that is identically zero.

8.5 (Operational amplifier) Consider the operational amplifier introduced in Section 3.3 and analyzed in Example 8.3. A PI controller can be constructed using an op amp by replacing the resistor R_2 with a resistor and capacitor in series, as shown in Figure 3.10. The resulting transfer function of the circuit is given by

$$G(s) = -\left(R_2 + \frac{1}{Cs}\right) \cdot \left(\frac{kCs}{(kR_1 C + R_2 C)s + 1}\right),$$

where k is the gain of the op amp, R_1 and R_2 are the resistances in the compensation network and C is the capacitance.

(a) Sketch the Bode plot for the system under the assumption that $k \gg R_2 > R_1$. You should label the key features in your plot, including the gain and phase at low frequency, the slopes of the gain curve, the frequencies at which the gain changes slope, etc.

(b) Suppose now that we include some dynamics in the amplifier, as outlined in Example 8.1. This would involve replacing the gain k with the transfer function

$$G(s) = \frac{k}{1 + sT}.$$

Compute the resulting transfer function for the system (i.e., replace k with $G(s)$) and find the poles and zeros assuming the following parameter values

$$\frac{R_2}{R_1} = 100, \quad k = 10^6, \quad R_2 C = 1, \quad T = 0.01.$$

(c) Sketch the Bode plot for the transfer function in part (b) using straight line approximations and compare this to the exact plot of the transfer function (using MATLAB). Make sure to label the important features in your plot.

8.6 (Transfer function for state space system) Consider the linear state space system

$$\frac{dx}{dt} = Ax + Bu, \qquad y = Cx.$$

Show that the transfer function is

$$G(s) = \frac{b_1 s^{n-1} + b_2 s^{n-2} + \cdots + b_n}{s^n + a_1 s^{n-1} + \cdots + a_n},$$

where

$$b_1 = CB, \quad b_2 = CAB + a_1 CB, \quad \ldots, \quad b_n = CA^{n-1}B + a_1 CA^{n-1}B + \cdots + a_{n-1}CB$$

and $\lambda(s) = s^n + a_1 s^{n-1} + \cdots + a_n$ is the characteristic polynomial for A.

8.7 (Kalman decomposition) Show that the transfer function of a system depends only on the dynamics in the reachable and observable subspace of the Kalman decomposition. (Hint: Consider the representation given by equation (7.27).)

8.8 Using block diagram algebra, show that the transfer functions from d to y and n to y in Figure 8.7 are given by

$$G_{yd} = \frac{P}{1 + PC} \qquad G_{yn} = \frac{1}{1 + PC}.$$

8.9 (Bode plot for a simple zero) Show that the Bode plot for transfer function $G(s) = (s + a)/a$ can be approximated by

$$\log |G(i\omega)| \approx \begin{cases} 0 & \text{if } \omega < a \\ \log \omega - \log a & \text{if } \omega > a, \end{cases}$$

$$\angle G(i\omega) \approx \begin{cases} 0 & \text{if } \omega < a/10 \\ 45 + 45(\log \omega - \log a) & a/10 < \omega < 10a \\ 90 & \text{if } \omega > 10a. \end{cases}$$

8.10 (Vectored thrust aircraft) Consider the lateral dynamics of a vectored thrust aircraft as described in Example 2.9. Show that the dynamics can be described using the following block diagram:

Use this block diagram to compute the transfer functions from u_1 to θ and x and show that they satisfy

$$H_{\theta u_1} = \frac{r}{Js^2}, \qquad H_{xu_1} = \frac{Js^2 - mgr}{Js^2(ms^2 + cs)}.$$

8.11 (Common poles) Consider a closed loop system of the form of Figure 8.7, with $F = 1$ and P and C having a common pole. Show that if each system is written in state space form, the resulting closed loop system is not reachable and not observable.

8.12 (Congestion control) Consider the congestion control model described in Section 3.4. Let w represent the individual window size for a set of N identical sources, q represent the end-to-end probability of a dropped packet, b represent the number of packets in the router's buffer and p represent the probability that that a packet is dropped by the router. We write $\bar{w} = Nw$ to represent the total number of packets being received from all N sources. Show that the linearized model can be described by the transfer functions

$$G_{b\bar{w}}(s) = \frac{e^{-\tau_f s}}{\tau_e s + e^{-\tau_f s}}, \qquad G_{\bar{w}q}(s) = -\frac{N}{q_e(\tau_e s + q_e w_e)}, \qquad G_{pb}(s) = \rho,$$

where (w_e, b_e) is the equilibrium point for the system, τ_e is the steady-state round-trip time and τ_f is the forward propagation time.

8.13 (Inverted pendulum with PD control) Consider the normalized inverted pendulum system, whose transfer function is given by $P(s) = 1/(s^2 - 1)$ (Exercise 8.3). A proportional-derivative control law for this system has transfer function $C(s) = k_p + k_d s$ (see Table 8.1). Suppose that we choose $C(s) = \alpha(s - 1)$. Compute the closed loop dynamics and show that the system has good tracking of reference signals but does not have good disturbance rejection properties.

8.14 (Vehicle suspension [96]) Active and passive damping are used in cars to give a smooth ride on a bumpy road. A schematic diagram of a car with a damping system in shown in the figure below.

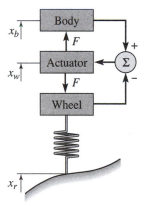

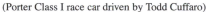

(Porter Class I race car driven by Todd Cuffaro)

This model is called a *quarter car model*, and the car is approximated with two masses, one representing one fourth of the car body and the other a wheel. The actuator exerts a force F between the wheel and the body based on feedback from the distance between the body and the center of the wheel (the *rattle space*).

Let x_b, x_w and x_r represent the heights of body, wheel and road measured from their equilibria. A simple model of the system is given by Newton's equations for the body and the wheel,

$$m_b \ddot{x}_b = F, \qquad m_w \ddot{x}_w = -F + k_t(x_r - x_w),$$

where m_b is a quarter of the body mass, m_w is the effective mass of the wheel including brakes and part of the suspension system (the *unsprung mass*) and k_t is the tire stiffness. For a conventional damper consisting of a spring and a damper, we have $F = k(x_w - x_b) + c(\dot{x}_w - \dot{x}_b)$. For an active damper the force F can be more general and can also depend on riding conditions. Rider comfort can be characterized by the transfer function G_{ax_r} from road height x_r to body acceleration $a = \ddot{x}_b$. Show that this transfer function has the property $G_{ax_r}(i\omega_t) = k_t/m_b$, where $\omega_t = \sqrt{k_t/m_w}$ (the *tire hop frequency*). The equation implies that there are fundamental limitations to the comfort that can be achieved with any damper.

8.15 (Vibration absorber) Damping vibrations is a common engineering problem. A schematic diagram of a damper is shown below:

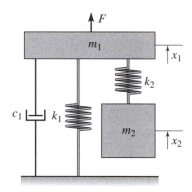

The disturbing vibration is a sinusoidal force acting on mass m_1, and the damper consists of the mass m_2 and the spring k_2. Show that the transfer function from disturbance force to height x_1 of the mass m_1 is

$$G_{x_1 F} = \frac{m_2 s^2 + k_2}{m_1 m_2 s^4 + m_2 c_1 s^3 + (m_1 k_2 + m_2(k_1 + k_2))s^2 + k_2 c_1 s + k_1 k_2}.$$

How should the mass m_2 and the stiffness k_2 be chosen to eliminate a sinusoidal oscillation with frequency ω_0. (More details are vibration absorbers is given in the classic text by Den Hartog [57, pp. 87–93].)

Chapter Nine
Frequency Domain Analysis

> *Mr. Black proposed a negative feedback repeater and proved by tests that it possessed the advantages which he had predicted for it. In particular, its gain was constant to a high degree, and it was linear enough so that spurious signals caused by the interaction of the various channels could be kept within permissible limits. For best results the feedback factor $\mu\beta$ had to be numerically much larger than unity. The possibility of stability with a feedback factor larger than unity was puzzling.*

Harry Nyquist, "The Regeneration Theory," 1956 [161].

In this chapter we study how the stability and robustness of closed loop systems can be determined by investigating how sinusoidal signals of different frequencies propagate around the feedback loop. This technique allows us to reason about the closed loop behavior of a system through the frequency domain properties of the open loop transfer function. The Nyquist stability theorem is a key result that provides a way to analyze stability and introduce measures of degrees of stability.

9.1 The Loop Transfer Function

Determining the stability of systems interconnected by feedback can be tricky because each system influences the other, leading to potentially circular reasoning. Indeed, as the quote from Nyquist above illustrates, the behavior of feedback systems can often be puzzling. However, using the mathematical framework of transfer functions provides an elegant way to reason about such systems, which we call *loop analysis*.

The basic idea of loop analysis is to trace how a sinusoidal signal propagates in the feedback loop and explore the resulting stability by investigating if the propagated signal grows or decays. This is easy to do because the transmission of sinusoidal signals through a linear dynamical system is characterized by the frequency response of the system. The key result is the Nyquist stability theorem, which provides a great deal of insight regarding the stability of a system. Unlike proving stability with Lyapunov functions, studied in Chapter 4, the Nyquist criterion allows us to determine more than just whether a system is stable or unstable. It provides a measure of the degree of stability through the definition of stability margins. The Nyquist theorem also indicates how an unstable system should be changed to make it stable, which we shall study in detail in Chapters 10–12.

Consider the system in Figure 9.1a. The traditional way to determine if the closed loop system is stable is to investigate if the closed loop characteristic polynomial has all its roots in the left half-plane. If the process and the controller have rational

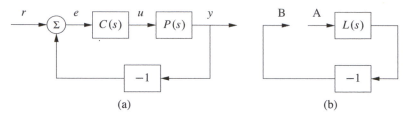

Figure 9.1: The loop transfer function. The stability of the feedback system (a) can be determined by tracing signals around the loop. Letting $L = PC$ represent the loop transfer function, we break the loop in (b) and ask whether a signal injected at the point A has the same magnitude and phase when it reaches point B.

transfer functions $P(s) = n_p(s)/d_p(s)$ and $C(s) = n_c(s)/d_c(s)$, then the closed loop system has the transfer function

$$G_{yr}(s) = \frac{PC}{1 + PC} = \frac{n_p(s)n_c(s)}{d_p(s)d_c(s) + n_p(s)n_c(s)},$$

and the characteristic polynomial is

$$\lambda(s) = d_p(s)d_c(s) + n_p(s)n_c(s).$$

To check stability, we simply compute the roots of the characteristic polynomial and verify that they each have negative real part. This approach is straightforward but it gives little guidance for design: it is not easy to tell how the controller should be modified to make an unstable system stable.

Nyquist's idea was to investigate conditions under which oscillations can occur in a feedback loop. To study this, we introduce the *loop transfer function* $L(s) = P(s)C(s)$, which is the transfer function obtained by breaking the feedback loop, as shown in Figure 9.1b. The loop transfer function is simply the transfer function from the input at position A to the output at position B multiplied by -1 (to account for the usual convention of negative feedback).

We will first determine conditions for having a periodic oscillation in the loop. Assume that a sinusoid of frequency ω_0 is injected at point A. In steady state the signal at point B will also be a sinusoid with the frequency ω_0. It seems reasonable that an oscillation can be maintained if the signal at B has the same amplitude and phase as the injected signal because we can then disconnect the injected signal and connect A to B. Tracing signals around the loop, we find that the signals at A and B are identical if

$$L(i\omega_0) = -1, \tag{9.1}$$

which then provides a condition for maintaining an oscillation. The key idea of the Nyquist stability criterion is to understand when this can happen in a general setting. As we shall see, this basic argument becomes more subtle when the loop transfer function has poles in the right half-plane.

Example 9.1 Operational amplifier circuit
Consider the op amp circuit in Figure 9.2a, where Z_1 and Z_2 are the transfer func-

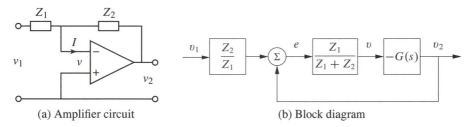

(a) Amplifier circuit   (b) Block diagram

Figure 9.2: Loop transfer function for an op amp. The op amp circuit (a) has a nominal transfer function $v_2/v_1 = Z_2(s)/Z_1(s)$, where Z_1 and Z_2 are the impedances of the circuit elements. The system can be represented by its block diagram (b), where we now include the op amp dynamics $G(s)$. The loop transfer function is $L = Z_1 G/(Z_1 + Z_2)$.

tions of the feedback elements from voltage to current. There is feedback because voltage v_2 is related to voltage v through the transfer function $-G$ describing the op amp dynamics and voltage v is related to voltage v_2 through the transfer function $Z_1/(Z_1 + Z_2)$. The loop transfer function is thus

$$L = \frac{GZ_1}{Z_1 + Z_2}. \tag{9.2}$$

Assuming that the current I is zero, the current through the elements Z_1 and Z_2 is the same, which implies

$$\frac{v_1 - v}{Z_1} = \frac{v - v_2}{Z_2}.$$

Solving for v gives

$$v = \frac{Z_2 v_1 + Z_1 v_2}{Z_1 + Z_2} = \frac{Z_2 v_1 - Z_1 G v}{Z_1 + Z_2} = \frac{Z_2}{Z_1} \frac{L}{G} v_1 - L v.$$

Since $v_2 = -Gv$ the input/output relation for the circuit becomes

$$G_{v_2 v_1} = -\frac{Z_2}{Z_1} \frac{L}{1 + L}.$$

A block diagram is shown in Figure 9.2b. It follows from (9.1) that the condition for oscillation of the op amp circuit is

$$L(i\omega) = \frac{Z_1(i\omega)G(i\omega)}{Z_1(i\omega) + Z_2(i\omega)} = -1 \tag{9.3}$$

$$\triangledown$$

One of the powerful concepts embedded in Nyquist's approach to stability analysis is that it allows us to study the stability of the feedback system by looking at properties of the loop transfer function. The advantage of doing this is that it is easy to see how the controller should be chosen to obtain a desired loop transfer function. For example, if we change the gain of the controller, the loop transfer function will be scaled accordingly. A simple way to stabilize an unstable system is then to reduce the gain so that the -1 point is avoided. Another way is to introduce

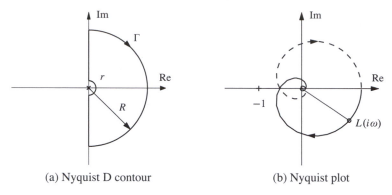

(a) Nyquist D contour   (b) Nyquist plot

Figure 9.3: The Nyquist contour Γ and the Nyquist plot. The Nyquist contour (a) encloses the right half-plane, with a small semicircle around any poles of $L(s)$ on the imaginary axis (illustrated here at the origin) and an arc at infinity, represented by $R \to \infty$. The Nyquist plot (b) is the image of the loop transfer function $L(s)$ when s traverses Γ in the clockwise direction. The solid line corresponds to $\omega > 0$, and the dashed line to $\omega < 0$. The gain and phase at the frequency ω are $g = |L(i\omega)|$ and $\varphi = \angle L(i\omega)$. The curve is generated for $L(s) = 1.4e^{-s}/(s+1)^2$.

a controller with the property that it bends the loop transfer function away from the critical point, as we shall see in the next section. Different ways to do this, called loop shaping, will be developed and will be discussed in Chapter 11.

9.2 The Nyquist Criterion

In this section we present Nyquist's criterion for determining the stability of a feedback system through analysis of the loop transfer function. We begin by introducing a convenient graphical tool, the Nyquist plot, and show how it can be used to ascertain stability.

The Nyquist Plot

We saw in the last chapter that the dynamics of a linear system can be represented by its frequency response and graphically illustrated by a Bode plot. To study the stability of a system, we will make use of a different representation of the frequency response called a *Nyquist plot*. The Nyquist plot of the loop transfer function $L(s)$ is formed by tracing $s \in \mathbb{C}$ around the Nyquist "D contour," consisting of the imaginary axis combined with an arc at infinity connecting the endpoints of the imaginary axis. The contour, denoted as $\Gamma \in \mathbb{C}$, is illustrated in Figure 9.3a. The image of $L(s)$ when s traverses Γ gives a closed curve in the complex plane and is referred to as the Nyquist plot for $L(s)$, as shown in Figure 9.3b. Note that if the transfer function $L(s)$ goes to zero as s gets large (the usual case), then the portion of the contour "at infinity" maps to the origin. Furthermore, the portion of the plot corresponding to $\omega < 0$ is the mirror image of the portion with $\omega > 0$.

There is a subtlety in the Nyquist plot when the loop transfer function has poles on the imaginary axis because the gain is infinite at the poles. To solve this problem, we modify the contour Γ to include small deviations that avoid any poles on the imaginary axis, as illustrated in Figure 9.3a (assuming a pole of $L(s)$ at the origin). The deviation consists of a small semicircle to the right of the imaginary axis pole location.

The condition for oscillation given in equation (9.1) implies that the Nyquist plot of the loop transfer function go through the point $L = -1$, which is called the *critical point*. Let ω_c represent a frequency at which $\angle L(i\omega_c) = 180°$, corresponding to the Nyquist curve crossing the negative real axis. Intuitively it seems reasonable that the system is stable if $|L(i\omega_c)| < 1$, which means that the critical point -1 is on the left-hand side of the Nyquist curve, as indicated in Figure 9.3b. This means that the signal at point B will have smaller amplitude than the injected signal. This is essentially true, but there are several subtleties that require a proper mathematical analysis to clear up. We defer the details for now and state the Nyquist condition for the special case where $L(s)$ is a stable transfer function.

Theorem 9.1 (Simplified Nyquist criterion). *Let $L(s)$ be the loop transfer function for a negative feedback system (as shown in Figure 9.1a) and assume that L has no poles in the closed right half-plane (Re $s \geq 0$) except for single poles on the imaginary axis. Then the closed loop system is stable if and only if the closed contour given by $\Omega = \{L(i\omega) : -\infty < \omega < \infty\} \subset \mathbb{C}$ has no net encirclements of the critical point $s = -1$.*

The following conceptual procedure can be used to determine that there are no encirclements. Fix a pin at the critical point $s = -1$, orthogonal to the plane. Attach a string with one end at the critical point and the other on the Nyquist plot. Let the end of the string attached to the Nyquist curve traverse the whole curve. There are no encirclements if the string does not wind up on the pin when the curve is encircled.

Example 9.2 Third-order system
Consider a third-order transfer function

$$L(s) = \frac{1}{(s + a)^3}.$$

To compute the Nyquist plot we start by evaluating points on the imaginary axis $s = i\omega$, which yields

$$L(i\omega) = \frac{1}{(i\omega + a)^3} = \frac{(a - i\omega)^3}{(a^2 + \omega^2)^3} = \frac{a^3 - 3a\omega^2}{(a^2 + \omega^2)^3} + i\frac{\omega^3 - 3a^2\omega}{(a^2 + \omega^2)^3}.$$

This is plotted in the complex plane in Figure 9.4, with the points corresponding to $\omega > 0$ drawn as a solid line and $\omega < 0$ as a dashed line. Notice that these curves are mirror images of each other.

To complete the Nyquist plot, we compute $L(s)$ for s on the outer arc of the

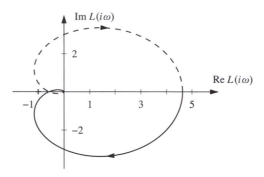

Figure 9.4: Nyquist plot for a third-order transfer function. The Nyquist plot consists of a trace of the loop transfer function $L(s) = 1/(s + a)^3$. The solid line represents the portion of the transfer function along the positive imaginary axis, and the dashed line the negative imaginary axis. The outer arc of the D contour maps to the origin.

Nyquist D contour. This arc has the form $s = Re^{i\theta}$ for $R \to \infty$. This gives

$$L(Re^{i\theta}) = \frac{1}{(Re^{i\theta} + a)^3} \to 0 \quad \text{as} \quad R \to \infty.$$

Thus the outer arc of the D contour maps to the origin on the Nyquist plot.   ∇

An alternative to computing the Nyquist plot explicitly is to determine the plot from the frequency response (Bode plot), which gives the Nyquist curve for $s = i\omega$, $\omega > 0$. We start by plotting $G(i\omega)$ from $\omega = 0$ to $\omega = \infty$, which can be read off from the magnitude and phase of the transfer function. We then plot $G(Re^{i\theta})$ with $\theta \in [-\pi/2, \pi/2]$ and $R \to \infty$, which almost always maps to zero. The remaining parts of the plot can be determined by taking the mirror image of the curve thus far (normally plotted using a dashed line). The plot can then be labeled with arrows corresponding to a clockwise traversal around the D contour (the same direction in which the first portion of the curve was plotted).

Example 9.3 Third-order system with a pole at the origin
Consider the transfer function

$$L(s) = \frac{k}{s(s + 1)^2},$$

where the gain has the nominal value $k = 1$. The Bode plot is shown in Figure 9.5a. The system has a single pole at $s = 0$ and a double pole at $s = -1$. The gain curve of the Bode plot thus has the slope -1 for low frequencies, and at the double pole $s = 1$ the slope changes to -3. For small s we have $L \approx k/s$, which means that the low-frequency asymptote intersects the unit gain line at $\omega = k$. The phase curve starts at $-90°$ for low frequencies, it is $-180°$ at the breakpoint $\omega = 1$ and it is $-270°$ at high frequencies.

Having obtained the Bode plot, we can now sketch the Nyquist plot, shown in Figure 9.5b. It starts with a phase of $-90°$ for low frequencies, intersects the negative real axis at the breakpoint $\omega = 1$ where $L(i) = 0.5$ and goes to zero along

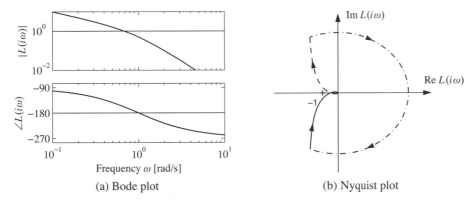

(a) Bode plot   (b) Nyquist plot

Figure 9.5: Sketching Nyquist and Bode plots. The loop transfer function is $L(s) = 1/(s(s + 1)^2)$. The large semicircle is the map of the small semicircle of the Γ contour around the pole at the origin. The closed loop is stable because the Nyquist curve does not encircle the critical point. The point where the phase is $-180°$ is marked with a circle in the Bode plot.

the imaginary axis for high frequencies. The small half-circle of the Γ contour at the origin is mapped on a large circle enclosing the right half-plane. The Nyquist curve does not encircle the critical point, and it follows from the simplified Nyquist theorem that the closed loop is stable. Since $L(i) = -k/2$, we find the system becomes unstable if the gain is increased to $k = 2$ or beyond. ∇

The Nyquist criterion does not require that $|L(i\omega_c)| < 1$ for all ω_c corresponding to a crossing of the negative real axis. Rather, it says that the number of encirclements must be zero, allowing for the possibility that the Nyquist curve could cross the negative real axis and cross back at magnitudes greater than 1. The fact that it was possible to have high feedback gains surprised the early designers of feedback amplifiers, as mentioned in the quote in the beginning of this chapter.

One advantage of the Nyquist criterion is that it tells us how a system is influenced by changes of the controller parameters. For example, it is very easy to visualize what happens when the gain is changed since this just scales the Nyquist curve.

Example 9.4 Congestion control

Consider the Internet congestion control system described in Section 3.4. Suppose we have N identical sources and a disturbance d representing an external data source, as shown in Figure 9.6a. We let w represent the individual window size for a source, q represent the end-to-end probability of a dropped packet, b represent the number of packets in the router's buffer and p represent the probability that that a packet is dropped by the router. We write $\bar{w}$ for the total number of packets being received from all N sources. We also include a time delay between the router and the senders, representing the time delays between the sender and receiver.

To analyze the stability of the system, we use the transfer functions computed

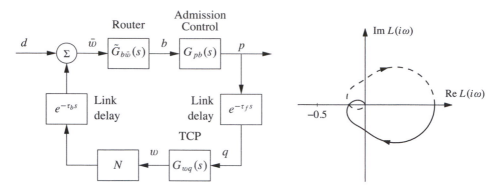

Figure 9.6: Internet congestion control. A set of N sources using TCP/Reno send messages through a single router with admission control (left). Link delays are included for the forward and backward directions. The Nyquist plot for the loop transfer function is shown on the right.

in Exercise 8.12:

$$\tilde{G}_{b\bar{w}}(s) = \frac{1}{\tau_e s + e^{-\tau_f s}}, \qquad G_{wq}(s) = -\frac{1}{q_e(\tau_e s + q_e w_e)}, \qquad G_{pb}(s) = \rho,$$

where (w_e, b_e) is the equilibrium point for the system, N is the number of sources, τ_e is the steady-state round-trip time and τ_f is the forward propagation time. We use $\tilde{G}_{b\bar{w}}$ to represent the transfer function with the forward time delay removed since this is accounted for as a separate block in Figure 9.6a. Similarly, $G_{wq} = G_{\bar{w}q}/N$ since we have pulled out the multiplier N as a separate block as well.

The loop transfer function is given by

$$L(s) = \rho \cdot \frac{N}{\tau_e s + e^{-\tau_f s}} \cdot \frac{1}{q_e(\tau_e s + q_e w_e)} e^{-\tau_e s}.$$

Using the fact that $q_e \approx 2N/w_e^2 = 2N^3/(\tau_e c)^2$ and $w_e = b_e/N = \tau_e c/N$ from equation (3.22), we can show that

$$L(s) = \rho \cdot \frac{N}{\tau_e s + e^{-\tau_f s}} \cdot \frac{c^3 \tau_e^3}{2N^3(c\tau_e^2 s + 2N^2)} e^{-\tau_e s}.$$

Note that we have chosen the sign of $L(s)$ to use the same sign convention as in Figure 9.1b. The exponential term representing the time delay gives significant phase above $\omega = 1/\tau_e$, and the gain at the crossover frequency will determine stability.

To check stability, we require that the gain be sufficiently small at crossover. If we assume that the pole due to the queue dynamics is sufficiently fast that the TCP dynamics are dominant, the gain at the crossover frequency ω_c is given by

$$|L(i\omega_c)| = \rho \cdot N \cdot \frac{c^3 \tau_e^3}{2N^3 c \tau_e^2 \omega_c} = \frac{\rho c^2 \tau_e}{2N\omega_c}.$$

Using the Nyquist criterion, the closed loop system will be unstable if this quantity is

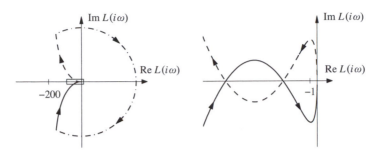

Figure 9.7: Nyquist curve for the loop transfer function $L(s) = \frac{3(s+1)^2}{s(s+6)^2}$. The plot on the right is an enlargement of the box around the origin of the plot on the left. The Nyquist curve intersections the negative real axis twice but has no net encirclements of -1.

greater than 1. In particular, for a fixed time delay, the system will become unstable as the link capacity c is increased. This indicates that the TCP protocol may not be scalable to high-capacity networks, as pointed out by Low et al. [137]. Exercise 9.7 provides some ideas of how this might be overcome. ∇

Conditional Stability

Normally, we find that unstable systems can be stabilized simply by reducing the loop gain. There are, however, situations where a system can be stabilized by increasing the gain. This was first encountered by electrical engineers in the design of feedback amplifiers, who coined the term *conditional stability*. The problem was actually a strong motivation for Nyquist to develop his theory. We will illustrate by an example.

Example 9.5 Third-order system
Consider a feedback system with the loop transfer function

$$L(s) = \frac{3(s+6)^2}{s(s+1)^2}. \tag{9.4}$$

The Nyquist plot of the loop transfer function is shown in Figure 9.7. Notice that the Nyquist curve intersects the negative real axis twice. The first intersection occurs at $L = -12$ for $\omega = 2$, and the second at $L = -4.5$ for $\omega = 3$. The intuitive argument based on signal tracing around the loop in Figure 9.1b is strongly misleading in this case. Injection of a sinusoid with frequency 2 rad/s and amplitude 1 at A gives, in steady state, an oscillation at B that is in phase with the input and has amplitude 12. Intuitively it is seems unlikely that closing of the loop will result in a stable system. However, it follows from Nyquist's stability criterion that the system is stable because there are no net encirclements of the critical point. Note, however, that if we *decrease* the gain, then we can get an encirclement, implying that the gain must be sufficiently large for stability. ∇

General Nyquist Criterion

Theorem 9.1 requires that $L(s)$ have no poles in the closed right half-plane. In some situations this is not the case and a more general result is required. Nyquist originally considered this general case, which we summarize as a theorem.

Theorem 9.2 (Nyquist's stability theorem). *Consider a closed loop system with the loop transfer function $L(s)$ that has P poles in the region enclosed by the Nyquist contour. Let N be the net number of clockwise encirclements of -1 by $L(s)$ when s encircles the Nyquist contour Γ in the clockwise direction. The closed loop system then has $Z = N + P$ poles in the right half-plane.*

The full Nyquist criterion states that if $L(s)$ has P poles in the right half-plane, then the Nyquist curve for $L(s)$ should have P counterclockwise encirclements of -1 (so that $N = -P$). In particular, this *requires* that $|L(i\omega_c)| > 1$ for some ω_c corresponding to a crossing of the negative real axis. Care has to be taken to get the right sign of the encirclements. The Nyquist contour has to be traversed clockwise, which means that ω moves from $-\infty$ to ∞ and N is positive if the Nyquist curve winds clockwise. If the Nyquist curve winds counterclockwise, then N will be negative (the desired case if $P \neq 0$).

As in the case of the simplified Nyquist criterion, we use small semicircles of radius r to avoid any poles on the imaginary axis. By letting $r \to 0$, we can use Theorem 9.2 to reason about stability. Note that the image of the small semicircles generates a section of the Nyquist curve whose magnitude approaches infinity, requiring care in computing the winding number. When plotting Nyquist curves on the computer, one must be careful to see that such poles are properly handled, and often one must sketch those portions of the Nyquist plot by hand, being careful to loop the right way around the poles.

Example 9.6 Stabilized inverted pendulum

The linearized dynamics of a normalized inverted pendulum can be represented by the transfer function $P(s) = 1/(s^2 - 1)$, where the input is acceleration of the pivot and the output is the pendulum angle θ, as shown in Figure 9.8 (Exercise 8.3). We attempt to stabilize the pendulum with a proportional-derivative (PD) controller having the transfer function $C(s) = k(s + 2)$. The loop transfer function is

$$L(s) = \frac{k(s+2)}{s^2 - 1}.$$

The Nyquist plot of the loop transfer function is shown in Figure 9.8b. We have $L(0) = -2k$ and $L(\infty) = 0$. If $k > 0.5$, the Nyquist curve encircles the critical point $s = -1$ in the counterclockwise direction when the Nyquist contour γ is encircled in the clockwise direction. The number of encirclements is thus $N = -1$. Since the loop transfer function has one pole in the right half-plane ($P = 1$), we find that $Z = N + P = 0$ and the system is thus stable for $k > 0.5$. If $k < 0.5$, there is no encirclement and the closed loop will have one pole in the right half-plane.

$$\nabla$$

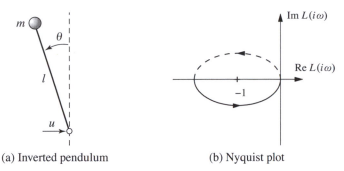

(a) Inverted pendulum   (b) Nyquist plot

Figure 9.8: PD control of an inverted pendulum. (a) The system consists of a mass that is balanced by applying a force at the pivot point. A proportional-derivative controller with transfer function $C(s) = k(s + 2)$ is used to command u based on θ. (b) A Nyquist plot of the loop transfer function for gain $k = 1$. There is one counterclockwise encirclement of the critical point, giving $N = -1$ clockwise encirclements.

Derivation of Nyquist's Stability Theorem

We will now prove the Nyquist stability theorem for a general loop transfer function $L(s)$. This requires some results from the theory of complex variables, for which the reader can consult Ahlfors [6]. Since some precision is needed in stating Nyquist's criterion properly, we will use a more mathematical style of presentation. We also follow the mathematical convention of counting encirclements in the counterclockwise direction for the remainder of this section. The key result is the following theorem about functions of complex variables.

Theorem 9.3 (Principle of variation of the argument). *Let D be a closed region in the complex plane and let Γ be the boundary of the region. Assume the function $f : \mathbb{C} \to \mathbb{C}$ is analytic in D and on Γ, except at a finite number of poles and zeros. Then the winding number w_n is given by*

$$w_n = \frac{1}{2\pi}\Delta_\Gamma \arg f(z) = \frac{1}{2\pi i}\int_\Gamma \frac{f'(z)}{f(z)}dz = Z - P,$$

where Δ_Γ is the net variation in the angle when z traverses the contour Γ in the counterclockwise direction, Z is the number of zeros in D and P is the number of poles in D. Poles and zeros of multiplicity m are counted m times.

Proof. Assume that $z = a$ is a zero of multiplicity m. In the neighborhood of $z = a$ we have

$$f(z) = (z - a)^m g(z),$$

where the function g is analytic and different from zero. The ratio of the derivative of f to itself is then given by

$$\frac{f'(z)}{f(z)} = \frac{m}{z - a} + \frac{g'(z)}{g(z)},$$

and the second term is analytic at $z = a$. The function f'/f thus has a single pole at $z = a$ with the residue m. The sum of the residues at the zeros of the function is Z. Similarly, we find that the sum of the residues for the poles is $-P$, and hence

$$Z - P = \frac{1}{2\pi i} \int_\Gamma \frac{f'(z)}{f(z)} \, dz = \frac{1}{2\pi i} \int_\Gamma \frac{d}{dz} \log f(z) \, dz = \frac{1}{2\pi i} \Delta_\Gamma \log f(z),$$

where Δ_Γ again denotes the variation along the contour Γ. We have

$$\log f(z) = \log |f(z)| + i \arg f(z),$$

and since the variation of $|f(z)|$ around a closed contour is zero it follows that

$$\Delta_\Gamma \log f(z) = i \Delta_\Gamma \arg f(z),$$

and the theorem is proved.   □

This theorem is useful in determining the number of poles and zeros of a function of complex variables in a given region. By choosing an appropriate closed region D with boundary Γ, we can determine the difference between the number of poles and zeros through computation of the winding number.

Theorem 9.3 can be used to prove Nyquist's stability theorem by choosing Γ as the Nyquist contour shown in Figure 9.3a, which encloses the right half-plane. To construct the contour, we start with part of the imaginary axis $-jR \leq s \leq jR$ and a semicircle to the right with radius R. If the function f has poles on the imaginary axis, we introduce small semicircles with radii r to the right of the poles as shown in the figure. The Nyquist contour is obtained by letting $R \to \infty$ and $r \to 0$. Note that Γ has orientation *opposite* that shown in Figure 9.3a. (The convention in engineering is to traverse the Nyquist contour in the clockwise direction since this corresponds to moving upwards along the imaginary axis, which makes it easy to sketch the Nyquist contour from a Bode plot.)

To see how we use the principle of variation of the argument to compute stability, consider a closed loop system with the loop transfer function $L(s)$. The closed loop poles of the system are the zeros of the function $f(s) = 1 + L(s)$. To find the number of zeros in the right half-plane, we investigate the winding number of the function $f(s) = 1 + L(s)$ as s moves along the Nyquist contour Γ in the *counterclockwise* direction. The winding number can conveniently be determined from the Nyquist plot. A direct application of Theorem 9.3 gives the Nyquist criterion, taking care to flip the orientation. Since the image of $1 + L(s)$ is a shifted version of $L(s)$, we usually state the Nyquist criterion as net encirclements of the -1 point by the image of $L(s)$.

9.3 Stability Margins

In practice it is not enough that a system is stable. There must also be some margins of stability that describe how stable the system is and its robustness to perturbations. There are many ways to express this, but one of the most common is the use of gain and phase margins, inspired by Nyquist's stability criterion. The key idea is that it

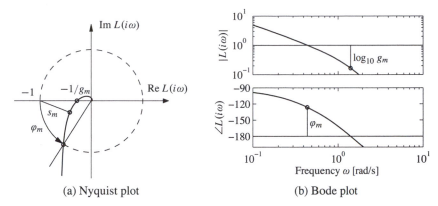

(a) Nyquist plot   (b) Bode plot

Figure 9.9: Stability margins. The gain margin g_m and phase margin φ_m are shown on the the Nyquist plot (a) and the Bode plot (b). The gain margin corresponds to the smallest increase in gain that creates an encirclement, and the phase margin is the smallest change in phase that creates an encirclement. The Nyquist plot also shows the stability margin s_m, which is the shortest distance to the critical point -1.

is easy to plot the loop transfer function $L(s)$. An increase in controller gain simply expands the Nyquist plot radially. An increase in the phase of the controller twists the Nyquist plot. Hence from the Nyquist plot we can easily pick off the amount of gain or phase that can be added without causing the system to become unstable.

Formally, the *gain margin g_m* of a system is defined as the smallest amount that the open loop gain can be increased before the closed loop system goes unstable. For a system whose phase decreases monotonically as a function of frequency starting at $0°$, the gain margin can be computed based on the smallest frequency where the phase of the loop transfer function $L(s)$ is $-180°$. Let ω_{pc} represent this frequency, called the *phase crossover frequency*. Then the gain margin for the system is given by

$$g_m = \frac{1}{|L(i\omega_{pc})|}. \tag{9.5}$$

Similarly, the *phase margin* is the amount of phase lag required to reach the stability limit. Let ω_{gc} be the *gain crossover frequency*, the smallest frequency where the loop transfer function $L(s)$ has unit magnitude. Then for a system with monotonically decreasing gain, the phase margin is given by

$$\varphi_m = \pi + \arg L(i\omega_{gc}). \tag{9.6}$$

These margins have simple geometric interpretations on the Nyquist diagram of the loop transfer function, as shown in Figure 9.9a, where we have plotted the portion of the curve corresponding to $\omega > 0$. The gain margin is given by the inverse of the distance to the nearest point between -1 and 0 where the loop transfer function crosses the negative real axis. The phase margin is given by the smallest angle on the unit circle between -1 and the loop transfer function. When the gain or phase is monotonic, this geometric interpretation agrees with the formulas above.

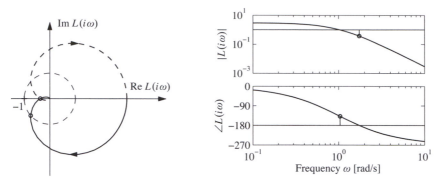

Figure 9.10: Stability margins for a third-order transfer function. The Nyquist plot on the left allows the gain, phase and stability margins to be determined by measuring the distances of relevant features. The gain and phase margins can also be read off of the Bode plot on the right.

A drawback with gain and phase margins is that it is necessary to give both of them in order to guarantee that the Nyquist curve is not close to the critical point. An alternative way to express margins is by a single number, the *stability margin* s_m, which is the shortest distance from the Nyquist curve to the critical point. This number is related to disturbance attenuation, as will be discussed in Section 11.3.

For many systems, the gain and phase margins can be determined from the Bode plot of the loop transfer function. To find the gain margin we first find the phase crossover frequency ω_{pc} where the phase is $-180°$. The gain margin is the inverse of the gain at that frequency. To determine the phase margin we first determine the gain crossover frequency ω_{gc}, i.e., the frequency where the gain of the loop transfer function is 1. The phase margin is the phase of the loop transfer function at that frequency plus $180°$. Figure 9.9b illustrates how the margins are found in the Bode plot of the loop transfer function. Note that the Bode plot interpretation of the gain and phase margins can be incorrect if there are multiple frequencies at which the gain is equal to 1 or the phase is equal to $-180°$.

Example 9.7 Third-order system
Consider a loop transfer function $L(s) = 3/(s+1)^3$. The Nyquist and Bode plots are shown in Figure 9.10. To compute the gain, phase and stability margins, we can use the Nyquist plot shown in Figure 9.10. This yields the following values:

$$g_m = 2.67, \qquad \varphi_m = 41.7°, \qquad s_m = 0.464.$$

The gain and phase margins can also be determined from the Bode plot. ∇

The gain and phase margins are classical robustness measures that have been used for a long time in control system design. The gain margin is well defined if the Nyquist curve intersects the negative real axis once. Analogously, the phase margin is well defined if the Nyquist curve intersects the unit circle at only one point. Other more general robustness measures will be introduced in Chapter 12.

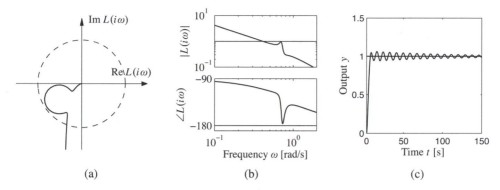

Figure 9.11: System with good gain and phase margins but a poor stability margin. Nyquist (a) and Bode (b) plots of the loop transfer function and step response (c) for a system with good gain and phase margins but with a poor stability margin. The Nyquist plot shows on the portion of the curve corresponding to $\omega > 0$.

Even if both the gain and phase margins are reasonable, the system may still not be robust, as is illustrated by the following example.

Example 9.8 Good gain and phase margins but poor stability margins
Consider a system with the loop transfer function

$$L(s) = \frac{0.38(s^2 + 0.1s + 0.55)}{s(s+1)(s^2 + 0.06s + 0.5)}.$$

A numerical calculation gives the gain margin as $g_m = 266$, and the phase margin is $70°$. These values indicate that the system is robust, but the Nyquist curve is still close to the critical point, as shown in Figure 9.11. The stability margin is $s_m = 0.27$, which is very low. The closed loop system has two resonant modes, one with damping ratio $\zeta = 0.81$ and the other with $\zeta = 0.014$. The step response of the system is highly oscillatory, as shown in Figure 9.11c.   ▽

The stability margin cannot easily be found from the Bode plot of the loop transfer function. There are, however, other Bode plots that will give s_m; these will be discussed in Chapter 12. In general, it is best to use the Nyquist plot to check stability since this provides more complete information than the Bode plot.

When designing feedback systems, it will often be useful to define the robustness of the system using gain, phase and stability margins. These numbers tell us how much the system can vary from our nominal model and still be stable. Reasonable values of the margins are phase margin $\varphi_m = 30°$–$60°$, gain margin $g_m = 2$–5 and stability margin $s_m = 0.5$–0.8.

There are also other stability measures, such as the *delay margin*, which is the smallest time delay required to make the system unstable. For loop transfer functions that decay quickly, the delay margin is closely related to the phase margin, but for systems where the gain curve of the loop transfer function has several peaks at high frequencies, the delay margin is a more relevant measure.

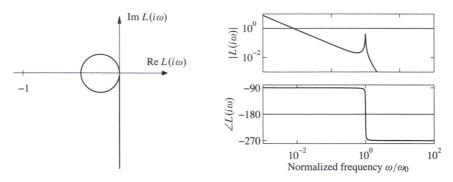

Figure 9.12: Nyquist and Bode plots of the loop transfer function for the AFM system (9.7) with an integral controller. The frequency in the Bode plot is normalized by a. The parameters are $\zeta = 0.01$ and $k_i = 0.008$.

Example 9.9 Nanopositioning system for an atomic force microscope

Consider the system for horizontal positioning of the sample in an atomic force microscope. The system has oscillatory dynamics, and a simple model is a spring–mass system with low damping. The normalized transfer function is given by

$$P(s) = \frac{\omega_0^2}{s^2 + 2\zeta\omega_0 s + \omega_0^2}, \tag{9.7}$$

where the damping ratio typically is a very small number, e.g., $\zeta = 0.1$.

We will start with a controller that has only integral action. The resulting loop transfer function is

$$L(s) = \frac{k_i \omega_0^2}{s(s^2 + 2\zeta\omega_0 s + \omega_0^2)},$$

where k_i is the gain of the controller. Nyquist and Bode plots of the loop transfer function are shown in Figure 9.12. Notice that the part of the Nyquist curve that is close to the critical point -1 is approximately circular.

From the Bode plot in Figure 9.12b, we see that the phase crossover frequency is $\omega_{pc} = a$, which will be independent of the gain k_i. Evaluating the loop transfer function at this frequency, we have $L(i\omega_0) = -k_i/(2\zeta\omega_0)$, which means that the gain margin is $g_m = 1 - k_i/(2\zeta\omega_0)$. To have a desired gain margin of g_m the integral gain should be chosen as

$$k_i = 2\omega_0\zeta(1 - g_m).$$

Figure 9.12 shows Nyquist and Bode plots for the system with gain margin $g_m = 1.67$ and stability margin $s_m = 0.597$. The gain curve in the Bode plot is almost a straight line for low frequencies and has a resonant peak at $\omega = \omega_0$. The gain crossover frequency is approximately equal to k_i. The phase decreases monotonically from $-90°$ to $-270°$: it is equal to $-180°$ at $\omega = \omega_0$. The curve can be shifted vertically by changing k_i: increasing k_i shifts the gain curve upward and increases the gain crossover frequency. Since the phase is $-180°$ at the resonant peak, it is necessary that the peak not touch the line $|L(i\omega)| = 1$. ∇

9.4 Bode's Relations and Minimum Phase Systems

An analysis of Bode plots reveals that there appears to be a relation between the gain curve and the phase curve. Consider, for example, the Bode plots for the differentiator and the integrator (shown in Figure 8.12). For the differentiator the slope is $+1$ and the phase is a constant $\pi/2$ radians. For the integrator the slope is -1 and the phase is $-\pi/2$. For the first-order system $G(s) = s + a$, the amplitude curve has the slope 0 for small frequencies and the slope $+1$ for high frequencies, and the phase is 0 for low frequencies and $\pi/2$ for high frequencies.

Bode investigated the relations between the curves for systems with no poles and zeros in the right half-plane. He found that the phase was uniquely given by the shape of the gain curve, and vice versa:

$$\arg G(i\omega_0) = \frac{\pi}{2} \int_0^\infty f(\omega) \frac{d \log |G(i\omega)|}{d \log \omega} d \log \omega \approx \frac{\pi}{2} \frac{d \log |G(i\omega)|}{d \log \omega}, \quad (9.8)$$

where f is the weighting kernel

$$f(\omega) = \frac{2}{\pi^2} \log \left| \frac{\omega + \omega_0}{\omega - \omega_0} \right|.$$

The phase curve is thus a weighted average of the derivative of the gain curve. If the gain curve has constant slope n, the phase curve has constant value $n\pi/2$.

Bode's relations (9.8) hold for systems that do not have poles and zeros in the right half-plane. Such systems are called *minimum phase systems* because systems with poles and zeros in the right half-plane have a larger phase lag. The distinction is important in practice because minimum phase systems are easier to control than systems with a larger phase lag. We will now give a few examples of nonminimum phase transfer functions.

The transfer function of a time delay of τ units is $G(s) = e^{-s\tau}$. This transfer function has unit gain $|G(i\omega)| = 1$, and the phase is $\arg G(i\omega) = -\omega\tau$. The corresponding minimum phase system with unit gain has the transfer function $G(s) = 1$. The time delay thus has an additional phase lag of $\omega\tau$. Notice that the phase lag increases linearly with frequency. Figure 9.13a shows the Bode plot of the transfer function. (Because we use a log scale for frequency, the phase falls off exponentially in the plot.)

Consider a system with the transfer function $G(s) = (a - s)/(a + s)$ with $a > 0$, which has a zero $s = a$ in the right half-plane. The transfer function has unit gain $|G(i\omega)| = 1$, and the phase is $\arg G(i\omega) = -2 \arctan (\omega/a)$. The corresponding minimum phase system with unit gain has the transfer function $G(s) = 1$. Figure 9.13b shows the Bode plot of the transfer function. A similar analysis of the transfer function $G(s) = (s + a)/s - a)$ with $a > 0$, which has a pole in the right half-plane, shows that its phase is $\arg G(i\omega) = -2 \arctan(a/\omega)$. The Bode plot is shown in Figure 9.13c.

The presence of poles and zeros in the right half-plane imposes severe limitations on the achievable performance. Dynamics of this type should be avoided by redesign of the system whenever possible. While the poles are intrinsic properties of the

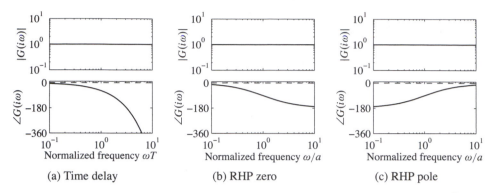

Figure 9.13: Bode plots of systems that are not minimum phase. (a) Time delay $G(s) = e^{-sT}$, (b) system with a right half-plane (RHP) zero $G(s) = (a - s)/(a + s)$ and (c) system with right half-plane pole. The corresponding minimum phase system has the transfer function $G(s) = 1$ in all cases, the phase curves for that system are shown as dashed lines.

system and they do not depend on sensors and actuators, the zeros depend on how inputs and outputs of a system are coupled to the states. Zeros can thus be changed by moving sensors and actuators or by introducing new sensors and actuators. Nonminimum phase systems are unfortunately quite common in practice.

The following example gives a system theoretic interpretation of the common experience that it is more difficult to drive in reverse gear and illustrates some of the properties of transfer functions in terms of their poles and zeros.

Example 9.10 Vehicle steering
The nonnormalized transfer function from steering angle to lateral velocity for the simple vehicle model is

$$G(s) = \frac{av_0 s + v_0^2}{bs},$$

where v_0 is the velocity of the vehicle and $a, b > 0$ (see Example 5.12). The transfer function has a zero at $s = v_0/a$. In normal driving this zero is in the left half-plane, but it is in the right half-plane when driving in reverse, $v_0 < 0$. The unit step response is

$$y(t) = \frac{av_0}{b} + \frac{v_0^2 t}{b}.$$

The lateral velocity thus responds immediately to a steering command. For reverse steering v_0 is negative and the initial response is in the wrong direction, a behavior that is representative for nonminimum phase systems (called an *inverse response*).

Figure 9.14 shows the step response for forward and reverse driving. In this simulation we have added an extra pole with the time constant T to approximately account for the dynamics in the steering system. The parameters are $a = b = 1$, $T = 0.1$, $v_0 = 1$ for forward driving and $v_0 = -1$ for reverse driving. Notice that for $t > t_0 = a/v_0$, where t_0 is the time required to drive the distance a, the step response for reverse driving is that of forward driving with the time delay t_0. The

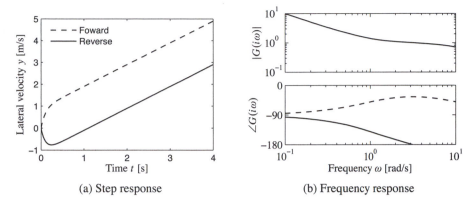

(a) Step response (b) Frequency response

Figure 9.14: Vehicle steering for driving in reverse. (a) Step responses from steering angle to lateral translation for a simple kinematics model when driving forward (dashed) and reverse (solid). With rear-wheel steering the center of mass first moves in the wrong direction and that the overall response with rear-wheel steering is significantly delayed compared with that for front-wheel steering. (b) Frequency response for driving forward (dashed) and reverse (solid). Notice that the gain curves are identical, but the phase curve for driving in reverse has nonminimum phase.

position of the zero v_0/a depends on the location of the sensor. In our calculation we have assumed that the sensor is at the center of mass. The zero in the transfer function disappears if the sensor is located at the rear wheel. The difficulty with zeros in the right half-plane can thus be visualized by a thought experiment where we drive a car in forward and reverse and observe the lateral position through a hole in the floor of the car. ∇

9.5 Generalized Notions of Gain and Phase

A key idea in frequency domain analysis is to trace the behavior of sinusoidal signals through a system. The concepts of gain and phase represented by the transfer function are strongly intuitive because they describe amplitude and phase relations between input and output. In this section we will see how to extend the concepts of gain and phase to more general systems, including some nonlinear systems. We will also show that there are analogs of Nyquist's stability criterion if signals are approximately sinusoidal.

System Gain

We begin by considering the case of a static linear system $y = Au$, where A is a matrix whose elements are complex numbers. The matrix does not have to be square. Let the inputs and outputs be vectors whose elements are complex numbers and use the Euclidean norm

$$\|u\| = \sqrt{\Sigma |u_i|^2}. \tag{9.9}$$

The norm of the output is

$$\|y\|^2 = u^* A^* A u,$$

where $*$ denotes the complex conjugate transpose. The matrix $A^* A$ is symmetric and positive semidefinite, and the right-hand side is a quadratic form. The square root of eigenvalues of the matrix $A^* A$ are all real, and we have

$$\|y\|^2 \leq \lambda_{\max}(A^* A) \|u\|^2.$$

The gain of the system can then be defined as the maximum ratio of the output to the input over all possible inputs:

$$\gamma = \max_u \frac{\|y\|}{\|u\|} = \sqrt{\lambda_{\max}(A^* A)}. \tag{9.10}$$

The square root of the eigenvalues of the matrix $A^* A$ are called the *singular values* of the matrix A, and the largest singular value is denoted $\bar{\sigma}(A)$.

To generalize this to the case of an input/output dynamical system, we need to think of the inputs and outputs not as vectors of real numbers but as vectors of *signals*. For simplicity, consider first the case of scalar signals and let the signal space L_2 be square-integrable functions with the norm

$$\|u\|_2 = \sqrt{\int_0^\infty |u|^2(\tau)\, d\tau}.$$

This definition can be generalized to vector signals by replacing the absolute value with the vector norm (9.9). We can now formally define the gain of a system taking inputs $u \in L_2$ and producing outputs $y \in L_2$ as

$$\gamma = \sup_{u \in L_2} \frac{\|y\|}{\|u\|}, \tag{9.11}$$

where sup is the *supremum,* defined as the smallest number that is larger than its argument. The reason for using the supremum is that the maximum may not be defined for $u \in L_2$. This definition of the system gain is quite general and can even be used for some classes of nonlinear systems, though one needs to be careful about how initial conditions and global nonlinearities are handled.

The norm (9.11) has some nice properties in the case of linear systems. In particular, given a single-input, single-output stable linear system with transfer function $G(s)$, it can be shown that the norm of the system is given by

$$\gamma = \sup_\omega |G(i\omega)| =: \|G\|_\infty. \tag{9.12}$$

In other words, the gain of the system corresponds to the peak value of the frequency response. This corresponds to our intuition that an input produces the largest output when we are at the resonant frequencies of the system. $\|G\|_\infty$ is called the *infinity norm* of the transfer function $G(s)$.

This notion of gain can be generalized to the multi-input, multi-output case as well. For a linear multivariable system with a real rational transfer function matrix

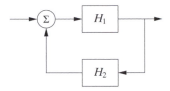

Figure 9.15: A feedback connection of two general nonlinear systems H_1 and H_2. The stability of the system can be explored using the small gain theorem.

$G(s)$ we can define the gain as

$$\gamma = \|G\|_\infty = \sup_\omega \bar{\sigma}\left(G(i\omega)\right). \tag{9.13}$$

Thus we can combine the idea of the gain of a matrix with the idea of the gain of a linear system by looking at the maximum singular value over all frequencies.

Small Gain and Passivity

For linear systems it follows from Nyquist's theorem that the closed loop is stable if the gain of the loop transfer function is less than 1 for all frequencies. This result can be extended to a larger class of systems by using the concept of the system gain defined in equation (9.11).

Theorem 9.4 (Small gain theorem). *Consider the closed loop system shown in Figure 9.15, where H_1 and H_2 are stable systems and the signal spaces are properly defined. Let the gains of the systems H_1 and H_2 be γ_1 and γ_2. Then the closed loop system is input/output stable if $\gamma_1\gamma_2 < 1$, and the gain of the closed loop system is*

$$\gamma = \frac{\gamma_1}{1 - \gamma_1\gamma_2}.$$

Notice that if systems H_1 and H_2 are linear, it follows from the Nyquist stability theorem that the closed loop is stable because if $\gamma_1\gamma_2 < 1$, the Nyquist curve is always inside the unit circle. The small gain theorem is thus an extension of the Nyquist stability theorem.

Although we have focused on linear systems, the small gain theorem also holds for nonlinear input/output systems. The definition of gain in equation (9.11) holds for nonlinear systems as well, with some care needed in handling the initial condition.

The main limitation of the small gain theorem is that it does not consider the phasing of signals around the loop, so it can be very conservative. To define the notion of phase we require that there be a scalar product. For square-integrable functions this can be defined as

$$\langle u, y \rangle = \int_0^\infty u(\tau)y(\tau)\,d\tau.$$

The phase φ between two signals can now be defined as

$$\langle u, y \rangle = \|u\|\|y\|\cos(\varphi).$$

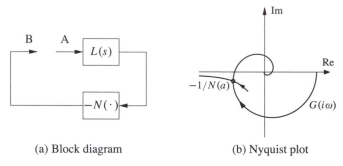

(a) Block diagram   (b) Nyquist plot

Figure 9.16: Describing function analysis. A feedback connection between a static nonlinearity and a linear system is shown in (a). The linear system is characterized by its transfer function $L(s)$, which depends on frequency, and the nonlinearity by its describing function $N(a)$, which depends on the amplitude a of its input. The Nyquist plot of $L(i\omega)$ and the plot of the $-1/N(a)$ are shown in (b). The intersection of the curves represents a possible limit cycle.

Systems where the phase between inputs and outputs is 90° or less for all inputs are called *passive systems*. It follows from the Nyquist stability theorem that a closed loop linear system is stable if the phase of the loop transfer function is between $-\pi$ and π. This result can be extended to nonlinear systems as well. It is called the *passivity theorem* and is closely related to the small gain theorem. See Khalil [123] for a more detailed description.

Additional applications of the small gain theorem and its application to robust stability are given in Chapter 12.

 Describing Functions

For special nonlinear systems like the one shown in Figure 9.16a, which consists of a feedback connection between a linear system and a static nonlinearity, it is possible to obtain a generalization of Nyquist's stability criterion based on the idea of *describing functions*. Following the approach of the Nyquist stability condition, we will investigate the conditions for maintaining an oscillation in the system. If the linear subsystem has low-pass character, its output is approximately sinusoidal even if its input is highly irregular. The condition for oscillation can then be found by exploring the propagation of a sinusoid that corresponds to the first harmonic.

To carry out this analysis, we have to analyze how a sinusoidal signal propagates through a static nonlinear system. In particular we investigate how the first harmonic of the output of the nonlinearity is related to its (sinusoidal) input. Letting F represent the nonlinear function, we expand $F(e^{i\omega t})$ in terms of its harmonics:

$$F(ae^{i\omega t}) = \sum_{n=0}^{\infty} M_n(a)e^{i(n\omega t + \varphi_n(a))},$$

where $M_n(a)$ and $\varphi_n(a)$ represent the gain and phase of the nth harmonic, which depend on the input amplitude since the function F is nonlinear. We define the

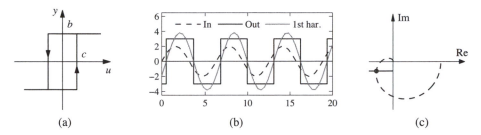

Figure 9.17: Describing function analysis for a relay with hysteresis. The input/output relation of the hysteresis is shown in (a) and the input with amplitude $a = 2$, the output and its first harmonic are shown in (b). The Nyquist plots of the transfer function $L(s) = (s + 1)^{-4}$ and the negative of the inverse describing function for the relay with $b = 3$ and $c = 1$ are shown in (c).

describing function to be the complex gain of the first harmonic:

$$N(a) = M_1(a)e^{i\varphi_n(a)}. \tag{9.14}$$

The function can also be computed by assuming that the input is a sinusoid and using the first term in the Fourier series of the resulting output.

Arguing as we did when deriving Nyquist's stability criterion, we find that an oscillation can be maintained if

$$L(i\omega)N(a) = -1. \tag{9.15}$$

This equation means that if we inject a sinusoid at A in Figure 9.16, the same signal will appear at B and an oscillation can be maintained by connecting the points. Equation (9.15) gives two conditions for finding the frequency ω of the oscillation and its amplitude a: the phase must be $180°$, and the magnitude must be unity. A convenient way to solve the equation is to plot $L(i\omega)$ and $-1/N(a)$ on the same diagram as shown in Figure 9.16b. The diagram is similar to the Nyquist plot where the critical point -1 is replaced by the curve $-1/N(a)$ and a ranges from 0 to ∞.

It is possible to define describing functions for types of inputs other than sinusoids. Describing function analysis is a simple method, but it is approximate because it assumes that higher harmonics can be neglected. Excellent treatments of describing function techniques can be found in the texts by Atherton [20] and Graham and McRuer [89].

Example 9.11 Relay with hysteresis
Consider a linear system with a nonlinearity consisting of a relay with hysteresis. The output has amplitude b and the relay switches when the input is $\pm c$, as shown in Figure 9.17a. Assuming that the input is $u = a\sin(\omega t)$, we find that the output is zero if $a \leq c$, and if $a > c$, the output is a square wave with amplitude b that switches at times $\omega t = \arcsin(c/a) + n\pi$. The first harmonic is then $y(t) = (4b/\pi)\sin(\omega t - \alpha)$, where $\sin\alpha = c/a$. For $a > c$ the describing function and its inverse are

$$N(a) = \frac{4b}{a\pi}\left(\sqrt{1 - \frac{c^2}{a^2}} - i\frac{c}{a}\right), \qquad \frac{1}{N(a)} = \frac{\pi\sqrt{a^2 - c^2}}{4b} + i\frac{\pi c}{4b},$$

where the inverse is obtained after simple calculations. Figure 9.17b shows the response of the relay to a sinusoidal input with the first harmonic of the output shown as a dashed line. Describing function analysis is illustrated in Figure 9.17c, which shows the Nyquist plot of the transfer function $L(s) = 2/(s + 1)^4$ (dashed line) and the negative inverse describing function of a relay with $b = 1$ and $c = 0.5$. The curves intersect for $a = 1$ and $\omega = 0.77$ rad/s, indicating the amplitude and frequency for a possible oscillation if the process and the relay are connected in a a feedback loop. ∇

9.6 Further Reading

Nyquist's original paper giving his now famous stability criterion was published in the *Bell Systems Technical Journal* in 1932 [160]. More accessible versions are found in the book [27], which also includes other interesting early papers on control. Nyquist's paper is also reprinted in an IEEE collection of seminal papers on control [23]. Nyquist used +1 as the critical point, but Bode changed it to −1, which is now the standard notation. Interesting perspectives on early developments are given by Black [36], Bode [41] and Bennett [29]. Nyquist did a direct calculation based on his insight into the propagation of sinusoidal signals through systems; he did not use results from the theory of complex functions. The idea that a short proof can be given by using the principle of variation of the argument is presented in the delightful book by MacColl [140]. Bode made extensive use of complex function theory in his book [40], which laid the foundation for frequency response analysis where the notion of minimum phase was treated in detail. A good source for complex function theory is the classic by Ahlfors [6]. Frequency response analysis was a key element in the emergence of control theory as described in the early texts by James et al. [110], Brown and Campbell [46] and Oldenburger [163], and it became one of the cornerstones of early control theory. Frequency response methods underwent a resurgence when robust control emerged in the 1980s, as will be discussed in Chapter 12.

Exercises

9.1 (Operational amplifier) Consider an op amp circuit with $Z_1 = Z_2$ that gives a closed loop system with nominally unit gain. Let the transfer function of the operational amplifier be

$$G(s) = \frac{ka_1a_2}{(s + a)(s + a_1)(s + a_2)},$$

where $a_1, a_2 \gg a$. Show that the condition for oscillation is $k < a_1 + a_2$ and compute the gain margin of the system. Hint: Assume $a = 0$.

9.2 (Atomic force microscope) The dynamics of the tapping mode of an atomic force microscope are dominated by the damping of the cantilever vibrations and the system that averages the vibrations. Modeling the cantilever as a spring–mass system with low damping, we find that the amplitude of the vibrations decays as $\exp(-\zeta\omega t)$, where ζ is the damping ratio and ω is the undamped natural frequency of the cantilever. The cantilever dynamics can thus be modeled by the transfer function

$$G(s) = \frac{b}{s + a},$$

where $a = \zeta\omega_0$. The averaging process can be modeled by the input/output relation

$$y(t) = \frac{1}{\tau}\int_{t-\tau}^{t} u(v)dv,$$

where the averaging time is a multiple n of the period of the oscillation $2\pi/\omega$. The dynamics of the piezo scanner can be neglected in the first approximation because it is typically much faster than a. A simple model for the complete system is thus given by the transfer function

$$P(s) = \frac{a(1 - e^{-s\tau})}{s\tau(s + a)}.$$

Plot the Nyquist curve of the system and determine the gain of a proportional controller that brings the system to the boundary of stability.

9.3 (Heat conduction) A simple model for heat conduction in a solid is given by the transfer function

$$P(s) = ke^{-\sqrt{s}}.$$

Sketch the Nyquist plot of the system. Determine the frequency where the phase of the process is $-180°$ and the gain at that frequency. Show that the gain required to bring the system to the stability boundary is $k = e^{\pi}$.

9.4 (Vectored thrust aircraft) Consider the state space controller designed for the vectored thrust aircraft in Examples 6.8 and 7.5. The controller consists of two components: an optimal estimator to compute the state of the system from the output and a state feedback compensator that computes the input given the (estimated) state. Compute the loop transfer function for the system and determine the gain, phase and stability margins for the closed loop dynamics.

9.5 (Vehicle steering) Consider the linearized model for vehicle steering with a controller based on state feedback discussed in Example 7.4. The transfer functions for the process and controller are given by

$$P(s) = \frac{\gamma s + 1}{s^2}, \quad C(s) = \frac{s(k_1 l_1 + k_2 l_2) + k_1 l_2}{s^2 + s(\gamma k_1 + k_2 + l_1) + k_1 + l_2 + k_2 l_1 - \gamma k_2 l_2},$$

as computed in Example 8.6. Let the process parameter be $\gamma = 0.5$ and assume that the state feedback gains are $k_1 = 1$ and $k_2 = 0.914$ and that the observer gains are $l_1 = 2.828$ and $l_2 = 4$. Compute the stability margins numerically.

9.6 (Stability margins for second-order systems) A process whose dynamics is described by a double integrator is controlled by an ideal PD controller with the transfer function $C(s) = k_d s + k_p$, where the gains are $k_d = 2\zeta\omega_0$ and $k_p = \omega_0^2$. Calculate and plot the gain, phase and stability margins as a function ζ.

9.7 (Congestion control in overload conditions) A strongly simplified flow model of a TCP loop under overload conditions is given by the loop transfer function

$$L(s) = \frac{k}{s} e^{-s\tau},$$

where the queuing dynamics are modeled by an integrator, the TCP window control is a time delay τ and the controller is simply a proportional controller. A major difficulty is that the time delay may change significantly during the operation of the system. Show that if we can measure the time delay, it is possible to choose a gain that gives a stability margin of $s_n \geq 0.6$ for all time delays τ.

9.8 (Bode's formula) Consider Bode's formula (9.8) for the relation between gain and phase for a transfer function that has all its singularities in the left half-plane. Plot the weighting function and make an assessment of the frequencies where the approximation $\arg G \approx (\pi/2)d \log |G|/d \log \omega$ is valid.

9.9 (Padé approximation to a time delay) Consider the transfer functions

$$G_1(s) = e^{-s\tau}, \qquad G_2(s) = e^{-s\tau} \approx \frac{1 - s\tau/2}{1 + s\tau/2}. \tag{9.16}$$

Show that the minimum phase properties of the transfer functions are similar for frequencies $\omega < 1/\tau$. A long time delay τ is thus equivalent to a small right half-plane zero. The approximation (9.16) is called a first-order *Padé approximation*.

9.10 (Inverse response) Consider a system whose input/output response is modeled by $G(s) = 6(-s + 1)/(s^2 + 5s + 6)$, which has a zero in the right half-plane. Compute the step response for the system, and show that the output goes in the wrong direction initially, which is also referred to as an *inverse response*. Compare the response to a minimum phase system by replacing the zero at $s = 1$ with a zero at $s = -1$.

9.11 (Describing function analysis) . Consider the system with the block diagram shown on the left below.

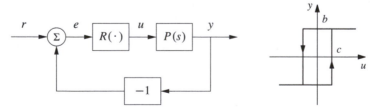

The block R is a relay with hysteresis whose input/output response is shown on the right and the process transfer function is $P(s) = e^{-s\tau}/s$. Use describing function analysis to determine frequency and amplitude of possible limit cycles. Simulate the system and compare with the results of the describing function analysis.

Chapter Ten
PID Control

Based on a survey of over eleven thousand controllers in the refining, chemicals and pulp and paper industries, 97% of regulatory controllers utilize PID feedback.

L. Desborough and R. Miller, 2002 [58].

This chapter treats the basic properties of proportional-integral-derivative (PID) control and the methods for choosing the parameters of the controllers. We also analyze the effects of actuator saturation and time delay, two important features of many feedback systems, and describe methods for compensating for these effects. Finally, we will discuss the implementation of PID controllers as an example of how to implement feedback control systems using analog or digital computation.

10.1 Basic Control Functions

PID control, which was introduced in Section 1.5 and has been used in several examples, is by far the most common way of using feedback in engineering systems. It appears in simple devices and in large factories with thousands of controllers. PID controllers appear in many different forms: as stand-alone controllers, as part of hierarchical, distributed control systems and built into embedded components. Most PID controllers do not use derivative action, so they should strictly speaking be called PI controllers; we will, however, use PID as a generic term for this class of controller. There is also growing evidence that PID control appears in biological systems [206].

Block diagrams of closed loop systems with PID controllers are shown in Figure 10.1. The control signal u for the system in Figure 10.1a is formed entirely from the error e; there is no feedforward term (which would correspond to $k_r r$ in the state feedback case). A common alternative in which proportional and derivative action do not act on the reference is shown in Figure 10.1b; combinations of the schemes will be discussed in Section 10.5. The command signal r is called the reference signal in regulation problems, or the *setpoint* in the literature of PID control. The input/output relation for an ideal PID controller with error feedback is

$$u = k_p e + k_i \int_0^t e(\tau)\, d\tau + k_d \frac{de}{dt} = k_p \Big(e + \frac{1}{T_i} \int_0^t e(\tau)\, d\tau + T_d \frac{de}{dt} \Big). \quad (10.1)$$

The control action is thus the sum of three terms: proportional feedback, the integral term and derivative action. For this reason PID controllers were originally called *three-term controllers*. The controller parameters are the proportional gain k_p, the

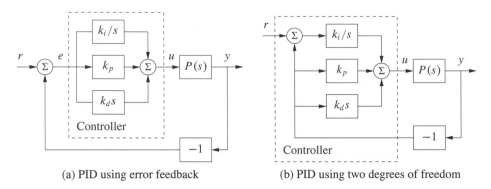

(a) PID using error feedback   (b) PID using two degrees of freedom

Figure 10.1: Block diagrams of closed loop systems with ideal PID controllers. Both controllers have one output, the control signal u. The controller in (a), which is based on error feedback, has one input, the control error $e = r - y$. For this controller proportional, integral and derivative action acts on the error $e = r - y$. The two degree-of-freedom controller in (b) has two inputs, the reference r and the process output y. Integral action acts on the error, but proportional and derivative action act on the process output y.

integral gain k_i and the derivative gain k_d. The time constants T_i and T_d, called integral time (constant) and derivative time (constant), are sometimes used instead of the integral and derivative gains.

The controller (10.1) represents an idealized controller. It is a useful abstraction for understanding the PID controller, but several modifications must be made to obtain a controller that is practically useful. Before discussing these practical issues we will develop some intuition about PID control.

We start by considering pure proportional feedback. Figure 10.2a shows the responses of the process output to a unit step in the reference value for a system with pure proportional control at different gain settings. In the absence of a feedforward term, the output never reaches the reference, and hence we are left with nonzero steady-state error. Letting the process and the controller have transfer functions $P(s)$ and $C(s)$, the transfer function from reference to output is

$$G_{yr} = \frac{PC}{1 + PC}, \tag{10.2}$$

and thus the steady-state error for a unit step is

$$1 - G_{yr}(0) = \frac{1}{1 + k_p P(0)}.$$

For the system in Figure 10.2a with gains $k_p = 1, 2$ and 5, the steady-state error is $0.5, 0.33$ and 0.17. The error decreases with increasing gain, but the system also becomes more oscillatory. Notice in the figure that the initial value of the control signal equals the controller gain.

To avoid having a steady-state error, the proportional term can be changed to

$$u(t) = k_p e(t) + u_{\text{ff}}, \tag{10.3}$$

where u_{ff} is a feedforward term that is adjusted to give the desired steady-state

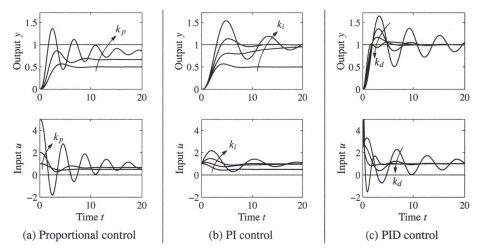

(a) Proportional control (b) PI control (c) PID control

Figure 10.2: Responses to step changes in the reference value for a system with a proportional controller (a), PI controller (b) and PID controller (c). The process has the transfer function $P(s) = 1/(s+1)^3$, the proportional controller has parameters $k_p = 1, 2$ and 5, the PI controller has parameters $k_p = 1, k_i = 0, 0.2, 0.5$ and 1, and the PID controller has parameters $k_p = 2.5$, $k_i = 1.5$ and $k_d = 0, 1, 2$ and 4.

value. If we choose $u_{\text{ff}} = r/P(0) = k_r r$, then the output will be exactly equal to the reference value, as it was in the state space case, provided that there are no disturbances. However, this requires exact knowledge of the process dynamics, which is usually not available. The parameter u_{ff}, called *reset* in the PID literature, must therefore be adjusted manually.

As we saw in Section 6.4, integral action guarantees that the process output agrees with the reference in steady state and provides an alternative to the feedforward term. Since this result is so important, we will provide a general proof. Consider the controller given by equation (10.1). Assume that there exists a steady state with $u = u_0$ and $e = e_0$. It then follows from equation (10.1) that

$$u_0 = k_p e_0 + k_i e_0 t,$$

which is a contradiction unless e_0 or k_i is zero. We can thus conclude that with integral action the error will be zero if it reaches a steady state. Notice that we have not made any assumptions about the linearity of the process or the disturbances. We have, however assumed that an equilibrium exists. Using integral action to achieve zero steady-state error is much better than using feedforward, which requires a precise knowledge of process parameters.

The effect of integral action can also be understood from frequency domain analysis. The transfer function of the PID controller is

$$C(s) = k_p + \frac{k_i}{s} + k_d s. \tag{10.4}$$

The controller has infinite gain at zero frequency ($C(0) = \infty$), and it then follows from equation (10.2) that $G_{yr}(0) = 1$, which implies that there is no steady-state

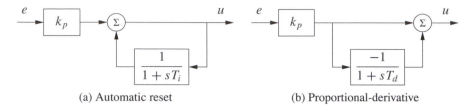

(a) Automatic reset   (b) Proportional-derivative

Figure 10.3: Implementation of PI and PD controllers. The block diagram in (a) shows how integral action is implemented using *positive feedback* with a first-order system, sometimes called automatic reset. The block diagram in (b) shows how derivative action can be implemented by taking differences between a static system and a first-order system.

error for a step input.

Integral action can also be viewed as a method for generating the feedforward term u_{ff} in the proportional controller (10.3) automatically. One way to do this is shown in Figure 10.3a, where the controller output is low-pass-filtered and fed back with positive gain. This implementation, called *automatic reset*, was one of the early inventions of integral control. The transfer function of the system in Figure 10.3a is obtained by block diagram algebra; we have

$$G_{ue} = k_p \frac{1 + sT_i}{sT_i} = k_p + \frac{k_p}{sT_i},$$

which is the transfer function for a PI controller.

The properties of integral action are illustrated in Figure 10.2b for a step input. The proportional gain is constant, $k_p = 1$, and the integral gains are $k_i = 0, 0.2, 0.5$ and 1. The case $k_i = 0$ corresponds to pure proportional control, with a steady-state error of 50%. The steady-state error is eliminated when integral gain action is used. The response creeps slowly toward the reference for small values of k_i and goes faster for larger integral gains, but the system also becomes more oscillatory.

The integral gain k_i is a useful measure for attenuation of load disturbances. Consider a closed loop system under PID control and assume that the system is stable and initially at rest with all signals being zero. Apply a unit step disturbance at the process input. After a transient the process output goes to zero and the controller output settles at a value that compensates for the disturbance. It follows from (10.1) that

$$u(\infty) = k_i \int_0^\infty e(t)dt.$$

The integrated error is thus inversely proportional to the integral gain k_i. The integral gain is thus a measure of the effectiveness of disturbance attenuation. A large gain k_i attenuates disturbances effectively, but too large a gain gives oscillatory behavior, poor robustness and possibly instability.

We now return to the general PID controller and consider the effect of the derivative term k_d. Recall that the original motivation for derivative feedback was to provide predictive or anticipatory action. Notice that the combination of the proportional and the derivative terms can be written as

$$u = k_p e + k_d \frac{de}{dt} = k_p \left(e + T_d \frac{de}{dt} \right) = k_p e_p,$$

where $e_p(t)$ can be interpreted as a prediction of the error at time $t + T_d$ by linear extrapolation. The prediction time $T_d = k_d/k_p$ is the derivative time constant of the controller.

Derivative action can be implemented by taking the difference between the signal and its low-pass filtered version as shown in Figure 10.3b. The transfer function for the system is

$$G_{ue}(s) = k_p \left(1 - \frac{1}{1 + sT_d} \right) = k_p \frac{sT_d}{1 + sT_d}. \tag{10.5}$$

The system thus has the transfer function $G(s) = sT_d/(1 + sT_d)$, which approximates a derivative for low frequencies ($|s| < T_d$).

Figure 10.2c illustrates the effect of derivative action: the system is oscillatory when no derivative action is used, and it becomes more damped as the derivative gain is increased. Performance deteriorates if the derivative gain is too high. When the input is a step, the controller output generated by the derivative term will be an impulse. This is clearly visible in Figure 10.2c. The impulse can be avoided by using the controller configuration shown in Figure 10.1b.

Although PID control was developed in the context of engineering applications, it also appears in nature. Disturbance attenuation by feedback in biological systems is often called *adaptation*. A typical example is the pupillary reflex discussed in Example 8.11, where it is said that the eye adapts to changing light intensity. Analogously, feedback with integral action is called perfect adaptation [206]. In biological systems proportional, integral and derivative action is generated by combining subsystems with dynamical behavior similarly to what is done in engineering systems. For example, PI action can be generated by the interaction of several hormones [68].

Example 10.1 PD action in the retina
The response of cone photoreceptors in the retina is an example where proportional and derivative action is generated by a combination of cones and horizontal cells. The cones are the primary receptors stimulated by light, which in turn stimulate the horizontal cells, and the horizontal cells give inhibitory (negative) feedback to the cones. A schematic diagram of the system is shown in Figure 10.4a. The system can be modeled by ordinary differential equations by representing neuron signals as continuous variables representing the average pulse rate. In [203] it is shown that the system can be represented by the differential equations

$$\frac{dx_1}{dt} = \frac{1}{T_c}(-x_1 - kx_2 + u), \qquad \frac{dx_2}{dt} = \frac{1}{T_h}(x_1 - x_2),$$

where u is the light intensity and x_1 and x_2 are the average pulse rates from the cones and the horizontal cells. A block diagram of the system is shown in Figure 10.4b. The step response of the system shown in Figure 10.4c shows that the system has

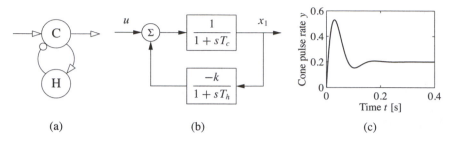

Figure 10.4: Schematic diagram of cone photoreceptors (C) and horizontal cells (H) in the retina. In the schematic diagram in (a), excitatory feedback is indicated by arrows and inhibitory feedback by circles. A block diagram is shown in (b) and the step response in (c).

a large initial response followed by a lower, constant steady-state response typical of proportional and derivative action. The parameters used in the simulation are $k = 4$, $T_c = 0.025$ and $T_h = 0.08$.   ∇

10.2 Simple Controllers for Complex Systems

Many of the design methods discussed in previous chapters have the property that the complexity of the controller is directly reflected by the complexity of the model. When designing controllers by output feedback in Chapter 7, we found for single-input, single-output systems that the order of the controller was the same as the order of the model, possibly one order higher if integral action was required. Applying similar design methods for PID control will require that we have low-order models of the processes to be able to easily analyze the results.

Low-order models can be obtained from first principles. Any stable system can be modeled by a static system if its inputs are sufficiently slow. Similarly a first-order model is sufficient if the storage of mass, momentum or energy can be captured by only one variable; typical examples are the velocity of a car on a road, angular velocity of a stiff rotational system, the level in a tank and the concentration in a volume with good mixing. System dynamics are of second order if the storage of mass, energy and momentum can be captured by two state variables; typical examples are the position of a car on the road, the stabilization of stiff satellites, the levels in two connected tanks and two-compartment models. A wide range of techniques for model reduction are also available. In this chapter we will focus on design techniques where we simplify the models to capture the essential properties that are needed for PID design.

We begin by analyzing the case of integral control. A stable system can be controlled by an integral controller provided that the requirements on the closed loop system are modest. To design the controller we assume that the transfer function of the process is a constant $K = P(0)$. The loop transfer function under integral control then becomes Kk_i/s, and the closed loop characteristic polynomial is simply $s + Kk_i$. Specifying performance by the desired time constant T_{cl} of the closed loop

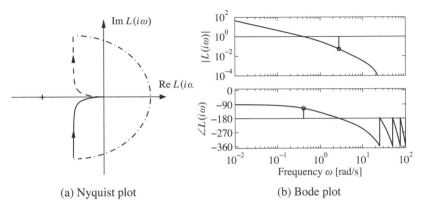

(a) Nyquist plot   (b) Bode plot

Figure 10.5: Integral control for AFM in tapping mode. An integral controller is designed based on the slope of the process transfer function at 0. The controller gives good robustness properties based on a very simple analysis.

system, we find that the integral gain is given by

$$k_i = 1/(T_{cl}P(0)).$$

The analysis requires that T_{cl} be sufficiently large that the process transfer function can be approximated by a constant.

For systems that are not well represented by a constant gain, we can obtain a better approximation by using the Taylor series expansion of the loop transfer function:

$$L(s) = \frac{k_i P(s)}{s} \approx \frac{k_i(P(0) + sP'(0))}{s} = k_i P'(0) + \frac{k_i P(0)}{s}.$$

Choosing $k_i P'(0) = -0.5$ gives a system with good robustness, as will be discussed in Section 12.5. The controller gain is then given by

$$k_i = -\frac{1}{2P'(0)}, \tag{10.6}$$

and the expected closed loop time constant is $T_{cl} \approx -2P'(0)/P(0)$.

Example 10.2 Integral control of AFM in tapping mode
A simplified model of the dynamics of the vertical motion of an atomic force microscope in tapping mode was discussed in Exercise 9.2. The transfer function for the system dynamics is

$$P(s) = \frac{a(1 - e^{-s\tau})}{s\tau(s + a)},$$

where $a = \zeta\omega_0$, $\tau = 2\pi n/\omega_0$ and the gain has been normalized to 1. We have $P(0) = 1$ and $P'(0) = -\tau/2 - 1/a$, and it follows from (10.6) that the integral gain can be chosen as $k_i = a/(2 + a\tau)$. Nyquist and Bode plots for the resulting loop transfer function are shown in Figure 10.5. $\triangledown$

A first-order system has the transfer function

$$P(s) = \frac{b}{s+a}.$$

With a PI controller the closed loop system has the characteristic polynomial

$$s(s+a) + bk_p s + bk_i s = s^2 + (a + bk_p)s + bk_i.$$

The closed loop poles can thus be assigned arbitrary values by proper choice of the controller gains. Requiring that the closed loop system have the characteristic polynomial

$$p(s) = s^2 + a_1 s + a_2^2,$$

we find that the controller parameters are

$$k_p = \frac{a_1 - a}{b}, \qquad k_i = \frac{a_2}{b}. \tag{10.7}$$

If we require a response of the closed loop system that is slower than that of the open loop system, a reasonable choice is $a_1 = a + \alpha$ and $a_2 = \alpha a$. If a response faster than that of the open loop system is required, it is reasonable to choose $a_1 = 2\zeta\omega_0$ and $a_2 = \omega_0^2$, where ω_0 and ζ are undamped natural frequency and damping ratio of the dominant mode. These choices have significant impact on the robustness of the system and will be discussed in Section 12.4. An upper limit to ω_0 is given by the validity of the model. Large values of ω_0 will require fast control actions, and actuators may saturate if the value is too large. A first-order model is unlikely to represent the true dynamics for high frequencies. We illustrate the design by an example.

Example 10.3 Cruise control using PI feedback
Consider the problem of maintaining the speed of a car as it goes up a hill. In Example 5.14 we found that there was little difference between the linear and nonlinear models when investigating PI control, provided that the throttle did not reach the saturation limits. A simple linear model of a car was given in Example 5.11:

$$\frac{d(v - v_e)}{dt} = -a(v - v_e) + b(u - u_e) - g\theta, \tag{10.8}$$

where v is the velocity of the car, u is the input from the engine and θ is the slope of the hill. The parameters were $a = 0.0101$, $b = 1.3203$, $g = 9.8$, $v_e = 20$ and $u_e = 0.1616$. This model will be used to find suitable parameters of a vehicle speed controller. The transfer function from throttle to velocity is a first-order system. Since the open loop dynamics is so slow, it is natural to specify a faster closed loop system by requiring that the closed loop system be of second-order with damping ratio ζ and undamped natural frequency ω_0. The controller gains are given by (10.7).

Figure 10.6 shows the velocity and the throttle for a car that initially moves on a horizontal road and encounters a hill with a slope of $4°$ at time $t = 6$ s. To design a PI controller we choose $\zeta = 1$ to obtain a response without overshoot, as shown in Figure 10.6a. The choice of ω_0 is a compromise between response speed

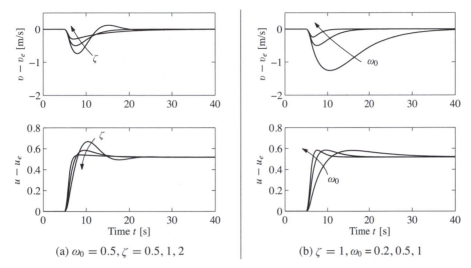

(a) $\omega_0 = 0.5, \zeta = 0.5, 1, 2$ (b) $\zeta = 1, \omega_0 = 0.2, 0.5, 1$

Figure 10.6: Cruise control using PI feedback. The step responses for the error and input illustrate the effect of parameters $\zeta = 1$ and ω_0 on the response of a car with cruise control. A change in road slope from $0°$ to $4°$ is applied between $t = 5$ and 6 s. (a) Responses for $\omega_0 = 0.5$ and $\zeta = 0.5, 1$ and 2. Choosing $\zeta = 1$ gives no overshoot. (b) Responses for $\zeta = 1$ and $\omega_0 = 0.2, 0.5$ and 1.0.

and control actions: a large value gives a fast response, but it requires fast control action. The trade-off is is illustrated in Figure 10.6b. The largest velocity error decreases with increasing ω_0, but the control signal also changes more rapidly. In the simple model (10.8) it was assumed that the force responds instantaneously to throttle commands. For rapid changes there may be additional dynamics that have to be accounted for. There are also physical limitations to the rate of change of the force, which also restricts the admissible value of ω_0. A reasonable choice of ω_0 is in the range 0.5–1.0. Notice in Figure 10.6 that even with $\omega_0 = 0.2$ the largest velocity error is only 1 m/s. ∇

A PI controller can also be used for a process with second-order dynamics, but there will be restrictions on the possible locations of the closed loop poles. Using a PID controller, it is possible to control a system of second order in such a way that the closed loop poles have arbitrary locations; see Exercise 10.2.

Instead of finding a low-order model and designing controllers for them, we can also use a high-order model and attempt to place only a few dominant poles. An integral controller has one parameter, and it is possible to position one pole. Consider a process with the transfer function $P(s)$. The loop transfer function with an integral controller is $L(s) = k_i P(s)/s$. The roots of the closed loop characteristic polynomial are the roots of $s + k_i P(s) = 0$. Requiring that $s = -a$ be a root, we find that the controller gain should be chosen as

$$k_i = \frac{a}{P(-a)}. \tag{10.9}$$

The pole $s = -a$ will be dominant if a is small. A similar approach can be applied

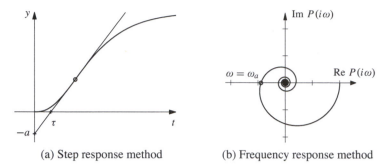

(a) Step response method   (b) Frequency response method

Figure 10.7: Ziegler–Nichols step and frequency response experiments. The unit step response in (a) is characterized by the parameters a and τ. The frequency response method (b) characterizes process dynamics by the point where the Nyquist curve of the process transfer function first intersects the negative real axis and the frequency ω_c where this occurs.

to PI and PID controllers.

10.3 PID Tuning

Users of control systems are frequently faced with the task of adjusting the controller parameters to obtain a desired behavior. There are many different ways to do this. One approach is to go through the conventional steps of modeling and control design as described in the previous section. Since the PID controller has so few parameters, a number of special empirical methods have also been developed for direct adjustment of the controller parameters. The first tuning rules were developed by Ziegler and Nichols [210]. Their idea was to perform a simple experiment, extract some features of process dynamics from the experiment and determine the controller parameters from the features.

Ziegler–Nichols' Tuning

In the 1940s, Ziegler and Nichols developed two methods for controller tuning based on simple characterization of process dynamics in the time and frequency domains.

The time domain method is based on a measurement of part of the open loop unit step response of the process, as shown in Figure 10.7a. The step response is measured by applying a unit step input to the process and recording the response. The response is characterized by parameters a and τ, which are the intercepts of the steepest tangent of the step response with the coordinate axes. The parameter τ is an approximation of the time delay of the system and a/τ is the steepest slope of the step response. Notice that it is not necessary to wait until steady state is reached to find the parameters, it suffices to wait until the response has had an inflection point. The controller parameters are given in Table 10.1. The parameters were obtained by extensive simulation of a range of representative processes. A controller was

Table 10.1: Ziegler–Nichols tuning rules. (a) The step response methods give the parameters in terms of the intercept a and the apparent time delay τ. (b) The frequency response method gives controller parameters in terms of *critical gain k_c* and *critical period T_c*.

Type	k_p	T_i	T_d		Type	k_p	T_i	T_d
P	1/a				P	$0.5k_c$		
PI	0.9/a	3τ			PI	$0.4k_c$	$0.8T_c$	
PID	1.2/a	2τ	0.5τ		PID	$0.6k_c$	$0.5T_c$	$0.125T_c$

(a) Step response method	(b) Frequency response method

tuned manually for each process, and an attempt was then made to correlate the controller parameters with a and τ.

In the frequency domain method, a controller is connected to the process, the integral and derivative gains are set to zero and the proportional gain is increased until the system starts to oscillate. The critical value of the proportional gain k_c is observed together with the period of oscillation T_c. It follows from Nyquist's stability criterion that the loop transfer function $L = k_c P(s)$ intersects the critical point at the frequency $\omega_c = 2\pi/T_c$. The experiment thus gives the point on the Nyquist curve of the process transfer function where the phase lag is 180°, as shown in Figure 10.7b.

The Ziegler–Nichols methods had a huge impact when they were introduced in the 1940s. The rules were simple to use and gave initial conditions for manual tuning. The ideas were adopted by manufacturers of controllers for routine use. The Ziegler–Nichols tuning rules unfortunately have two severe drawbacks: too little process information is used, and the closed loop systems that are obtained lack robustness.

The step response method can be improved significantly by characterizing the unit step response by parameters K, τ and T in the model

$$P(s) = \frac{K}{1 + sT} e^{-\tau s}. \tag{10.10}$$

The parameters can be obtained by fitting the model to a measured step response. Notice that the experiment takes a longer time than the experiment in Figure 10.7a because to determine K it is necessary to wait until the steady state has been reached. Also notice that the intercept a in the Ziegler–Nichols rule is given by $a = K\tau/T$.

The frequency response method can be improved by measuring more points on the Nyquist curve, e.g., the zero frequency gain K or the point where the process has a 90° phase lag. This latter point can be obtained by connecting an integral controller and increasing its gain until the system reaches the stability limit. The experiment can also be automated by using relay feedback, as will be discussed later in this section.

There are many versions of improved tuning rules. As an illustration we give

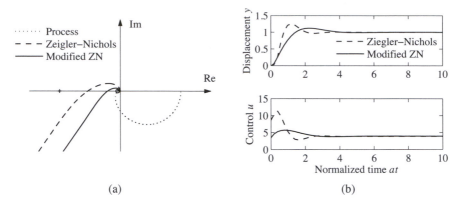

Figure 10.8: PI control of an AFM in tapping mode. Nyquist plots (a) and step responses (b) for PI control of the vertical motion of an atomic force microscope in tapping mode. The averaging parameter is $n = 20$. Results with Ziegler–Nichols tuning are shown by dashed lines, and modified Ziegler–Nichols tuning is shown by solid lines. The Nyquist plot of the process transfer function is shown by dotted lines.

the following rules for PI control, based on [16]:

$$
k_p = \frac{0.15\tau + 0.35T}{K\tau} \quad \left(\frac{0.9T}{K\tau}\right), \quad k_i = \frac{0.46\tau + 0.02T}{K\tau^2} \square \left(\frac{0.3T}{K\tau^2}\right),
$$

$$
k_p = 0.22k_c - \frac{0.07}{K} \quad \left(0.4k_c\right), \quad k_i = \frac{0.16k_c}{T_c} + \frac{0.62}{KT_c} \square \left(\frac{0.5k_c}{T_c}\right).
$$

(10.11)

The values for the Ziegler–Nichols rule are given in parentheses. Notice that the improved formulas typically give lower controller gains than the Ziegler–Nichols method. The integral gain is higher for systems where the dynamics are delay-dominated, $\tau \gg T$.

Example 10.4 Atomic force microscope in tapping mode

A simplified model of the dynamics of the vertical motion of an atomic force microscope in tapping mode was discussed in Example 10.2. The transfer function is normalized by choosing $1/a$ as the time unit. The normalized transfer function is

$$
P(s) = \frac{1 - e^{-sT_n}}{sT_n(s + 1)},
$$

where $T_n = 2n\pi a/\omega_0 = 2n\pi \zeta$. The Nyquist plot of the transfer function is shown in Figure 10.8a for $z = 0.002$ and $n = 20$. The leftmost intersection of the Nyquist curve with the real axis occurs at $\text{Re } s = -0.0461$ for $\omega = 13.1$. The critical gain is thus $k_c = 21.7$ and the critical period is $T_c = 0.48$. Using the Ziegler–Nichols tuning rule, we find the parameters $k_p = 8.87$ and $k_i = 22.6$ ($T_i = 0.384$) for a PI controller. With this controller the stability margin is $s_m = 0.31$, which is quite small. The step response of the controller is shown in Figure 10.8. Notice in particular that there is a large overshoot in the control signal.

The modified Ziegler–Nichols rule (10.11) gives the controller parameters $k =$

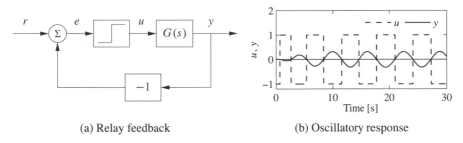

(a) Relay feedback   (b) Oscillatory response

Figure 10.9: Block diagram of a process with relay feedback (a) and typical signals (b). The process output y is a solid line, and the relay output u is a dashed line. Notice that the signals u and y have opposite phases.

3.47 and $k_i = 8.73$ ($T_i = 0.459$) and the stability margin becomes $s_m = 0.61$. The step response with this controller is shown in Figure 10.8. A comparison of the responses obtained with the original Ziegler–Nichols rule shows that the overshoot has been reduced. Notice that the control signal reaches its steady-state value almost instantaneously. It follows from Example 10.2 that a pure integral controller has the normalized gain $k_i = 1/(2 + T_n) = 0.44$. Comparing this with the gains of a PI controller, we can conclude that a PI controller gives much better performance than a pure integral controller.   ▽

Relay Feedback

The Ziegler–Nichols frequency response method increases the gain of a proportional controller until oscillation to determine the critical gain k_c and the corresponding critical period T_c or, equivalently, the point where the Nyquist curve intersects the negative real axis. One way to obtain this information automatically is to connect the process in a feedback loop with a nonlinear element having a relay function as shown in Figure 10.9a. For many systems there will then be an oscillation, as shown in Figure 10.9b, where the relay output u is a square wave and the process output y is close to a sinusoid. Moreover the input and the output are out of phase, which means that the system oscillates with the critical period T_c, where the process has a phase lag of 180°. Notice that an oscillation with constant period is established quickly.

The critical period is simply the period of the oscillation. To determine the critical gain we expand the square wave relay output in a Fourier series. Notice in the figure that the process output is practically sinusoidal because the process attenuates higher harmonics effectively. It is then sufficient to consider only the first harmonic component of the input. Letting d be the relay amplitude, the first harmonic of the square wave input has amplitude $4d/\pi$. If a is the amplitude of the process output, the process gain at the critical frequency $\omega_c = 2\pi/T_c$ is $|P(i\omega_c)| = \pi a/(4d)$ and the critical gain is

$$K_c = \frac{4d}{a\pi}. \tag{10.12}$$

Having obtained the critical gain K_c and the critical period T_c, the controller parameters can then be determined using the Ziegler–Nichols rules. Improved tuning can be obtained by fitting a model to the data obtained from the relay experiment.

The relay experiment can be automated. Since the amplitude of the oscillation is proportional to the relay output, it is easy to control it by adjusting the relay output. *Automatic tuning* based on relay feedback is used in many commercial PID controllers. Tuning is accomplished simply by pushing a button that activates relay feedback. The relay amplitude is automatically adjusted to keep the oscillations sufficiently small, and the relay feedback is switched to a PID controller as soon as the tuning is finished.

10.4 Integrator Windup

Many aspects of a control system can be understood from linear models. There are, however, some nonlinear phenomena that must be taken into account. These are typically limitations in the actuators: a motor has limited speed, a valve cannot be more than fully opened or fully closed, etc. For a system that operates over a wide range of conditions, it may happen that the control variable reaches the actuator limits. When this happens, the feedback loop is broken and the system runs in open loop because the actuator remains at its limit independently of the process output as long as the actuator remains saturated. The integral term will also build up since the error is typically nonzero. The integral term and the controller output may then become very large. The control signal will then remain saturated even when the error changes, and it may take a long time before the integrator and the controller output come inside the saturation range. The consequence is that there are large transients. This situation is referred to as *integrator windup*, illustrated in the following example.

Example 10.5 Cruise control
The windup effect is illustrated in Figure 10.10a, which shows what happens when a car encounters a hill that is so steep ($6°$) that the throttle saturates when the cruise controller attempts to maintain speed. When encountering the slope at time $t = 5$, the velocity decreases and the throttle increases to generate more torque. However, the torque required is so large that the throttle saturates. The error decreases slowly because the torque generated by the engine is just a little larger than the torque required to compensate for gravity. The error is large and the integral continues to build up until the error reaches zero at time 30, but the controller output is still larger than the saturation limit and the actuator remains saturated. The integral term starts to decrease, and at time 45 and the velocity settles quickly to the desired value. Notice that it takes considerable time before the controller output comes into the range where it does not saturate, resulting in a large overshoot.   ∇

There are many methods to avoid windup. One method is illustrated in Figure 10.11: the system has an extra feedback path that is generated by measuring the actual actuator output, or the output of a mathematical model of the saturating

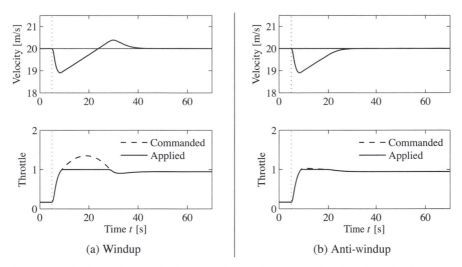

Figure 10.10: Simulation of PI cruise control with windup (a) and anti-windup (b). The figure shows the speed v and the throttle u for a car that encounters a slope that is so steep that the throttle saturates. The controller output is a dashed line. The controller parameters are $k_p = 0.5$ and $k_i = 0.1$. The anti-windup compensator eliminates the overshoot by preventing the error for building up in the integral term of the controller.

actuator, and forming an error signal e_s as the difference between the output of the controller v and the actuator output u. The signal e_s is fed to the input of the integrator through gain k_t. The signal e_s is zero when there is no saturation and the extra feedback loop has no effect on the system. When the actuator saturates, the signal e_s is fed back to the integrator in such a way that e_s goes toward zero. This implies that controller output is kept close to the saturation limit. The controller output will then change as soon as the error changes sign and integral windup is avoided.

The rate at which the controller output is reset is governed by the feedback gain k_t; a large value of k_t gives a short reset time. The parameter k_t cannot be too large because measurement noise can then cause an undesirable reset. A reasonable choice is to choose k_t as a fraction of $1/T_i$. We illustrate how integral windup can be avoided by investigating the cruise control system.

Example 10.6 Cruise control with anti-windup
Figure 10.10b shows what happens when a controller with anti-windup is applied to the system simulated in Figure 10.10a. Because of the feedback from the actuator model, the output of the integrator is quickly reset to a value such that the controller output is at the saturation limit. The behavior is drastically different from that in Figure 10.10a and the large overshoot is avoided. The tracking gain is $k_t = 2$ in the simulation. ∇

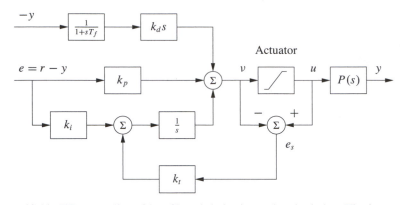

Figure 10.11: PID controller with a filtered derivative and anti-windup. The input to the integrator ($1/s$) consists of the error term plus a "reset" based on input saturation. If the actuator is not saturated, then $e_s = u - v$, otherwise e_s will decrease the integrator input to prevent windup.

10.5 Implementation

There are many practical issues that have to be considered when implementing PID controllers. They have been developed over time based on practical experience. In this section we consider some of the most common. Similar considerations also apply to other types of controllers.

Filtering the Derivative

A drawback with derivative action is that an ideal derivative has high gain for high-frequency signals. This means that high-frequency measurement noise will generate large variations in the control signal. The effect of measurement noise may be reduced by replacing the term $k_d s$ by $k_d s/(1 + sT_f)$, which can be interpreted as an ideal derivative of a low-pass filtered signal. For small s the transfer function is approximately $k_d s$ and for large s it is equal to k_d/T_f. The approximation acts as a derivative for low-frequency signals and as a constant gain for high-frequency signals. The filtering time is chosen as $T_f = (k_d/k)/N$, with N in the range 2–20. Filtering is obtained automatically if the derivative is implemented by taking the difference between the signal and its filtered version as shown in Figure 10.3b (see equation (10.5)).

Instead of filtering just the derivative, it is also possible to use an ideal controller and filter the measured signal. The transfer function of such a controller with a filter is then

$$C(s) = k_p \left(1 + \frac{1}{sT_i} + sT_d\right) \frac{1}{1 + sT_f + (sT_f)^2/2}, \qquad (10.13)$$

where a second-order filter is used.

Setpoint Weighting

Figure 10.1 shows two configurations of a PID controller. The system in Figure 10.1a has a controller with *error feedback* where proportional, integral and derivative action acts on the error. In the simulation of PID controllers in Figure 10.2c there is a large initial peak in the control signal, which is caused by the derivative of the reference signal. The peak can be avoided by using the controller in Figure 10.1b, where proportional and derivative action acts only on the process output. An intermediate form is given by

$$ u = k_p \big(\beta r - y\big) + k_i \int_0^\infty \big(r(\tau) - y(\tau)\big)\, d\tau + k_d \Big(\gamma \frac{dr}{dt} - \frac{dy}{dt}\Big), \qquad (10.14) $$

where the proportional and derivative actions act on fractions β and γ of the reference. Integral action has to act on the error to make sure that the error goes to zero in steady state. The closed loop systems obtained for different values of β and γ respond to load disturbances and measurement noise in the same way. The response to reference signals is different because it depends on the values of β and γ, which are called *reference weights* or *setpoint weights*. We illustrate the effect of setpoint weighting by an example.

Example 10.7 Cruise control with setpoint weighting

Consider the PI controller for the cruise control system derived in Example 10.3. Figure 10.12 shows the effect of setpoint weighting on the response of the system to a reference signal. With $\beta = 1$ (error feedback) there is an overshoot in velocity and the control signal (throttle) is initially close to the saturation limit. There is no overshoot with $\beta = 0$ and the control signal is much smaller, clearly a much better drive comfort. The frequency responses gives another view of the same effect. The parameter β is typically in the range 0–1, and γ is normally zero to avoid large transients in the control signal when the reference is changed. ∇

The controller given by equation (10.14) is a special case of the general controller structure having two degrees of freedom, which was discussed in Section 7.5.

Implementation Based on Operational Amplifiers

PID controllers have been implemented in different technologies. Figure 10.13 shows how PI and PID controllers can be implemented by feedback around operational amplifiers.

To show that the circuit in Figure 10.13b is a PID controller we will use the approximate relation between the input voltage e and the output voltage u of the operational amplifier derived in Example 8.3,

$$ u = -\frac{Z_2}{Z_1} e. $$

In this equation Z_1 is the impedance between the negative input of the amplifier and the input voltage e, and Z_2 is the impedance between the zero input of the amplifier

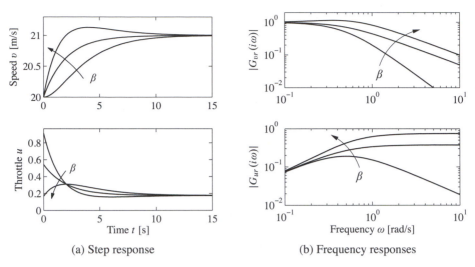

Figure 10.12: Time and frequency responses for PI cruise control with setpoint weighting. Step responses are shown in (a), and the gain curves of the frequency responses in (b). The controller gains are $k_p = 0.74$ and $k_i = 0.19$. The setpoint weights are $\beta = 0, 0.5$ and 1, and $\gamma = 0$.

and the output voltage u. The impedances are given by

$$Z_1(s) = \frac{R_1}{1 + R_1 C_1 s}, \qquad Z_2(s) = R_2 + \frac{1}{C_2 s},$$

and we find the following relation between the input voltage e and the output voltage u:

$$u = -\frac{Z_2}{Z_1} e = -\frac{R_2}{R_1} \frac{(1 + R_1 C_1 s)(1 + R_2 C_2 s)}{R_2 C_2 s} e.$$

This is the input/output relation for a PID controller of the form (10.1) with parameters

$$k_p = \frac{R_2}{R_1}, \qquad T_i = R_2 C_1, \qquad T_d = R_1 C_1.$$

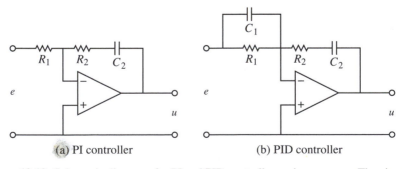

Figure 10.13: Schematic diagrams for PI and PID controllers using op amps. The circuit in (a) uses a capacitor in the feedback path to store the integral of the error. The circuit in (b) adds a filter on the input to provide derivative action.

The corresponding results for a PI controller are obtained by setting $C_1 = 0$ (removing the capacitor).

Computer Implementation

In this section we briefly describe how a PID controller may be implemented using a computer. The computer typically operates periodically, with signals from the sensors sampled and converted to digital form by the A/D converter, and the control signal computed and then converted to analog form for the actuators. The sequence of operation is as follows:

1. Wait for clock interrupt
2. Read input from sensor
3. Compute control signal
4. Send output to the actuator
5. Update controller variables
6. Repeat

Notice that an output is sent to the actuators as soon as it is available. The time delay is minimized by making the calculations in step 3 as short as possible and performing all updates after the output is commanded. This simple way of reducing the latency is, unfortunately, seldom used in commercial systems.

As an illustration we consider the PID controller in Figure 10.11, which has a filtered derivative, setpoint weighting and protection against integral windup. The controller is a continuous-time dynamical system. To implement it using a computer, the continuous-time system has to be approximated by a discrete-time system.

A block diagram of a PID controller with anti-windup is shown in Figure 10.11. The signal v is the sum of the proportional, integral and derivative terms, and the controller output is $u = \text{sat}(v)$, where sat is the saturation function that models the actuator. The proportional term $k_p(\beta r - y)$ is implemented simply by replacing the continuous variables with their sampled versions. Hence

$$P(t_k) = k_p \left(\beta r(t_k) - y(t_k) \right), \tag{10.15}$$

where $\{t_k\}$ denotes the sampling instants, i.e., the times when the computer reads its input. We let h represent the sampling time, so that $t_{k+1} = t_k + h$. The integral term is obtained by approximating the integral with a sum,

$$I(t_{k+1}) = I(t_k) + k_i h\, e(t_k) + \frac{h}{T_t} \left(\text{sat}(v) - v \right), \tag{10.16}$$

where $T_t = h/k_t$ represents the anti-windup term. The filtered derivative term D is given by the differential equation

$$T_f \frac{dD}{dt} + D = -k_d \dot{y}.$$

Approximating the derivative with a backward difference gives

$$T_f \frac{D(t_k) - D(t_{k-1})}{h} + D(t_k) = -k_d \frac{y(t_k) - y(t_{k-1})}{h},$$

which can be rewritten as

$$D(t_k) = \frac{T_f}{T_f + h} D(t_{k-1}) - \frac{k_d}{T_f + h} \left(y(t_k) - y(t_{k-1}) \right). \tag{10.17}$$

The advantage of using a backward difference is that the parameter $T_f/(T_f + h)$ is nonnegative and less than 1 for all $h > 0$, which guarantees that the difference equation is stable. Reorganizing equations (10.15)–(10.17), the PID controller can be described by the following pseudocode:

```
% Precompute controller coefficients
bi=ki*h
ad=Tf/(Tf+h)
bd=kd/(Tf+h)
br=h/Tt

% Control algorithm - main loop
while (running) {
   r=adin(ch1)              % read setpoint from ch1
   y=adin(ch2)              % read process variable from ch2
   P=kp*(b*r-y)             % compute proportional part
   D=ad*D-bd*(y-yold)       % update derivative part
   v=P+I+D                  % compute temporary output
   u=sat(v,ulow,uhigh)      % simulate actuator saturation
   daout(ch1)              % set analog output ch1
   I=I+bi*(r-y)+br*(u-v)   % update integral
   yold=y                  % update old process output
   sleep(h)               % wait until next update interval
}
```

Precomputation of the coefficients `bi`, `ad`, `bd` and `br` saves computer time in the main loop. These calculations have to be done only when controller parameters are changed. The main loop is executed once every sampling period. The program has three states: `yold`, `I`, and `D`. One state variable can be eliminated at the cost of less readable code. The latency between reading the analog input and setting the analog output consists of four multiplications, four additions and evaluation of the `sat` function. All computations can be done using fixed-point calculations if necessary. Notice that the code computes the filtered derivative of the process output and that it has setpoint weighting and anti-windup protection.

10.6 Further Reading

The history of PID control is very rich and stretches back to the beginning of the foundation of control theory. Very readable treatments are given by Bennett [28, 29] and Mindel [152]. The Ziegler–Nichols rules for tuning PID controllers, first presented in 1942 [210], were developed based on extensive experiments with pneumatic simulators and Vannevar Bush's differential analyzer at MIT. An interesting view of the development of the Ziegler–Nichols rules is given in an interview with Ziegler [39]. An industrial perspective on PID control is given in [33], [180] and

[205] and in the paper [58] cited in the beginning of this chapter. A comprehensive presentation of PID control is given in [16]. Interactive learning tools for PID control can be downloaded from http://www.calerga.com/contrib.

Exercises

10.1 (Ideal PID controllers) Consider the systems represented by the block diagrams in Figure 10.1. Assume that the process has the transfer function $P(s) = b/(s+a)$ and show that the transfer functions from r to y are

$$(a) \quad G_{yr}(s) = \frac{bk_d s^2 + bk_p s + bk_i}{(1 + bk_d)s^2 + (a + bk_p)s + bk_i},$$

$$(b) \quad G_{yr}(s) = \frac{bk_i}{(1 + bk_d)s^2 + (a + bk_p)s + bk_i}.$$

Pick some parameters and compare the step responses of the systems.

10.2 Consider a second-order process with the transfer function

$$P(s) = \frac{b}{s^2 + a_1 s + a_2}.$$

The closed loop system with a PI controller is a third-order system. Show that it is possible to position the closed loop poles as long as the sum of the poles is $-a_1$. Give equations for the parameters that give the closed loop characteristic polynomial

$$(s + \alpha_0)(s^2 + 2\zeta_0\omega_0 s + \omega_0^2).$$

10.3 Consider a system with the transfer function $P(s) = (s + 1)^{-2}$. Find an integral controller that gives a closed loop pole at $s = -a$ and determine the value of a that maximizes the integral gain. Determine the other poles of the system and judge if the pole can be considered dominant. Compare with the value of the integral gain given by equation (10.6).

10.4 (Ziegler–Nichols tuning) Consider a system with transfer function $P(s) = e^{-s}/s$. Determine the parameters of P, PI and PID controllers using Ziegler–Nichols step and frequency response methods. Compare the parameter values obtained by the different rules and discuss the results.

10.5 (Vehicle steering) Design a proportional-integral controller for the vehicle steering system that gives the closed loop characteristic polynomial

$$s^3 + 2\omega_0 s^2 + 2\omega_0 s + \omega_0^3.$$

10.6 (Congestion control) A simplified flow model for TCP transmission is derived in [101, 137]. The linearized dynamics are modeled by the transfer function

$$G_{qp}(s) = \frac{b}{(s + a_1)(s + a_2)} e^{-s\tau_e},$$

which describes the dynamics relating the expected queue length q to the expected packet drop p. The parameters are given by $a_1 = 2N^2/(c\tau_e^2)$, $a_2 = 1/\tau_e$ and $b = c^2/(2N)$. The parameter c is the bottleneck capacity, N is the number of sources feeding the link and τ_e is the round-trip delay time. Use the parameter values $N = 75$ sources, $C = 1250$ packets/s and $\tau_e = 0.15$ and find the parameters of a PI controller using one of the Ziegler–Nichols rules and the corresponding improved rule. Simulate the responses of the closed loop systems obtained with the PI controllers.

10.7 (Motor drive) Consider the model of the motor drive in Exercise 2.10. Develop an approximate second-order model of the system and use it to design an ideal PD controller that gives a closed loop system with eigenvalues in $\zeta\omega_0 \pm i\omega_0\sqrt{1 - \zeta^2}$. Add low-pass filtering as shown in equation (10.13) and explore how large ω_0 can be made while maintaining a good stability margin. Simulate the closed loop system with the chosen controller and compare the results with the controller based on state feedback in Exercise 6.11.

10.8 Consider the system in Exercise 10.7 investigate what happens if the second-order filtering of the derivative is replace by a first-order filter.

10.9 (Tuning rules) Apply the Ziegler–Nichols and the modified tuning rules to design PI controllers for systems with the transfer functions

$$P_1 = \frac{e^{-s}}{s}, \qquad P_2 = \frac{e^{-s}}{s + 1}, \qquad P_3 = e^{-s}.$$

Compute the stability margins and explore any patterns.

10.10 (Windup and anti-windup) Consider a PI controller of the form $C(s) = 1 + 1/s$ for a process with input that saturates when $|u| > 1$, and whose linear dynamics are given by the transfer function $P(s) = 1/s$. Simulate the response of the system to step changes in the reference signal of magnitude $1, 2$ and 3. Repeat the simulation when the windup protection scheme in Figure 10.11 is used.

10.11 (Windup protection by conditional integration) Many methods have been proposed to avoid integrator windup. One method called *conditional integration* is to update the integral only when the error is sufficiently small. To illustrate this method we consider a system with PI control described by

$$\frac{dx_1}{dt} = u, \qquad u = \mathrm{sat}_{u_0}(k_p e + k_i x_2), \qquad \frac{dx_2}{dt} = \begin{cases} e & \text{if } |e| < e_0 \\ 0 & \text{if } |e| \geq e_0, \end{cases}$$

where $e = r - x$. Plot the phase portrait of the system for the parameter values $k_p = 1$, $k_i = 1$, $u_0 = 1$ and $e_0 = 1$ and discuss the properties of the system. The example illustrates the difficulties of introducing ad hoc nonlinearities without careful analysis.

Chapter Eleven
Frequency Domain Design

Sensitivity improvements in one frequency range must be paid for with sensitivity deteriorations in another frequency range, and the price is higher if the plant is open-loop unstable. This applies to every controller, no matter how it was designed.

Gunter Stein in the inaugural IEEE Bode Lecture, 1989 [185].

In this chapter we continue to explore the use of frequency domain techniques with a focus on the design of feedback systems. We begin with a more thorough description of the performance specifications for control systems and then introduce the concept of "loop shaping" as a mechanism for designing controllers in the frequency domain. We also introduce some fundamental limitations to performance for systems with time delays and right half-plane poles and zeros.

11.1 Sensitivity Functions

In the previous chapter, we considered the use of proportional-integral-derivative (PID) feedback as a mechanism for designing a feedback controller for a given process. In this chapter we will expand our approach to include a richer repertoire of tools for shaping the frequency response of the closed loop system.

One of the key ideas in this chapter is that we can design the behavior of the closed loop system by focusing on the open loop transfer function. This same approach was used in studying stability using the Nyquist criterion: we plotted the Nyquist plot for the *open* loop transfer function to determine the stability of the *closed* loop system. From a design perspective, the use of loop analysis tools is very powerful: since the loop transfer function is $L = PC$, if we can specify the desired performance in terms of properties of L, we can directly see the impact of changes in the controller C. This is much easier, for example, than trying to reason directly about the tracking response of the closed loop system, whose transfer function is given by $G_{yr} = PC/(1 + PC)$.

We will start by investigating some key properties of the feedback loop. A block diagram of a basic feedback loop is shown in Figure 11.1. The system loop is composed of two components: the process and the controller. The controller itself has two blocks: the feedback block C and the feedforward block F. There are two disturbances acting on the process, the load disturbance d and the measurement noise n. The load disturbance represents disturbances that drive the process away from its desired behavior, while the measurement noise represents disturbances that corrupt information about the process given by the sensors. In the figure, the load

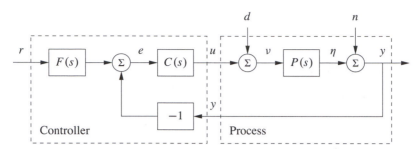

Figure 11.1: Block diagram of a basic feedback loop with two degrees of freedom. The controller has a feedback block C and a feedforward block F. The external signals are the reference signal r, the load disturbance d and the measurement noise n. The process output is η, and the control signal is u.

disturbance is assumed to act on the process input. This is a simplification since disturbances often enter the process in many different ways, but it allows us to streamline the presentation without significant loss of generality.

The process output η is the real variable that we want to control. Control is based on the measured signal y, where the measurements are corrupted by measurement noise n. The process is influenced by the controller via the control variable u. The process is thus a system with three inputs—the control variable u, the load disturbance d and the measurement noise n—and one output—the measured signal y. The controller is a system with two inputs and one output. The inputs are the measured signal y and the reference signal r, and the output is the control signal u. Note that the control signal u is an input to the process and the output of the controller, and that the measured signal y is the output of the process and an input to the controller.

The feedback loop in Figure 11.1 is influenced by three external signals, the reference r, the load disturbance d and the measurement noise n. Any of the remaining signals can be of interest in controller design, depending on the particular application. Since the system is linear, the relations between the inputs and the interesting signals can be expressed in terms of the transfer functions. The following relations are obtained from the block diagram in Figure 11.1:

$$
\begin{bmatrix} y \\ \eta \\ v \\ u \\ e \end{bmatrix} = \begin{bmatrix} \dfrac{PCF}{1+PC} & \dfrac{P}{1+PC} & \dfrac{1}{1+PC} \\[2ex] \dfrac{PCF}{1+PC} & \dfrac{P}{1+PC} & \dfrac{-PC}{1+PC} \\[2ex] \dfrac{CF}{1+PC} & \dfrac{1}{1+PC} & \dfrac{-C}{1+PC} \\[2ex] \dfrac{CF}{1+PC} & \dfrac{-PC}{1+PC} & \dfrac{-C}{1+PC} \\[2ex] \dfrac{F}{1+PC} & \dfrac{-P}{1+PC} & \dfrac{-1}{1+PC} \end{bmatrix} \begin{bmatrix} r \\ d \\ n \end{bmatrix}. \tag{11.1}
$$

In addition, we can write the transfer function for the error between the reference

r and the output η (not an explicit signal in the diagram), which satisfies

$$\epsilon = r - \eta = \left(1 - \frac{PCF}{1 + PC}\right)r + \frac{-P}{1 + PC}d + \frac{PC}{1 + PC}n.$$

There are several interesting conclusions we can draw from these equations. First we can observe that several transfer functions are the same and that the majority of the relations are given by the following set of six transfer functions, which we call the *Gang of Six*:

$$TF = \frac{PCF}{1 + PC}, \qquad T = \frac{PC}{1 + PC}, \qquad PS = \frac{P}{1 + PC},$$
$$CFS = \frac{CF}{1 + PC}, \qquad CS = \frac{C}{1 + PC}, \qquad S = \frac{1}{1 + PC}. \tag{11.2}$$

The transfer functions in the first column give the response of the process output and control signal to the reference signal. The second column gives the response of the control variable to the load disturbance and the noise, and the final column gives the response of the process output to those two inputs. Notice that only four transfer functions are required to describe how the system reacts to load disturbances and measurement noise, and that two additional transfer functions are required to describe how the system responds to reference signals.

The linear behavior of the system is determined by the six transfer functions in equation (11.2), and specifications can be expressed in terms of these transfer functions. The special case when $F = 1$ is called a system with (pure) error feedback. In this case all control actions are based on feedback from the error only and the system is completely characterized by four transfer functions, namely, the four rightmost transfer functions in equation (11.2), which have specific names:

$$S = \frac{1}{1 + PC} \quad \text{sensitivity function} \qquad PS = \frac{P}{1 + PC} \quad \text{load sensitivity function}$$

$$T = \frac{PC}{1 + PC} \quad \text{complementary sensitivity function} \qquad CS = \frac{C}{1 + PC} \quad \text{noise sensitivity function} \tag{11.3}$$

These transfer functions and their equivalent systems are called the *Gang of Four*. The load sensitivity function is sometimes called the input sensitivity function and the noise sensitivity function is sometimes called the output sensitivity function. These transfer functions have many interesting properties that will be discussed in detail in the rest of the chapter. Good insight into these properties is essential in understanding the performance of feedback systems for the purposes of both analysis and design.

Analyzing the Gang of Six, we find that the feedback controller C influences the effects of load disturbances and measurement noise. Notice that measurement noise enters the process via the feedback. In Section 12.2 it will be shown that the controller influences the sensitivity of the closed loop to process variations.

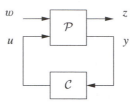

Figure 11.2: A more general representation of a feedback system. The process input u represents the control signal, which can be manipulated, and the process input w represents other signals that influence the process. The process output y is the vector of measured variables and z are other signals of interest.

The feedforward part F of the controller influences only the response to command signals.

In Chapter 9 we focused on the loop transfer function, and we found that its properties gave useful insights into the properties of a system. To make a proper assessment of a feedback system it is necessary to consider the properties of all the transfer functions (11.2) in the Gang of Six or the Gang of Four, as illustrated in the following example.

Example 11.1 The loop transfer function gives only limited insight
Consider a process with the transfer function $P(s) = 1/(s - a)$ controlled by a PI controller with error feedback having the transfer function $C(s) = k(s - a)/s$. The loop transfer function is $L = k/s$, and the sensitivity functions are

$$T = \frac{PC}{1 + PC} = \frac{k}{s + k}, \qquad PS = \frac{P}{1 + PC} = \frac{s}{(s - a)(s + k)},$$

$$CS = \frac{C}{1 + PC} = \frac{k(s - a)}{s + k}, \qquad S = \frac{1}{1 + PC} = \frac{s}{s + k}.$$

Notice that the factor $s - a$ is canceled when computing the loop transfer function and that this factor also does not appear in the sensitivity function or the complementary sensitivity function. However, cancellation of the factor is very serious if $a > 0$ since the transfer function PS relating load disturbances to process output is then unstable. In particular, a small disturbance d can lead to an unbounded output, which is clearly not desirable. ∇

The system in Figure 11.1 represents a special case because it is assumed that the load disturbance enters at the process input and that the measured output is the sum of the process variable and the measurement noise. Disturbances can enter in many different ways, and the sensors may have dynamics. A more abstract way to capture the general case is shown in Figure 11.2, which has only two blocks representing the process ($\mathcal{P}$) and the controller ($\mathcal{C}$). The process has two inputs, the control signal u and a vector of disturbances w, and two outputs, the measured signal y and a vector of signals z that is used to specify performance. The system in Figure 11.1 can be captured by choosing $w = (d, n)$ and $z = (\eta, v, e, \epsilon)$. The process transfer function $\mathcal{P}$ is a 4×3 matrix, and the controller transfer function $\mathcal{C}$ is a 1×2 matrix; compare with Exercise 11.3.

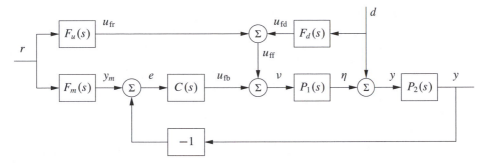

Figure 11.3: Block diagram of a system with feedforward compensation for improved response to reference signals and measured disturbances (2 DOF system). Three feedforward elements are present: $F_m(s)$ sets the desired output value, $F_u(s)$ generates the feedforward command u_{fr} and $F_d(s)$ attempts to cancel disturbances.

Processes with multiple inputs and outputs can also be considered by regarding u and y as vectors. Representations at these higher levels of abstraction are useful for the development of theory because they make it possible to focus on fundamentals and to solve general problems with a wide range of applications. However, care must be exercised to maintain the coupling to the real-world control problems we intend to solve.

11.2 Feedforward Design

Most of our analysis and design tools up to this point have focused on the role of feedback and its effect on the dynamics of the system. Feedforward is a simple and powerful technique that complements feedback. It can be used both to improve the response to reference signals and to reduce the effect of measurable disturbances. Feedforward compensation admits perfect elimination of disturbances, but it is much more sensitive to process variations than feedback compensation. A general scheme for feedforward was discussed in Section 7.5 using Figure 7.10. A simple form of feedforward for PID controllers was discussed in Section 10.5. The controller in Figure 11.1 also has a feedforward block to improve response to command signals. An alternative version of feedforward is shown in Figure 11.3, which we will use in this section to understand some of the trade-offs between feedforward and feedback.

Controllers with two degrees of freedom (feedforward and feedback) have the advantage that the response to reference signals can be designed independently of the design for disturbance attenuation and robustness. We will first consider the response to reference signals, and we will therefore initially assume that the load disturbance d is zero. Let F_m represent the ideal response of the system to reference signals. The feedforward compensator is characterized by the transfer functions F_u and F_m. When the reference is changed, the transfer function F_u generates the signal u_{fr}, which is chosen to give the desired output when applied as input to the process. Under ideal conditions the output y is then equal to y_m, the error signal is zero and there will be no feedback action. If there are disturbances or modeling errors, the signals y_m and y will differ. The feedback then attempts to bring the error to zero.

To make a formal analysis, we compute the transfer function from reference input to process output:

$$G_{yr}(s) = \frac{P(CF_m + F_u)}{1 + PC} = F_m + \frac{PF_u - F_m}{1 + PC}, \qquad (11.4)$$

where $P = P_2 P_1$. The first term represents the desired transfer function. The second term can be made small in two ways. Feedforward compensation can be used to make $PF_u - F_m$ small, or feedback compensation can be used to make $1 + PC$ large. Perfect feedforward compensation is obtained by choosing

$$F_u = \frac{F_m}{P}. \qquad (11.5)$$

Design of feedforward using transfer functions is thus a very simple task. Notice that the feedforward compensator F_u contains an inverse model of the process dynamics.

Feedback and feedforward have different properties. Feedforward action is obtained by matching two transfer functions, requiring precise knowledge of the process dynamics, while feedback attempts to make the error small by dividing it by a large quantity. For a controller having integral action, the loop gain is large for low frequencies, and it is thus sufficient to make sure that the condition for ideal feedforward holds at higher frequencies. This is easier than trying to satisfy the condition (11.5) for all frequencies.

We will now consider reduction of the effects of the load disturbance d in Figure 11.3 by feedforward control. We assume that the disturbance signal is measured and that the disturbance enters the process dynamics in a known way (captured by P_1 and P_2). The effect of the disturbance can be reduced by feeding the measured signal through a dynamical system with the transfer function F_d. Assuming that the reference r is zero, we can use block diagram algebra to find that the transfer function from the disturbance to the process output is

$$G_{yd} = \frac{P_2(1 + F_d P_1)}{1 + PC}, \qquad (11.6)$$

where $P = P_1 P_2$. The effect of the disturbance can be reduced by making $1 + F_d P_1$ small (feedforward) or by making $1 + PC$ large (feedback). Perfect compensation is obtained by choosing

$$F_d = -P_1^{-1}, \qquad (11.7)$$

requiring inversion of the transfer function P_1.

As in the case of reference tracking, disturbance attenuation can be accomplished by combining feedback and feedforward control. Since low-frequency disturbances can be eliminated by feedback, we require the use of feedforward only for high-frequency disturbances, and the transfer function F_d in equation (11.7) can then be computed using an approximation of P_1 for high frequencies.

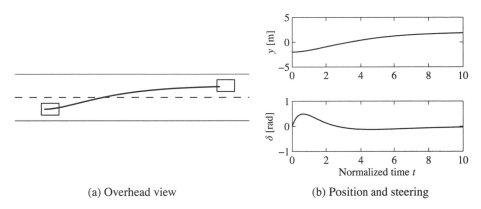

(a) Overhead view   (b) Position and steering

Figure 11.4: Feedforward control for vehicle steering. The plot on the left shows the trajectory generated by the controller for changing lanes. The plots on the right show the lateral deviation y (top) and the steering angle δ (bottom) for a smooth lane change control using feedforward (based on the linearized model).

Equations (11.5) and (11.7) give analytic expressions for the feedforward compensator. To obtain a transfer function that can be implemented without difficulties we require that the feedforward compensator be stable and that it does not require differentiation. Therefore there may be constraints on possible choices of the desired response F_m, and approximations are needed if the process has zeros in the right half-plane or time delays.

Example 11.2 Vehicle steering
A linearized model for vehicle steering was given in Example 6.4. The normalized transfer function from steering angle δ to lateral deviation y is $P(s) = (\gamma s + 1)/s^2$. For a lane transfer system we would like to have a nice response without overshoot, and we therefore choose the desired response as $F_m(s) = a^2/(s + a)^2$, where the response speed or aggressiveness of the steering is governed by the parameter a. Equation (11.5) gives

$$F_u = \frac{F_m}{P} = \frac{a^2 s^2}{(\gamma s + 1)(s + a)^2},$$

which is a stable transfer function as long as $\gamma > 0$. Figure 11.4 shows the responses of the system for $a = 0.5$. The figure shows that a lane change is accomplished in about 10 vehicle lengths with smooth steering angles. The largest steering angle is slightly larger than 0.1 rad (6°). Using the scaled variables, the curve showing lateral deviations (y as a function of t) can also be interpreted as the vehicle path (y as a function of x) with the vehicle length as the length unit.   ∇

A major advantage of controllers with two degrees of freedom that combine feedback and feedforward is that the control design problem can be split in two parts. The feedback controller C can be designed to give good robustness and effective disturbance attenuation, and the feedforward part can be designed independently to give the desired response to command signals.

11.3 Performance Specifications

A key element of the control design process is how we specify the desired performance of the system. It is also important for users to understand performance specifications so that they know what to ask for and how to test a system. Specifications are often given in terms of robustness to process variations and responses to reference signals and disturbances. They can be given in terms of both time and frequency responses. Specifications for the step response to reference signals were given in Figure 5.9 in Section 5.3 and in Section 6.3. Robustness specifications based on frequency domain concepts were provided in Section 9.3 and will be considered further in Chapter 12. The specifications discussed previously were based on the loop transfer function. Since we found in Section 11.1 that a single transfer function did not always characterize the properties of the closed loop completely, we will give a more complete discussion of specifications in this section, based on the full Gang of Six.

The transfer function gives a good characterization of the linear behavior of a system. To provide specifications it is desirable to capture the characteristic properties of a system with a few parameters. Common features for time responses are overshoot, rise time and settling time, as shown in Figure 5.9. Common features of frequency responses are resonant peak, peak frequency, gain crossover frequency and bandwidth. A *resonant peak* is a maximum of the gain, and the peak frequency is the corresponding frequency. The *gain crossover frequency* is the frequency where the open loop gain is equal one. The *bandwidth* is defined as the frequency range where the closed loop gain is $1/\sqrt{2}$ of the low-frequency gain (low-pass), mid-frequency gain (band-pass) or high-frequency gain (high-pass). There are interesting relations between specifications in the time and frequency domains. Roughly speaking, the behavior of time responses for short times is related to the behavior of frequency responses at high frequencies, and vice versa. The precise relations are not trivial to derive.

Response to Reference Signals

Consider the basic feedback loop in Figure 11.1. The response to reference signals is described by the transfer functions $G_{yr} = PCF/(1 + PC)$ and $G_{ur} = CF/(1 + PC)$ ($F = 1$ for systems with error feedback). Notice that it is useful to consider both the response of the output and that of the control signal. In particular, the control signal response allows us to judge the magnitude and rate of the control signal required to obtain the output response.

Example 11.3 Third-order system
Consider a process with the transfer function $P(s) = (s + 1)^{-3}$ and a PI controller with error feedback having the gains $k_p = 0.6$ and $k_i = 0.5$. The responses are illustrated in Figure 11.5. The solid lines show results for a proportional-integral (PI) controller with error feedback. The dashed lines show results for a controller with feedforward designed to give the transfer function $G_{yr} = (0.5s + 1)^{-3}$. Looking at the time responses, we find that the controller with feedforward gives a faster

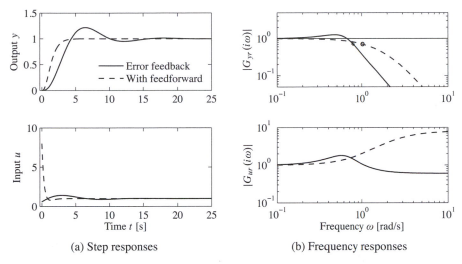

Figure 11.5: Reference signal responses. The responses in process output y and control signal u to a unit step in the reference signal r are shown in (a), and the gain curves of G_{yr} and G_{ur} are shown in (b). Results with PI control with error feedback are shown by solid lines, and the dashed lines show results for a controller with a feedforward compensator.

response with no overshoot. However, much larger control signals are required to obtain the fast response. The largest value of the control signal is 8, compared to 1.2 for the regular PI controller. The controller with feedforward has a larger bandwidth (marked with ○) and no resonant peak. The transfer function G_{ur} also has higher gain at high frequencies. ▽

Response to Load Disturbances and Measurement Noise

A simple criterion for disturbance attenuation is to compare the output of the closed loop system in Figure 11.1 with the output of the corresponding open loop system obtained by setting $C = 0$. If we let the disturbances for the open and closed loop systems be identical, the output of the closed loop system is then obtained simply by passing the open loop output through a system with the transfer function S. The sensitivity function tells how the variations in the output are influenced by feedback (Exercise 11.7). Disturbances with frequencies such that $|S(i\omega)| < 1$ are attenuated, but disturbances with frequencies such that $|S(i\omega)| > 1$ are amplified by feedback. The maximum sensitivity M_s, which occurs at the frequency ω_{ms}, is thus a measure of the largest amplification of the disturbances. The maximum magnitude of $1/(1 + L)$ is also the minimum of $|1 + L|$, which is precisely the stability margin s_m defined in Section 9.3, so that $M_s = 1/s_m$. The maximum sensitivity is therefore also a robustness measure.

If the sensitivity function is known, the potential improvements by feedback can be evaluated simply by recording a typical output and filtering it through the sensitivity function. A plot of the gain curve of the sensitivity function is a good way to make an assessment of the disturbance attenuation. Since the sensitivity function

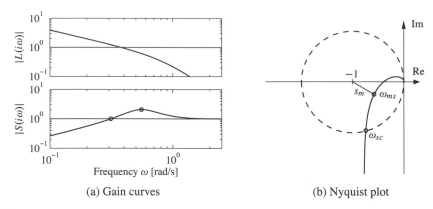

(a) Gain curves   (b) Nyquist plot

Figure 11.6: Graphical interpretation of the sensitivity function. Gain curves of the loop transfer function and the sensitivity function (a) can be used to calculate the properties of the sensitivity function through the relation $S = 1/(1 + L)$. The sensitivity crossover frequency ω_{sc} and the frequency ω_{ms} where the sensitivity has its largest value are indicated in the sensitivity plot. The Nyquist plot (b) shows the same information in a different form. All points inside the dashed circle have sensitivities greater than 1.

depends only on the loop transfer function, its properties can also be visualized graphically using the Nyquist plot of the loop transfer function. This is illustrated in Figure 11.6. The complex number $1 + L(i\omega)$ can be represented as the vector from the point -1 to the point $L(i\omega)$ on the Nyquist curve. The sensitivity is thus less than 1 for all points outside a circle with radius 1 and center at -1. Disturbances with frequencies in this range are attenuated by the feedback.

The transfer function G_{yd} from load disturbance d to process output y for the system in Figure 11.1 is

$$G_{yd} = \frac{P}{1 + PC} = PS = \frac{T}{C}. \tag{11.8}$$

Since load disturbances typically have low frequencies, it is natural to focus on the behavior of the transfer function at low frequencies. For a system with $P(0) \neq 0$ and a controller with integral action, the controller gain goes to infinity for small frequencies and we have the following approximation for small s:

$$G_{yd} = \frac{T}{C} \approx \frac{1}{C} \approx \frac{s}{k_i}, \tag{11.9}$$

where k_i is the integral gain. Since the sensitivity function S goes to 1 for large s, we have the approximation $G_{yd} \approx P$ for high frequencies.

Measurement noise, which typically has high frequencies, generates rapid variations in the control variable that are detrimental because they cause wear in many actuators and can even saturate an actuator. It is thus important to keep variations in the control signal due to measurement noise at reasonable levels—a typical requirement is that the variations are only a fraction of the span of the control signal. The variations can be influenced by filtering and by proper design of the high-frequency

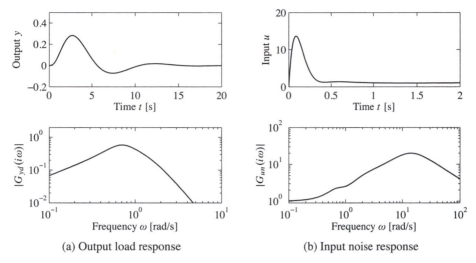

Figure 11.7: Disturbance responses. The time and frequency responses of process output y to load disturbance d are shown in (a) and the responses of control signal u to measurement noise n are shown in (b).

properties of the controller.

The effects of measurement noise are captured by the transfer function from the measurement noise to the control signal,

$$-G_{un} = \frac{C}{1 + PC} = CS = \frac{T}{P}. \tag{11.10}$$

The complementary sensitivity function is close to 1 for low frequencies ($\omega < \omega_{gc}$), and G_{un} can be approximated by $-1/P$. The sensitivity function is close to 1 for high frequencies ($\omega > \omega_{gc}$), and G_{un} can be approximated by $-C$.

Example 11.4 Third-order system

Consider a process with the transfer function $P(s) = (s+1)^{-3}$ and a proportional-integral-derivative (PID) controller with gains $k_p = 0.6$, $k_i = 0.5$ and $k_d = 2.0$. We augment the controller using a second-order noise filter with $T_f = 0.1$, so that its transfer function is

$$C(s) = \frac{k_d s^2 + k_p s + k_i}{s(s^2 T_f^2/2 + s T_f + 1)}.$$

The system responses are illustrated in Figure 11.7. The response of the output to a step in the load disturbance in the top part of Figure 11.7a has a peak of 0.28 at time $t = 2.73$ s. The frequency response in Figure 11.7a shows that the gain has a maximum of 0.58 at $\omega = 0.7$ rad/s.

The response of the control signal to a step in measurement noise is shown in Figure 11.7b. The high-frequency roll-off of the transfer function $G_{un}(i\omega)$ is due to filtering; without it the gain curve in Figure 11.7b would continue to rise after 20 rad/s. The step response has a peak of 13 at $t = 0.08$ s. The frequency response has its peak 20 at $\omega = 14$ rad/s. Notice that the peak occurs far above the peak of the response to load disturbances and far above the gain crossover frequency $\omega_{gc} = 0.78$ rad/s. An approximation derived in Exercise 11.9 gives $\max |CS(i\omega)| \approx k_d/T_f = 20$, which occurs at $\omega = \sqrt{2}/T_d = 14.1$ rad/s. ∇

11.4 Feedback Design via Loop Shaping

One advantage of the Nyquist stability theorem is that it is based on the loop transfer function, which is related to the controller transfer function through $L = PC$. It is thus easy to see how the controller influences the loop transfer function. To make an unstable system stable we simply have to bend the Nyquist curve away from the critical point.

This simple idea is the basis of several different design methods collectively called *loop shaping*. These methods are based on choosing a compensator that gives a loop transfer function with a desired shape. One possibility is to determine a loop transfer function that gives a closed loop system with the desired properties and to compute the controller as $C = L/P$. Another is to start with the process transfer function, change its gain and then add poles and zeros until the desired shape is obtained. In this section we will explore different loop-shaping methods for control law design.

Design Considerations

We will first discuss a suitable shape for the loop transfer function that gives good performance and good stability margins. Figure 11.8 shows a typical loop transfer function. Good robustness requires good stability margins (or good gain and phase margins), which imposes requirements on the loop transfer function around the crossover frequencies ω_{pc} and ω_{gc}. The gain of L at low frequencies must be large in order to have good tracking of command signals and good attenuation of low-frequency disturbances. Since $S = 1/(1 + L)$, it follows that for frequencies where $|L| > 101$ disturbances will be attenuated by a factor of 100 and the tracking error is less than 1%. It is therefore desirable to have a large crossover frequency and a steep (negative) slope of the gain curve. The gain at low frequencies can be increased by a controller with integral action, which is also called *lag compensation*. To avoid injecting too much measurement noise into the system, the loop transfer function should have low gain at high frequencies, which is called *high-frequency roll-off*. The choice of gain crossover frequency is a compromise among attenuation of load disturbances, injection of measurement noise and robustness.

Bode's relations (see Section 9.4) impose restrictions on the shape of the loop transfer function. Equation (9.8) implies that the slope of the gain curve at gain crossover cannot be too steep. If the gain curve has a constant slope, we have the following relation between slope n_{gc} and phase margin φ_m:

$$n_{gc} = -2 + \frac{2\varphi_m}{\pi} \text{ [rad]}. \qquad (11.11)$$

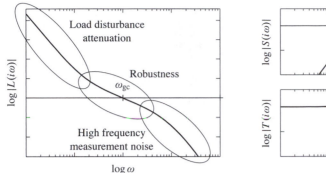
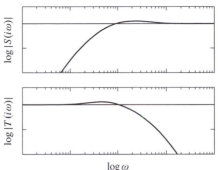

Figure 11.8: Gain curve and sensitivity functions for a typical loop transfer function. The plot on the left shows the gain curve and the plots on the right show the sensitivity function and complementary sensitivity function. The gain crossover frequency ω_{gc} and the slope n_{gc} of the gain curve at crossover are important parameters that determine the robustness of closed loop systems. At low frequency, a large magnitude for L provides good load disturbance rejection and reference tracking, while at high frequency a small loop gain is used to avoid amplifying measurement noise.

This formula is a reasonable approximation when the gain curve does not deviate too much from a straight line. It follows from equation (11.11) that the phase margins $30°, 45°$ and $60°$ correspond to the slopes $-5/3, -3/2$ and $-4/3$.

Loop shaping is a trial-and-error procedure. We typically start with a Bode plot of the process transfer function. We then attempt to shape the loop transfer function by changing the controller gain and adding poles and zeros to the controller transfer function. Different performance specifications are evaluated for each controller as we attempt to balance many different requirements by adjusting controller parameters and complexity. Loop shaping is straightforward to apply to single-input, single-output systems. It can also be applied to systems with one input and many outputs by closing the loops one at a time starting with the innermost loop. The only limitation for minimum phase systems is that large phase leads and high controller gains may be required to obtain closed loop systems with a fast response. Many specific procedures are available: they all require experience, but they also give good insight into the conflicting requirements. There are fundamental limitations to what can be achieved for systems that are not minimum phase; they will be discussed in the next section.

Lead and Lag Compensation

A simple way to do loop shaping is to start with the transfer function of the process and add simple compensators with the transfer function

$$C(s) = k \frac{s+a}{s+b}. \tag{11.12}$$

The compensator is called a *lead compensator* if $a < b$, and a *lag compensator* if $a > b$. The PI controller is a special case of a lag compensator with $b = 0$, and

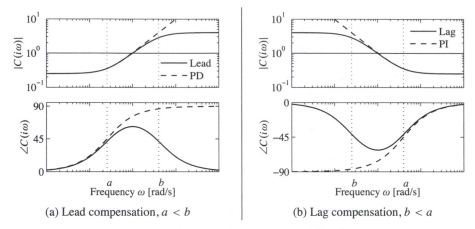

Figure 11.9: Frequency response for lead and lag compensators $C(s) = k(s+a)/(s+b)$. Lead compensation (a) occurs when $a < b$ and provides phase lead between $\omega = a$ and $\omega = b$. Lag compensation (b) corresponds to $a > b$ and provides low-frequency gain. PI control is a special case of lag compensation and PD control is a special case of lead compensation. PI/PD frequency responses are shown by dashed curves.

the ideal PD controller is a special case of a lead compensator with $a = 0$. Bode plots of lead and lag compensators are shown in Figure 11.9. Lag compensation, which increases the gain at low frequencies, is typically used to improve tracking performance and disturbance attenuation at low frequencies. Compensators that are tailored to specific disturbances can be also designed, as shown in Exercise 11.10. Lead compensation is typically used to improve phase margin. The following examples give illustrations.

Example 11.5 Atomic force microscope in tapping mode

A simple model of the dynamics of the vertical motion of an atomic force microscope in tapping mode was given in Exercise 9.2. The transfer function for the system dynamics is

$$P(s) = \frac{a(1 - e^{-s\tau})}{s\tau(s + a)},$$

where $a = \zeta\omega_0$, $\tau = 2\pi n/\omega_0$ and the gain has been normalized to 1. A Bode plot of this transfer function for the parameters $a = 1$ and $\tau - 0.25$ is shown in dashed curves in Figure 11.10a. To improve the attenuation of load disturbances we increase the low-frequency gain by introducing an integral controller. The loop transfer function then becomes $L = k_i P(s)/s$, and we adjust the gain so that the phase margin is zero, giving $k_i = 8.3$. Notice the increase of the gain at low frequencies. The Bode plot is shown by the dotted line in Figure 11.10a, where the critical point is indicated by $\circ$. To improve the phase margin we introduce proportional action and we increase the proportional gain k_p gradually until reasonable values of the sensitivities are obtained. The value $k_p = 3.5$ gives maximum sensitivity $M_s = 1.6$ and maximum complementary sensitivity $M_t = 1.3$. The loop transfer function is shown in solid lines in Figure 11.10a. Notice the significant increase of the phase

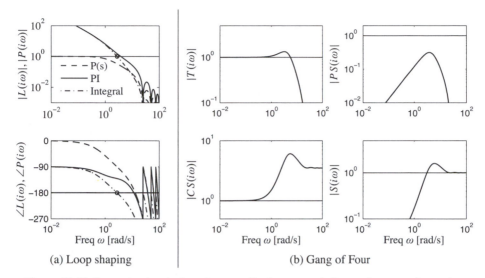

Figure 11.10: Loop-shaping design of a controller for an atomic force microscope in tapping mode. (a) Bode plots of the process (dashed), the loop transfer function for an integral controller with critical gain (dotted) and a PI controller (solid) adjusted to give reasonable robustness. (b) Gain curves for the Gang of Four for the system.

margin compared with the purely integral controller (dotted line).

To evaluate the design we also compute the gain curves of the transfer functions in the Gang of Four. They are shown in Figure 11.10b. The peaks of the sensitivity curves are reasonable, and the plot of PS shows that the largest value of PS is 0.3, which implies that the load disturbances are well attenuated. The plot of CS shows that the largest controller gain is 6. The controller has a gain of 3.5 at high frequencies, and hence we may consider adding high-frequency roll-off.   ∇

A common problem in the design of feedback systems is that the phase margin is too small, and phase *lead* must then be added to the system. If we set $a < b$ in equation (11.12), we add phase lead in the frequency range between the pole/zero pair (and extending approximately 10× in frequency in each direction). By appropriately choosing the location of this phase lead, we can provide additional phase margin at the gain crossover frequency.

Because the phase of a transfer function is related to the slope of the magnitude, increasing the phase requires increasing the gain of the loop transfer function over the frequency range in which the lead compensation is applied. In Exercise 11.11 it is shown that the gain increases exponentially with the amount of phase lead. We can also think of the lead compensator as changing the slope of the transfer function and thus shaping the loop transfer function in the crossover region (although it can be applied elsewhere as well).

Example 11.6 Roll control for a vectored thrust aircraft
Consider the control of the roll of a vectored thrust aircraft such as the one illustrated in Figure 11.11. Following Exercise 8.10, we model the system with a second-order

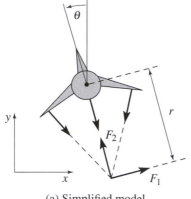

Symbol	Description	Value
m	Vehicle mass	4.0 kg
J	Vehicle inertia, φ_3 axis	0.0475 kg m^2
r	Force moment arm	25.0 cm
c	Damping coefficient	0.05 kg m/s
g	Gravitational constant	9.8 m/s^2

(a) Simplified model   (b) Parameter values

Figure 11.11: Roll control of a vectored thrust aircraft. (a) The roll angle θ is controlled by applying maneuvering thrusters, resulting in a moment generated by F_z. (b) The table lists the parameter values for a laboratory version of the system.

transfer function of the form

$$P(s) = \frac{r}{Js^2},$$

with the parameters given in Figure 11.11b. We take as our performance specification that we would like less than 1% error in steady state and less than 10% tracking error up to 10 rad/s.

The open loop transfer function is shown in Figure 11.12a. To achieve our performance specification, we would like to have a gain of at least 10 at a frequency of 10 rad/s, requiring the gain crossover frequency to be at a higher frequency. We see from the loop shape that in order to achieve the desired performance we cannot simply increase the gain since this would give a very low phase margin. Instead, we must increase the phase at the desired crossover frequency.

To accomplish this, we use a lead compensator (11.12) with $a = 2$ and $b = 50$. We then set the gain of the system to provide a large loop gain up to the desired bandwidth, as shown in Figure 11.12b. We see that this system has a gain of greater than 10 at all frequencies up to 10 rad/s and that it has more than 60° of phase margin.   ∇

The action of a lead compensator is essentially the same as that of the derivative portion of a PID controller. As described in Section 10.5, we often use a filter for the derivative action of a PID controller to limit the high-frequency gain. This same effect is present in a lead compensator through the pole at $s = b$.

Equation (11.12) is a first-order compensator and can provide up to 90° of phase lead. Larger phase lead can be obtained by using a higher-order lead compensator (Exercise 11.11):

$$C(s) = k\frac{(s + a)^n}{(s + b)^n}, \qquad a < b.$$

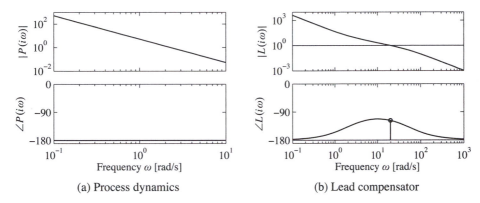

Figure 11.12: Control design for a vectored thrust aircraft using lead compensation. The Bode plot for the open loop process P is shown in (a) and the loop transfer function $L = PC$ using a lead compensator in (b). Note the phase lead in the crossover region near $\omega = 100$ rad/s.

11.5 Fundamental Limitations

Although loop shaping gives us a great deal of flexibility in designing the closed loop response of a system, there are certain fundamental limits on what can be achieved. We consider here some of the primary performance limitations that can occur because of difficult dynamics; additional limitations related to robustness are considered in the next chapter.

Right Half-Plane Poles and Zeros and Time Delays

There are linear systems that are inherently difficult to control. The limitations are related to poles and zeros in the right half-plane and time delays. To explore the limitations caused by poles and zeros in the right half-plane we factor the process transfer function as

$$P(s) = P_{mp}(s)P_{ap}(s), \tag{11.13}$$

where P_{mp} is the minimum phase part and P_{ap} is the nonminimum phase part. The factorization is normalized so that $|P_{ap}(i\omega)| = 1$, and the sign is chosen so that P_{ap} has negative phase. The transfer function P_{ap} is called an *all-pass system* because it has unit gain for all frequencies. Requiring that the phase margin be φ_m, we get

$$\arg L(i\omega_{gc}) = \arg P_{ap}(i\omega_{gc}) + \arg P_{mp}(i\omega_{gc}) + \arg C(i\omega_{gc}) \geq -\pi + \varphi_m, \tag{11.14}$$

where C is the controller transfer function. Let n_{gc} be the slope of the gain curve at the crossover frequency. Since $|P_{ap}(i\omega)| = 1$, it follows that

$$n_{gc} = \left.\frac{d \log |L(i\omega)|}{d \log \omega}\right|_{\omega=\omega_{gc}} = \left.\frac{d \log |P_{mp}(i\omega)C(i\omega)|}{d \log \omega}\right|_{\omega=\omega_{gc}}.$$

Assuming that the slope n_{gc} is negative, it has to be larger than -2 for the system to be stable. It follows from Bode's relations, equation (9.8), that

$$\arg P_{mp}(i\omega) + \arg C(i\omega) \approx n_{gc}\frac{\pi}{2}.$$

Combining this with equation (11.14) gives the following inequality for the allowable phase lag of the all-pass part at the gain crossover frequency:

$$-\arg P_{ap}(i\omega_{gc}) \le \pi - \varphi_m + n_{gc}\frac{\pi}{2} =: \varphi_l. \tag{11.15}$$

This condition, which we call the *gain crossover frequency inequality*, shows that the gain crossover frequency must be chosen so that the phase lag of the nonminimum phase component is not too large. For systems with high robustness requirements we may choose a phase margin of $60°$ ($\varphi_m = \pi/3$) and a slope $n_{gc} = -1$, which gives an admissible phase lag $\varphi_l = \pi/6 = 0.52$ rad ($30°$). For systems where we can accept a lower robustness we may choose a phase margin of $45°$ ($\varphi_m = \pi/4$) and the slope $n_{gc} = -1/2$, which gives an admissible phase lag $\varphi_l = \pi/2 = 1.57$ rad ($90°$).

The crossover frequency inequality shows that nonminimum phase components impose severe restrictions on possible crossover frequencies. It also means that there are systems that cannot be controlled with sufficient stability margins. We illustrate the limitations in a number of commonly encountered situations.

Example 11.7 Zero in the right half-plane
The nonminimum phase part of the process transfer function for a system with a right half-plane zero is

$$P_{ap}(s) = \frac{z - s}{z + s},$$

where $z > 0$. The phase lag of the nonminimum phase part is

$$-\arg P_{ap}(i\omega) = 2\arctan\frac{\omega}{z}.$$

Since the phase lag of P_{ap} increases with frequency, the inequality (11.15) gives the following bound on the crossover frequency:

$$\omega_{gc} < z\,\tan(\varphi_l/2). \tag{11.16}$$

With $\varphi_l = \pi/3$ we get $\omega_{gc} < 0.6\,z$. Slow right half-plane zeros (z small) therefore give tighter restrictions on possible gain crossover frequencies than fast right half-plane zeros. ∇

Time delays also impose limitations similar to those given by zeros in the right half-plane. We can understand this intuitively from the Padé approximation

$$e^{-s\tau} \approx \frac{1 - 0.5s\tau}{1 + 0.5s\tau} = \frac{2/\tau - s}{2/\tau + s}.$$

A long time delay is thus equivalent to a slow right half-plane zero $z = 2/\tau$.

Example 11.8 Pole in the right half-plane

The nonminimum phase part of the transfer function for a system with a pole in the right half-plane is

$$P_{ap}(s) = \frac{s+p}{s-p},$$

where $p > 0$. The phase lag of the nonminimum phase part is

$$-\arg P_{ap}(i\omega) = 2\arctan\frac{p}{\omega},$$

and the crossover frequency inequality becomes

$$\omega_{gc} > \frac{p}{\tan(\varphi_l/2)}. \tag{11.17}$$

Right half-plane poles thus require that the closed loop system have a sufficiently high bandwidth. With $\varphi_l = \pi/3$ we get $\omega_{gc} > 1.7p$. Fast right half-plane poles (p large) therefore give tighter restrictions on possible gain crossover frequencies than slow right half-plane poles. The control of unstable systems imposes minimum bandwidth requirements for process actuators and sensors. ∇

We will now consider systems with a right half-plane zero z and a right half-plane pole p. If $p = z$, there will be an unstable subsystem that is neither reachable nor observable, and the system cannot be stabilized (see Section 7.5). We can therefore expect that the system is difficult to control if the right half-plane pole and zero are close. A straightforward way to use the crossover frequency inequality is to plot the phase of the nonminimum phase factor P_{ap} of the process transfer function. Such a plot, which can be incorporated in an ordinary Bode plot, will immediately show the permissible gain crossover frequencies. An illustration is given in Figure 11.13, which shows the phase of P_{ap} for systems with a right half-plane pole/zero pair and systems with a right half-plane pole and a time delay. If we require that the phase lag φ_l of the nonminimum phase factor be less than 90°, we must require that the ratio z/p be larger than 6 or smaller than 1/6 for systems with right half-plane poles and zeros and that the product $p\tau$ be less than 0.3 for systems with a time delay and a right half-plane pole. Notice the symmetry in the problem for $z > p$ and $z < p$: in either case the zeros and the poles must be sufficiently far apart (Exercise 11.12). Also notice that possible values of the gain crossover frequency ω_{gc} are quite restricted.

Using the theory of functions of complex variables, it can be shown that for systems with a right half-plane pole p and a right half-plane zero z (or a time delay τ), any stabilizing controller gives sensitivity functions with the property

$$\sup_{\omega} |S(i\omega)| \geq \frac{p+z}{|p-z|}, \qquad \sup_{\omega} |T(i\omega)| \geq e^{p\tau}. \tag{11.18}$$

This result is proven in Exercise 11.13.

As the examples above show, right half-plane poles and zeros significantly limit the achievable performance of a system, hence one would like to avoid these whenever possible. The poles of a system depend on the intrinsic dynamics of the

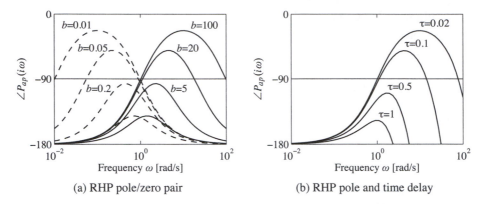

(a) RHP pole/zero pair   (b) RHP pole and time delay

Figure 11.13: Example limitations due to the gain crossover frequency inequality. The figures show the phase lag of the all-pass factor P_{ap} as a function of frequency. Since the phase lag of P_{ap} at the gain crossover frequency cannot be too large, it is necessary to choose the gain crossover frequency properly. All systems have a right half-plane pole at $s = 1$. The system in (a) has zeros at $s = 2, 5, 20$ and 100 (solid lines) and at $s = 0.5, 0.2, 0.05$ and 0.01 (dashed lines). The system in (b) has time delays $\tau = 0.02\ 0.1, 0.5$ and 1.

system and are given by the eigenvalues of the dynamics matrix A of a linear system. Sensors and actuators have no effect on the poles; the only way to change poles is to redesign the system. Notice that this does not imply that unstable systems should be avoided. Unstable system may actually have advantages; one example is high-performance supersonic aircraft.

The zeros of a system depend on how the sensors and actuators are coupled to the states. The zeros depend on all the matrices A, B, C and D in a linear system. The zeros can thus be influenced by moving the sensors and actuators or by adding sensors and actuators. Notice that a fully actuated system $B = I$ does not have any zeros.

Example 11.9 Balance system
As an example of a system with both right half-plane poles and zeros, consider the balance system with zero damping, whose dynamics are given by

$$H_{\theta F} = \frac{ml}{-(M_t J_t - m^2 l^2)s^2 + mgl M_t},$$

$$H_{pF} = \frac{-J_t s^2 + mgl}{s^2\left(-(M_t J_t - m^2 l^2)s^2 + mgl M_t\right)}.$$

Assume that we want to stabilize the pendulum by using the cart position as the measured signal. The transfer function from the input force F to the cart position p has poles $\{0, 0, \pm\sqrt{mgl M_t/(M_t J_t - m^2 l^2)}\}$ and zeros $\{\pm\sqrt{mgl/J_t}\}$. Using the parameters in Example 6.7, the right half-plane pole is at $p = 2.68$ and the zero is at $z = 2.09$. Equation (11.18) then gives $|S(i\omega)| \geq 8$, which shows that it is not possible to control the system robustly.

The right half-plane zero of the system can be eliminated by changing the output of the system. For example, if we choose the output to correspond to a position at a distance r along the pendulum, we have $y = p - r \sin \theta$ and the transfer function for the linearized output becomes

$$H_{y,F} = H_{pF} - r H_{\theta F} = \frac{(mlr - J_t)s^2 + mgl}{s^2 \left(-(M_t J_t - m^2 l^2)s^2 + mgl M_t \right)}.$$

If we choose r sufficiently large, then $mlr - J_t > 0$ and we eliminate the right half-plane zero, obtaining instead two pure imaginary zeros. The gain crossover frequency inequality is then based just on the right half-plane pole (Example 11.8). If our admissible phase lag for the nonminimum phase part is $\varphi_l = 45°$, then our gain crossover must satisfy

$$\omega_{gc} > \frac{p}{\tan(\varphi_l/2)} = 6.48 \text{ rad/s}.$$

If the actuators have sufficiently high bandwidth, e.g., a factor of 10 above ω_{gc} or roughly 10 Hz, then we can provide robust tracking up to this frequency. $\quad \nabla$

Bode's Integral Formula

In addition to providing adequate phase margin for robust stability, a typical control design will have to satisfy performance conditions on the sensitivity functions (Gang of Four). In particular, the sensitivity function $S = 1/(1 + PC)$ represents the disturbance attenuation and also relates the tracking error e to the reference signal: we usually want the sensitivity to be small over the range of frequencies where we want small tracking error and good disturbance attenuation. A basic problem is to investigate if S can be made small over a large frequency range. We will start by investigating an example.

Example 11.10 System that admits small sensitivities
Consider a closed loop system consisting of a first-order process and a proportional controller. Let the loop transfer function be

$$L(s) = PC = \frac{k}{s + 1},$$

where parameter k is the controller gain. The sensitivity function is

$$S(s) = \frac{s + 1}{s + 1 + k}$$

and we have

$$|S(i\omega)| = \sqrt{\frac{1 + \omega^2}{1 + 2k + k^2 + \omega^2}}.$$

This implies that $|S(i\omega)| < 1$ for all finite frequencies and that the sensitivity can be made arbitrarily small for any finite frequency by making k sufficiently large. $\quad \nabla$

The system in Example 11.10 is unfortunately an exception. The key feature of the system is that the Nyquist curve of the process is completely contained in the right half-plane. Such systems are called *passive*, and their transfer functions are *positive real*. For typical control systems there are severe constraints on the sensitivity function. The following theorem, due to Bode, provides insights into the limits of performance under feedback.

Theorem 11.1 (Bode's integral formula). *Assume that the loop transfer function $L(s)$ of a feedback system goes to zero faster than $1/s$ as $s \to \infty$, and let $S(s)$ be the sensitivity function. If the loop transfer function has poles p_k in the right half-plane, then the sensitivity function satisfies the following integral:*

$$\int_0^\infty \log |S(i\omega)| \, d\omega = \int_0^\infty \log \frac{1}{|1 + L(i\omega)|} \, d\omega = \pi \sum p_k. \tag{11.19}$$

Equation (11.19) implies that there are fundamental limitations to what can be achieved by control and that control design can be viewed as a redistribution of disturbance attenuation over different frequencies. In particular, this equation shows that if the sensitivity function is made smaller for some frequencies, it must increase at other frequencies so that the integral of $\log |S(i\omega)|$ remains constant. This means that if disturbance attenuation is improved in one frequency range, it will be worse in another, a property sometime referred to as the *waterbed effect*. It also follows that systems with open loop poles in the right half-plane have larger overall sensitivity than stable systems.

Equation (11.19) can be regarded as a *conservation law*: if the loop transfer function has no poles in the right half-plane, the equation simplifies to

$$\int_0^\infty \log |S(i\omega)| \, d\omega = 0.$$

This formula can be given a nice geometric interpretation as illustrated in Figure 11.14, which shows $\log |S(i\omega)|$ as a function of ω. The area over the horizontal axis must be equal to the area under the axis when the frequency is plotted on a *linear* scale. Thus if we wish to make the sensitivity smaller up to some frequency ω_{sc}, we must balance this by increased sensitivity above ω_{sc}. Control system design can be viewed as trading the disturbance attenuation at some frequencies for disturbance amplification at other frequencies. Notice that the system in Example 11.10 violates the condition that $\lim_{s \to \infty} sL(s) = 0$ and hence the integral formula does not apply.

There is result analogous to equation (11.19) for the complementary sensitivity function:

$$\int_0^\infty \frac{\log |T(i\omega)|}{\omega^2} \, d\omega = \pi \sum \frac{1}{z_i}, \tag{11.20}$$

where the summation is over all right half-plane zeros. Notice that slow right half-plane zeros are worse than fast ones and that fast right half-plane poles are worse than slow ones.

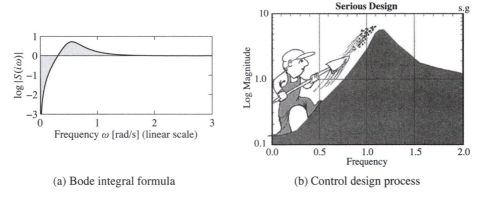

(a) Bode integral formula

(b) Control design process

Figure 11.14: Interpretation of the *waterbed effect*. The function $\log |S(i\omega)|$ is plotted versus ω in linear scales in (a). According to Bode's integral formula (11.19), the area of $\log |S(i\omega)|$ above zero must be equal to the area below zero. Gunter Stein's interpretation of design as a trade-off of sensitivities at different frequencies is shown in (b) (from [185]).

Example 11.11 X-29 aircraft

As an example of the application of Bode's integral formula, we present an analysis of the control system for the X-29 aircraft (see Figure 11.15a), which has an unusual configuration of aerodynamic surfaces that are designed to enhance its maneuverability. This analysis was originally carried out by Gunter Stein in his article "Respect the Unstable" [185], which is also the source of the quote at the beginning of this chapter.

To analyze this system, we make use of a small set of parameters that describe the key properties of the system. The X-29 has longitudinal dynamics that are very similar to inverted pendulum dynamics (Exercise 8.3) and, in particular, have a pair of poles at approximately $p = \pm 6$ and a zero at $z = 26$. The actuators that stabilize the pitch have a bandwidth of $\omega_a = 40$ rad/s and the desired bandwidth of the pitch control loop is $\omega_1 = 3$ rad/s. Since the ratio of the zero to the pole is only 4.3, we may expect that it may be difficult to achieve the specifications.

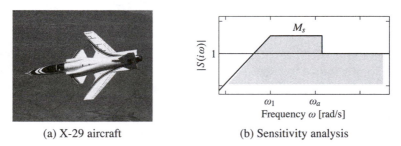

(a) X-29 aircraft

(b) Sensitivity analysis

Figure 11.15: X-29 flight control system. The aircraft makes use of forward swept wings and a set of canards on the fuselage to achieve high maneuverability (a). The desired sensitivity for the closed loop system is shown in (b). We seek to use our control authority to shape the sensitivity curve so that we have low sensitivity (good performance) up to frequency ω_1 by creating higher sensitivity up to our actuator bandwidth ω_a.

To evaluate the achievable performance, we search for a control law such that the sensitivity function is small up to the desired bandwidth and not greater than M_s beyond that frequency. Because of the Bode integral formula, we know that M_s must be greater than 1 at high frequencies to balance the small sensitivity at low frequency. We thus ask if we can find a controller that has the shape shown in Figure 11.15b with the smallest value of M_s. Note that the sensitivity above the frequency ω_a is not specified since we have no actuator authority at that frequency. However, assuming that the process dynamics fall off at high frequency, the sensitivity at high frequency will approach 1. Thus, we desire to design a closed loop system that has low sensitivity at frequencies below ω_1 and sensitivity that is not too large between ω_1 and ω_a.

From Bode's integral formula, we know that whatever controller we choose, equation (11.19) must hold. We will assume that the sensitivity function is given by

$$|S(i\omega)| = \begin{cases} \frac{\omega M_s}{\omega_1} & \omega \leq \omega_1 \\ M_s & \omega_1 \leq \omega \leq \omega_a, \end{cases}$$

corresponding to Figure 11.15b. If we further assume that $|L(s)| \leq \delta/\omega^2$ for frequencies larger than the actuator bandwidth, Bode's integral becomes

$$\int_0^\infty \log|S(i\omega)|\,d\omega = \int_0^{\omega_a} \log|S(i\omega)|\,d\omega$$
$$= \int_0^{\omega_1} \log\frac{\omega M_s}{\omega_1}\,d\omega + (\omega_a - \omega_1)\log M_s = \pi p.$$

Evaluation of the integral gives $-\omega_1 + \omega_a \log M_s = \pi p$ or

$$M_s = e^{(\pi p + \omega_1)/\omega_a}.$$

This formula tells us what the achievable value of M_s will be for the given control specifications. In particular, using $p = 6$, $\omega_1 = 3$ and $\omega_a = 40$ rad/s, we find that $M_s = 1.75$, which means that in the range of frequencies between ω_1 and ω_a, disturbances at the input to the process dynamics (such as wind) will be amplified by a factor of 1.75 in terms of their effect on the aircraft.

Another way to view these results is to compute the phase margin that corresponds to the given level of sensitivity. Since the peak sensitivity normally occurs at or near the crossover frequency, we can compute the phase margin corresponding to $M_s = 1.75$. As shown in Exercise 11.14, the maximum achievable phase margin for this system is approximately $35°$, which is below the usual design limit of $45°$ in aerospace systems. The zero at $s = 26$ limits the maximum gain crossover the can be achieved. ∇

Derivation of Bode's Formula

We now derive Bode's integral formula (Theorem 11.1). This is a technical section that requires some knowledge of the theory of complex variables, in particular contour integration. Assume that the loop transfer function has distinct poles at

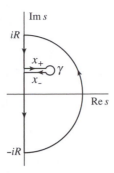

Figure 11.16: Contour used to prove Bode's theorem. For each right half-plane pole we create a path from the imaginary axis that encircles the pole as shown. To avoid clutter we have shown only one of the paths that enclose one right half-plane.

$s = p_k$ in the right half-plane and that $L(s)$ goes to zero faster than $1/s$ for large values of s.

Consider the integral of the logarithm of the sensitivity function $S(s) = 1/(1 + L(s))$ over the contour shown in Figure 11.16. The contour encloses the right half-plane except for the points $s = p_k$ where the loop transfer function $L(s) = P(s)C(s)$ has poles and the sensitivity function $S(s)$ has zeros. The direction of the contour is counterclockwise.

The integral of the log of the sensitivity function around this contour is given by

$$\int_\Gamma \log(S(s))\,ds = \int_{iR}^{-iR} \log(S(s))\,ds + \int_R \log(S(s))\,ds + \sum_k \int_\gamma \log(S(s))\,ds$$

$$= I_1 + I_2 + I_3 = 0,$$

where R is a large semicircle on the right and γ_k is the contour starting on the imaginary axis at $s = \operatorname{Im} p_k$ and a small circle enclosing the pole p_k. The integral is zero because the function $\log S(s)$ is analytic inside the contour. We have

$$I_1 = -i \int_{-iR}^{iR} \log(S(i\omega))d\omega = -2i \int_0^{iR} \log(|S(i\omega)|)d\omega$$

because the real part of $\log S(i\omega)$ is an even function and the imaginary part is an odd function. Furthermore we have

$$I_2 = \int_R \log(S(s))\,ds = -\int_R \log(1 + L(s))\,ds \approx -\int_R L(s)\,ds.$$

Since $L(s)$ goes to zero faster than $1/s$ for large s, the integral goes to zero when the radius of the circle goes to infinity.

Next we consider the integral I_3. For this purpose we split the contour into three parts X_+, γ and X_-, as indicated in Figure 11.16. We can then write the integral as

$$I_3 = \int_{X_+} \log S(s)\,ds + \int_\gamma \log S(s)\,ds + \int_{X_-} \log S(s)\,ds.$$

The contour γ is a small circle with radius r around the pole p_k. The magnitude of the integrand is of the order $\log r$, and the length of the path is $2\pi r$. The integral thus goes to zero as the radius r goes to zero. Furthermore, making use of the fact that X_- is oriented oppositely from X_+, we have

$$\int_{X_+} \log S(s)\, ds + \int_{X_-} \log S(s)\, ds = \int_{X_+} \big(\log S(s) - \log S(s - 2\pi i)\big)\, ds = 2\pi \operatorname{Re} p_k.$$

Since $|S(s)| = |S(s - 2\pi i)|$, we have

$$\log S(s) - \log S(s - 2\pi i) = \arg S(s) - \arg S(s - 2\pi i) = 2\pi\, i,$$

and we find that

$$I_3 = 2\pi\, i\, \sum_k \operatorname{Re} p_k.$$

Letting the small circles go to zero and the large circle go to infinity and adding the contributions from all right half-plane poles p_k gives

$$I_1 + I_2 + I_3 = -2i \int_0^R \log |S(i\omega)|\, d\omega + i \sum_k 2\pi \operatorname{Re} p_k = 0.$$

Since complex poles appear as complex conjugate pairs, $\sum_k \operatorname{Re} p_k = \sum_k p_k$, which gives Bode's formula (11.19).

11.6 Design Example

In this section we present a detailed example that illustrates the main design techniques described in this chapter.

Example 11.12 Lateral control of a vectored thrust aircraft
The problem of controlling the motion of a vertical takeoff and landing (VTOL) aircraft was introduced in Example 2.9 and in Example 11.6, where we designed a controller for the roll dynamics. We now wish to control the position of the aircraft, a problem that requires stabilization of both the attitude and the position.

To control the lateral dynamics of the vectored thrust aircraft, we make use of a "inner/outer" loop design methodology, as illustrated in Figure 11.17. This diagram shows the process dynamics and controller divided into two components: an *inner loop* consisting of the roll dynamics and control and an *outer loop* consisting of the lateral position dynamics and controller. This decomposition follows the block diagram representation of the dynamics given in Exercise 8.10.

The approach that we take is to design a controller C_i for the inner loop so that the resulting closed loop system H_i provides fast and accurate control of the roll angle for the aircraft. We then design a controller for the lateral position that uses the approximation that we can directly control the roll angle as an input to the dynamics controlling the position. Under the assumption that the dynamics of the roll controller are fast relative to the desired bandwidth of the lateral position control, we can then combine the inner and outer loop controllers to get a single

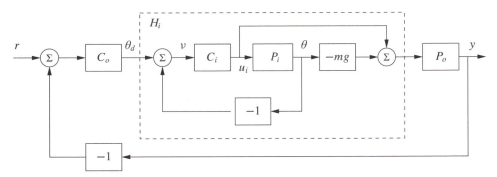

Figure 11.17: Inner/outer control design for a vectored thrust aircraft. The inner loop H_i controls the roll angle of the aircraft using the vectored thrust. The outer loop controller C_o commands the roll angle to regulate the lateral position. The process dynamics are decomposed into inner loop (P_i) and outer loop (P_o) dynamics, which combine to form the full dynamics for the aircraft.

controller for the entire system. As a performance specification for the entire system, we would like to have zero steady-state error in the lateral position, a bandwidth of approximately 1 rad/s and a phase margin of 45°.

For the inner loop, we choose our design specification to provide the outer loop with accurate and fast control of the roll. The inner loop dynamics are given by

$$P_i = H_{\theta u_1} = \frac{r}{Js^2 + cs}.$$

We choose the desired bandwidth to be 10 rad/s (10 times that of the outer loop) and the low-frequency error to be no more than 5%. This specification is satisfied using the lead compensator of Example 11.6 designed previously, so we choose

$$C_i(s) = k\frac{s+a}{s+b}, \qquad a = 2, \quad b = 50, \quad k = 1.$$

The closed loop dynamics for the system satisfy

$$H_i = \frac{C_i}{1 + C_i P_i} - mg\frac{C_i P_i}{1 + C_i P_i} = \frac{C_i(1 - mg P_i)}{1 + C_i P_i}.$$

A plot of the magnitude of this transfer function is shown in Figure 11.18, and we see that $H_i \approx -mg = 39.2$ is a good approximation up to 10 rad/s.

To design the outer loop controller, we assume the inner loop roll control is perfect, so that we can take θ_d as the input to our lateral dynamics. Following the diagram shown in Exercise 8.10, the outer loop dynamics can be written as

$$P(s) = H_i(0)P_o(s) = \frac{H_i(0)}{ms^2},$$

where we replace $H_i(s)$ with $H_i(0)$ to reflect our approximation that the inner loop will eventually track our commanded input. Of course, this approximation may not be valid, and so we must verify this when we complete our design.

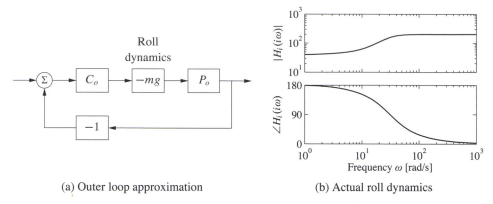

(a) Outer loop approximation   (b) Actual roll dynamics

Figure 11.18: Outer loop control design for a vectored thrust aircraft. (a) The outer loop approximates the roll dynamics as a state gain $-mg$. (b) The Bode plot for the roll dynamics, indicating that this approximation is accurate up to approximately 10 rad/s.

Our control goal is now to design a controller that gives zero steady-state error in x and has a bandwidth of 1 rad/s. The outer loop process dynamics are given by a second-order integrator, and we can again use a simple lead compensator to satisfy the specifications. We also choose the design such that the loop transfer function for the outer loop has $|L_o| < 0.1$ for $\omega > 10$ rad/s, so that the H_i dynamics can be neglected. We choose the controller to be of the form

$$C_o(s) = -k_o \frac{s + a_o}{s + b_o},$$

with the negative sign to cancel the negative sign in the process dynamics. To find the location of the poles, we note that the phase lead flattens out at approximately $b/10$. We desire phase lead at crossover, and we desire the crossover at $\omega_{gc} = 1$ rad/s, so this gives $b_o = 10$. To ensure that we have adequate phase lead, we must choose a_o such that $b_o/10 < 10a_o < b_o$, which implies that a_o should be between 0.1 and 1. We choose $a_o = 0.3$. Finally, we need to set the gain of the system such that at crossover the loop gain has magnitude 1. A simple calculation shows that $k_o = 2$ satisfies this objective. Thus, the final outer loop controller becomes

$$C_o(s) = 0.8 \frac{s + 0.3}{s + 10}.$$

Finally, we can combine the inner and outer loop controllers and verify that the system has the desired closed loop performance. The Bode and Nyquist plots corresponding to Figure 11.17 with inner and outer loop controllers are shown in Figure 11.19, and we see that the specifications are satisfied. In addition, we show the Gang of Four in Figure 11.20, and we see that the transfer functions between all inputs and outputs are reasonable. The sensitivity to load disturbances PS is large at low frequency because the controller does not have integral action.

The approach of splitting the dynamics into an inner and an outer loop is common in many control applications and can lead to simpler designs for complex systems.

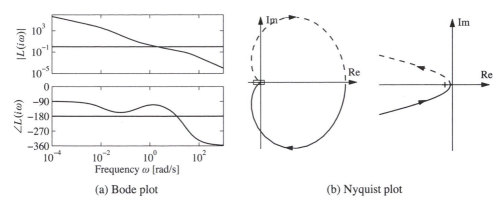

(a) Bode plot   (b) Nyquist plot

Figure 11.19: Inner/outer loop controller for a vectored thrust aircraft. The Bode plot (a) and Nyquist plot (b) for the transfer function for the combined inner and outer loop transfer functions are shown. The system has a phase margin of $68°$ and a gain margin of 6.2.

Indeed, for the aircraft dynamics studied in this example, it is very challenging to directly design a controller from the lateral position x to the input u_1. The use of the additional measurement of θ greatly simplifies the design because it can be broken up into simpler pieces.   ∇

11.7 Further Reading

Design by loop shaping was a key element in the early development of control, and systematic design methods were developed; see James, Nichols and Phillips [110], Chestnut and Mayer [51], Truxal [194] and Thaler [191]. Loop shaping is also treated in standard textbooks such as Franklin, Powell and Emami-Naeini [79], Dorf and Bishop [61], Kuo and Golnaraghi [133] and Ogata [162]. Systems with two degrees of freedom were developed by Horowitz [102], who also discussed the limitations of poles and zeros in the right half-plane. Fundamental results on limitations are given in Bode [40]; more recent presentations are found in Goodwin, Graebe and Salgado [88]. The treatment in Section 11.5 is based on [14]. Much of the early work was based on the loop transfer function; the importance of the sensitivity functions appeared in connection with the development in the 1980s that resulted in H_∞ design methods. A compact presentation is given in the texts by Doyle, Francis and Tannenbaum [64] and Zhou, Doyle and Glover [209]. Loop shaping was integrated with the robust control theory in McFarlane and Glover [150] and Vinnicombe [196]. Comprehensive treatments of control system design are given in Maciejowski [141] and Goodwin, Graebe and Salgado [88].

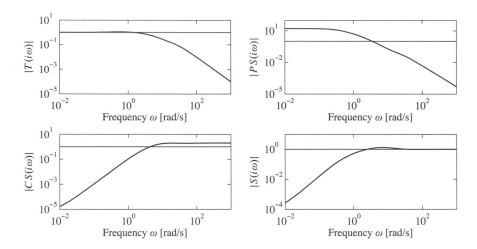

Figure 11.20: Gang of Four for vectored thrust aircraft system.

Exercises

11.1 Consider the system in Figure 11.1. Give all signal pairs that are related by the transfer functions $1/(1 + PC)$, $P/(1 + PC)$, $C/(1 + PC)$ and $PC/(1 + PC)$.

11.2 Consider the system in Example 11.1. Choose the parameters $a = -1$ and compute the time and frequency responses for all the transfer functions in the Gang of Four for controllers with $k = 0.2$ and $k = 5$.

11.3 (Equivalence of Figures 11.1 and 11.2) Consider the system in Figure 11.1 and let the outputs of interest be $z = (\eta, \nu)$ and the major disturbances be $w = (n, d)$. Show that the system can be represented by Figure 11.2 and give the matrix transfer functions $\mathcal{P}$ and $\mathcal{C}$. Verify that the closed loop transfer function H_{zw} gives the Gang of Four.

11.4 Consider the spring–mass system given by (2.14), which has the transfer function

$$P(s) = \frac{1}{ms^2 + cs + k}.$$

Design a feedforward compensator that gives a response with critical damping ($\zeta = 1$).

11.5 (Sensitivity of feedback and feedforward) Consider the system in Figure 11.1 and let G_{yr} be the transfer function relating the measured signal y to the reference r. Show that the sensitivities of G_{yr} with respect to the feedforward and feedback transfer functions F and C are given by $dG_{yr}/dF = CP/(1 + PC)$ and $dG_{yr}/dC = FP/(1 + PC)^2 = G_{yr}L/C$.

11.6 (Equivalence of controllers with two degrees of freedom) Show that the systems in Figures 11.1 and 11.3 give the same responses to command signals if $F_m C + F_u = CF$.

11.7 (Disturbance attenuation) Consider the feedback system shown in Figure 11.1. Assume that the reference signal is constant. Let y_{ol} be the measured output when there is no feedback and y_{cl} be the output with feedback. Show that $Y_{cl}(s) = S(s)Y_{ol}(s)$, where S is the sensitivity function.

11.8 (Disturbance reduction through feedback) Consider a problem in which an output variable has been measured to estimate the potential for disturbance attenuation by feedback. Suppose an analysis shows that it is possible to design a closed loop system with the sensitivity function

$$S(s) = \frac{s}{s^2 + s + 1}.$$

Estimate the possible disturbance reduction when the measured disturbance is

$$y(t) = 5\sin(0.1\,t) + 3\sin(0.17\,t) + 0.5\cos(0.9\,t) + 0.1\,t.$$

11.9 Show that the effect of high frequency measurement noise on the control signal for the system in Example 11.4 can be approximated by

$$CS \approx C = \frac{k_d s}{(sT_f)^2/2 + sT_f + 1},$$

and that the largest value of $|CS(i\omega)|$ is k_d/T_f which occurs for $\omega = \sqrt{2}/T_f$.

11.10 (Attenuation of low-frequency sinusoidal disturbances) Integral action eliminates constant disturbances and reduces low-frequency disturbances because the controller gain is infinite at zero frequency. A similar idea can be used to reduce the effects of sinusoidal disturbances of known frequency ω_0 by using the controller

$$C(s) = k_p + \frac{k_s s}{s^2 + 2\zeta\omega_0 s + \omega_0^2}.$$

This controller has the gain $C_s(i\omega) = k_p + k_s/(2\zeta)$ for the frequency ω_0, which can be large by choosing a small value of ζ. Assume that the process has the transfer function $P(s) = 1/s$. Determine the Bode plot of the loop transfer function and simulate the system. Compare the results with PI control.

11.11 Consider a lead compensator with the transfer function

$$C_n(s) = \left(\frac{s\sqrt[n]{k} + a}{s + a}\right)^n,$$

which has zero frequency gain $C(0) = 1$ and high-frequency gain $C(\infty) = k$. Show that the gain required to give a given phase lead φ is

$$k = \left(1 + 2\tan^2(\varphi/n) + 2\tan(\varphi/n)\sqrt{1 + \tan^2(\varphi/n)}\right)^n,$$

and that $\lim_{n\to\infty} k = e^{2\varphi}$.

11.12 Consider a process with the loop transfer function

$$L(s) = k\frac{z-s}{s-p},$$

with positive z and p. Show that the system is stable if $p/z < k < 1$ or $1 < k < p/z$, and that the largest stability margin is $s_m = |p-z|/(p+z)$ is obtained for $k = 2p/(p+z)$. Determine the pole/zero ratios that gives the stability margin $s_m = 2/3$.

 11.13 Prove the inequalities given by equation (11.18). (Hint: Use the maximum modulus theorem.)

11.14 (Phase margin formulas) Show that the relationship between the phase margin and the values of the sensitivity functions at gain crossover is given by

$$|S(i\omega_{gc})| = |T(i\omega_{gc})| = \frac{1}{2\sin(\varphi_m/2)}.$$

11.15 (Stabilization of an inverted pendulum with visual feedback) Consider stabilization of an inverted pendulum based on visual feedback using a video camera with a 50-Hz frame rate. Let the effective pendulum length be l. Use the gain crossover frequency inequality to determine the minimum length of the pendulum that can be stabilized if we desire a phase lag φ_l of no more than $90°$.

11.16 (Rear-steered bicycle) Consider the simple model of a bicycle in Equation (3.5), which has one pole in the right half-plane. The model is also valid for a bicycle with rear wheel steering, but the sign of the velocity is then reversed and the system also has a zero in the right half-plane. Use the results of Exercise 11.12 to give a condition on the physical parameters that admits a controller with the stability margin s_m.

 11.17 Prove the formula (11.20) for the complementary sensitivity.

Chapter Twelve
Robust Performance

> *However, by building an amplifier whose gain is deliberately made, say 40 decibels higher than necessary (10000 fold excess on energy basis), and then feeding the output back on the input in such a way as to throw away that excess gain, it has been found possible to effect extraordinary improvement in constancy of amplification and freedom from non-linearity.*
>
> Harold S. Black, "Stabilized Feedback Amplifiers," 1934 [35].

This chapter focuses on the analysis of robustness of feedback systems, a vast topic for which we provide only an introduction to some of the key concepts. We consider the stability and performance of systems whose process dynamics are uncertain and derive fundamental limits for robust stability and performance. To do this we develop ways to describe uncertainty, both in the form of parameter variations and in the form of neglected dynamics. We also briefly mention some methods for designing controllers to achieve robust performance.

12.1 Modeling Uncertainty

Harold Black's quote above illustrates that one of the key uses of feedback is to provide robustness to uncertainty ("constancy of amplification"). It is one of the most useful properties of feedback and is what makes it possible to design feedback systems based on strongly simplified models.

One form of uncertainty in dynamical systems is *parametric uncertainty* in which the parameters describing the system are unknown. A typical example is the variation of the mass of a car, which changes with the number of passengers and the weight of the baggage. When linearizing a nonlinear system, the parameters of the linearized model also depend on the operating conditions. It is straightforward to investigate the effects of parametric uncertainty simply by evaluating the performance criteria for a range of parameters. Such a calculation reveals the consequences of parameter variations. We illustrate by a simple example.

Example 12.1 Cruise control
The cruise control problem was described in Section 3.1, and a PI controller was designed in Example 10.3. To investigate the effect of parameter variations, we will choose a controller designed for a nominal operating condition corresponding to mass $m = 1600$ kg, fourth gear ($\alpha = 12$) and speed $v_e = 25$ m/s; the controller gains are $k_p = 0.72$ and $k_i = 0.18$. Figure 12.1a shows the velocity v and the throttle u when encountering a hill with a $3°$ slope with masses in the range $1600 < m < 2000$ kg, gear ratios 3–5 ($\alpha = 10$, 12 and 16) and velocity $10 \leq v \leq 40$ m/s.

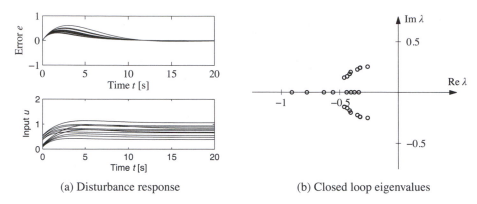

(a) Disturbance response   (b) Closed loop eigenvalues

Figure 12.1: Responses of the cruise control system to a slope increase of $3°$ (a) and the eigenvalues of the closed loop system (b). Model parameters are swept over a wide range.

The simulations were done using models that were linearized around the different operating conditions. The figure shows that there are variations in the response but that they are quite reasonable. The largest velocity error is in the range of 0.2–0.6 m/s, and the settling time is about 15 s. The control signal is marginally larger than 1 in some cases, which implies that the throttle is fully open. A full nonlinear simulation using a controller with windup protection is required if we want to explore these cases in more detail. Figure 12.1b shows the eigenvalues of the closed loop system for the different operating conditions. The figure shows that the closed loop system is well damped in all cases.   ∇

This example indicates that at least as far as parametric variations are concerned, the design based on a simple nominal model will give satisfactory control. The example also indicates that a controller with fixed parameters can be used in all cases. Notice that we have not considered operating conditions in low gear and at low speed, but cruise controllers are not typically used in these cases.

Unmodeled Dynamics

It is generally easy to investigate the effects of parametric variations. However, there are other uncertainties that also are important, as discussed at the end of Section 2.3. The simple model of the cruise control system captures only the dynamics of the forward motion of the vehicle and the torque characteristics of the engine and transmission. It does not, for example, include a detailed model of the engine dynamics (whose combustion processes are extremely complex) or the slight delays that can occur in modern electronically controlled engines (as a result of the processing time of the embedded computers). These neglected mechanisms are called *unmodeled dynamics*.

Unmodeled dynamics can be accounted for by developing a more complex model. Such models are commonly used for controller development, but substantial effort is required to develop them. An alternative is to investigate if the closed loop system is sensitive to generic forms of unmodeled dynamics. The basic idea is to

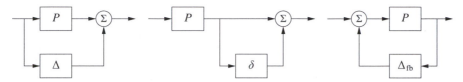

Figure 12.2: Unmodeled dynamics in linear systems. Uncertainty can be represented using additive perturbations (left), multiplicative perturbations (middle) or feedback perturbations (right). The nominal system is P, and Δ, $\delta = \Delta/P$ and Δ_{fb} represent unmodeled dynamics.

describe the unmodeled dynamics by including a transfer function in the system description whose frequency response is bounded but otherwise unspecified. For example, we might model the engine dynamics in the cruise control example as a system that quickly provides the torque that is requested through the throttle, giving a small deviation from the simplified model, which assumed the torque response was instantaneous. This technique can also be used in many instances to model parameter variations, allowing a quite general approach to uncertainty management.

In particular, we wish to explore if additional linear dynamics may cause difficulties. A simple way is to assume that the transfer function of the process is $P(s) + \Delta$, where $P(s)$ is the nominal simplified transfer function and Δ represents the unmodeled dynamics in terms of *additive uncertainty*. Different representations of uncertainty are shown in Figure 12.2.

When Are Two Systems Similar? The Vinnicombe Metric

A fundamental issue in describing robustness is to determine when two systems are close. Given such a characterization, we can then attempt to describe robustness according to how close the actual system must be to the model in order to still achieve the desired levels of performance. This seemingly innocent problem is not as simple as it may appear. A naive approach is to say that two systems are close if their open loop responses are close. Even if this appears natural, there are complications, as illustrated by the following examples.

Example 12.2 Similar in open loop but large differences in closed loop
The systems with the transfer functions

$$P_1(s) = \frac{k}{s+1}, \qquad P_2(s) = \frac{k}{(s+1)(sT+1)^2} \qquad (12.1)$$

have very similar open loop responses for small values of T, as illustrated in the top plot in Figure 12.3a, which is plotted for $T = 0.025$ and $k = 100$. The differences between the step responses are barely noticeable in the figure. The step responses with unit gain error feedback are shown in the bottom plot in Figure 12.3a. Notice that one closed loop system is stable and the other one is unstable. ∇

Example 12.3 Different in open loop but similar in closed loop

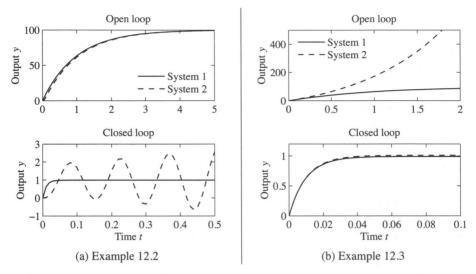

Figure 12.3: Determining when two systems are close. The plots in (a) show a situation when the open loop responses are almost identical, but the closed loop responses are very different. The processes are given by equation (12.1) with $k = 100$ and $T = 0.025$. The plots in (b) show the opposite situation: the systems are different in open loop but similar in closed loop. The processes are given by equation (12.2) with $k = 100$.

Consider the systems

$$P_1(s) = \frac{k}{s+1}, \qquad P_2(s) = \frac{k}{s-1}. \tag{12.2}$$

The open loop responses are very different because P_1 is stable and P_2 is unstable, as shown in the top plot in Figure 12.3b. Closing a feedback loop with unit gain around the systems, we find that the closed loop transfer functions are

$$T_1(s) = \frac{k}{s+k+1}, \qquad T_2(s) = \frac{k}{s+k-1},$$

which are very close for large k, as shown in Figure 12.3b.   ∇

These examples show that if our goal is to close a feedback loop, it may be very misleading to compare the open loop responses of the system.

Inspired by these examples we introduce the *Vinnicombe metric*, which is a distance measure that is appropriate for closed loop systems. Consider two systems with the transfer functions P_1 and P_2, and define

$$d(P_1, P_2) = \sup_\omega \frac{|P_1(i\omega) - P_2(i\omega)|}{\sqrt{(1 + |P_1(i\omega)|^2)(1 + |P_2(i\omega)|^2)}}, \tag{12.3}$$

which is a metric with the property $0 \le d(P_1, P_2) \le 1$. The number $d(P_1, P_2)$ can be interpreted as the difference between the complementary sensitivity functions for the closed loop systems that are obtained with unit feedback around P_1 and P_2; see Exercise 12.3. The metric also has a nice geometric interpretation, as shown in

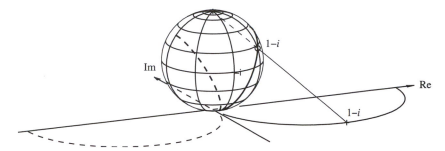

Figure 12.4: Geometric interpretation of $d(P_1, P_2)$. At each frequency, the points on the Nyquist curve for P_1 (solid) and P_2 (dashed) are projected onto a sphere of radius 1 sitting at the origin of the complex plane. The projection of the point $1 - i$ is shown. The distance between the two systems is defined as the maximum distance between the projections of $P_1(i\omega)$ and $P_2(i\omega)$ over all frequencies ω. The figure is plotted for the transfer functions $P_1(s) = 2/(s + 1)$ and $P_2(s) = 2/(s - 1)$. (Diagram courtesy G. Vinnicombe.)

Figure 12.4, where the Nyquist plots of P_1 and P_2 are projected onto a sphere with radius 1 at the origin of the complex plane (called the *Riemann sphere*). Points in the complex plane are projected onto the sphere by a line through the point and the north pole (Figure 12.4). The distance $d(P_1, P_2)$ is the longest chordal distance between the projections of $P_1(i\omega)$ and $P_2(i\omega)$. The distance is small when P_1 and P_2 are small or large, but it emphasizes the behavior around the gain crossover frequency.

The distance $d(P_1, P_2)$ has one drawback for the purpose of comparing the behavior of systems under feedback. If P_2 is perturbed continuously from P_1 to P_2, there can be intermediate transfer functions P where $d(P_1, P)$ is 1 even if $d(P_1, P_2)$ is small (see Exercise 12.4). To explore when this could happen, we observe that

$$1 - d^2(P_1, P) = \frac{(1 + P(i\omega)P_1(-i\omega))(1 + P(-i\omega)P_1(i\omega))}{(1 + |P_1(i\omega)|^2)(1 + |P(i\omega)|^2)}.$$

The right-hand side is zero, and hence $d(P_1, P) = 1$ if $1 + P(i\omega)P_1(-i\omega) = 0$ for some ω. To explore when this could occur, we investigate the behavior of the function $1 + P(s)P_1(-s)$ when P is perturbed from P_1 to P_2. If the functions $f_1(s) = 1 + P_1(s)P_1(-s)$ and $f_2(s) = 1 + P_2(s)P_1(-s)$ do not have the same number of zeros in the right half-plane, there is an intermediate P such that $1 + P(i\omega)P_1(-i\omega) = 0$ for some ω. To exclude this case we introduce the set $\mathcal{C}$ as all pairs (P_1, P_2) such that the functions $f_1 = 1 + P_1(s)P_1(-s)$ and $f_2 = 1 + P_2(s)P_1(-s)$ have the same number of zeros in the right half-plane.

The *Vinnicombe metric* or *ν-gap metric* is defined as

$$\delta_\nu(P_1, P_2) = \begin{cases} d(P_1, P_2), & \text{if } (P_1, P_2) \in \mathcal{C} \\ 1, & \text{otherwise.} \end{cases} \tag{12.4}$$

Vinnicombe [196] showed that $\delta_\nu(P_1, P_2)$ is a metric, he gave strong robustness results based on the metric and he developed the theory for systems with many inputs and many outputs. We illustrate its use by computing the metric for the systems in the previous examples.

Example 12.4 Vinnicombe metric for Examples 12.2 and 12.3

For the systems in Example 12.2 we have

$$f_1(s) = 1 + P_1(s)P_1(-s) = \frac{1 + k^2 - s^2}{1 - s^2},$$

$$f_2(s) = 1 + P_2(s)P_1(-s) = \frac{1 + k^2 + 2sT + (T^2 - 1)s^2 - 2s^3T - s^4T^2}{(1 - s^2)(1 + 2sT + s^2T^2)}.$$

The function f_1 has one zero in the right half-plane. A numerical calculation for $k = 100$ and $T = 0.025$ shows that the function f_2 has the roots 46.3, -86.3, $-20.0 \pm 60.0i$. Both functions have one zero in the right half-plane, allowing us to compute the norm (12.4). For $T = 0.025$ this gives $\delta_\nu(P_1, P_2) = 0.98$, which is a quite large value. To have reasonable robustness Vinnicombe recommended values less than 1/3.

For the system in Example 12.3 we have

$$1 + P_1(s)P_1(-s) = \frac{1 + k^2 - s^2}{1 - s^2}, \qquad 1 + P_2(s)P_1(-s) = \frac{1 - k^2 - 2s + s^2}{(s + 1)^2}$$

These functions have the same number of zeros in the right half-plane if $k > 1$. In this particular case the Vinnicombe metric is $d(P_1, P_2) = 2k/(1 + k^2)$ (Exercise 12.4) and with $k = 100$ we get $\delta_\nu(P_1, P_2) = 0.02$. Figure 12.4 shows the Nyquist curves and their projections for $k = 2$. Notice that $d(P_1, P_2)$ is very small for small k even though the closed loop systems are very different. It is therefore essential to consider the condition $(P_1, P_2) \in \mathcal{C}$, as discussed in Exercise 12.4. ∇

12.2 Stability in the Presence of Uncertainty

Having discussed how to describe uncertainty and the similarity between two systems, we now consider the problem of robust stability: When can we show that the stability of a system is robust with respect to process variations? This is an important question since the potential for instability is one of the main drawbacks of feedback. Hence we want to ensure that even if we have small inaccuracies in our model, we can still guarantee stability and performance.

Robust Stability Using Nyquist's Criterion

The Nyquist criterion provides a powerful and elegant way to study the effects of uncertainty for linear systems. A simple criterion is that the Nyquist curve be sufficiently far from the critical point -1. Recall that the shortest distance from the Nyquist curve to the critical point is $s_m = 1/M_s$, where M_s is the maximum of the sensitivity function and s_m is the stability margin introduced in Section 9.3.

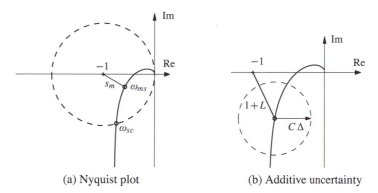

(a) Nyquist plot   (b) Additive uncertainty

Figure 12.5: Robust stability using the Nyquist criterion. (a) This plot shows that the shortest distance to the critical point s_m is a robustness measure. (b) This plot shows the Nyquist curve of a nominal loop transfer function and its uncertainty caused by additive process variations Δ.

The maximum sensitivity M_s or the stability margin s_m is thus a good robustness measure, as illustrated in Figure 12.5a.

We will now derive explicit conditions for permissible process uncertainties. Consider a stable feedback system with a process P and a controller C. If the process is changed from P to $P + \Delta$, the loop transfer function changes from PC to $PC + C\Delta$, as illustrated in Figure 12.5b. If we have a bound on the size of Δ (represented by the dashed circle in the figure), then the system remains stable as long as the process variations never overlap the -1 point, since this leaves the number of encirclements of -1 unchanged.

Some additional assumptions are required for the analysis to hold. Most importantly, we require that the process perturbations Δ be stable so that we do not introduce any new right half-plane poles that would require additional encirclements in the Nyquist criterion.

We will now compute an analytical bound on the allowable process disturbances. The distance from the critical point -1 to the loop transfer function L is $|1 + L|$. This means that the perturbed Nyquist curve will not reach the critical point -1 provided that $|C\Delta| < |1 + L|$, which implies

$$|\Delta| < \left| \frac{1 + PC}{C} \right| \qquad \text{or} \qquad |\delta| = \left| \frac{\Delta}{P} \right| < \frac{1}{|T|}. \tag{12.5}$$

This condition must be valid for all points on the Nyquist curve, i.e, pointwise for all frequencies. The condition for robust stability can thus be written as

$$|\delta(i\omega)| = \left| \frac{\Delta(i\omega)}{P(i\omega)} \right| < \frac{1}{|T(i\omega)|} \qquad \text{for all } \omega \geq 0. \tag{12.6}$$

Notice that the condition is conservative because it follows from Figure 12.5 that the critical perturbation is in the direction toward the critical point -1. Larger perturbations can be permitted in the other directions.

The condition in equation (12.6) allows us to reason about uncertainty without exact knowledge of the process perturbations. Namely, we can verify stability for *any* uncertainty Δ that satisfies the given bound. From an analysis perspective, this gives us a measure of the robustness for a given design. Conversely, if we require robustness of a given level, we can attempt to choose our controller C such that the desired level of robustness is available (by asking that T be small) in the appropriate frequency bands.

Equation (12.6) is one of the reasons why feedback systems work so well in practice. The mathematical models used to design control systems are often simplified, and the properties of a process may change during operation. Equation (12.6) implies that the closed loop system will at least be stable for substantial variations in the process dynamics.

It follows from equation (12.6) that the variations can be large for those frequencies where T is small and that smaller variations are allowed for frequencies where T is large. A conservative estimate of permissible process variations that will not cause instability is given by

$$|\delta(i\omega)| = \left|\frac{\Delta(i\omega)}{P(i\omega)}\right| < \frac{1}{M_t},$$

where M_t is the largest value of the complementary sensitivity

$$M_t = \sup_{\omega} |T(i\omega)| = \left\|\frac{PC}{1 + PC}\right\|_{\infty}. \tag{12.7}$$

The value of M_t is influenced by the design of the controller. For example, it is shown in Exercise 12.5 that if $M_t = 2$ then pure gain variations of 50% or pure phase variations of 30° are permitted without making the closed loop system unstable.

Example 12.5 Cruise control
Consider the cruise control system discussed in Section 3.1. The model of the car in fourth gear at speed 25 m/s is

$$P(s) = \frac{1.38}{s + 0.0142},$$

and the controller is a PI controller with gains $k_p = 0.72$ and $k_i = 0.18$. Figure 12.6 plots the allowable size of the process uncertainty using the bound in equation (12.6). At low frequencies, $T(0) = 1$ and so the perturbations can be as large as the original process ($|\delta| = |\Delta/P| < 1$). The complementary sensitivity has its maximum $M_t = 1.14$ at $\omega_{mt} = 0.35$, and hence this gives the minimum allowable process uncertainty, with $|\delta| < 0.87$ or $|\Delta| < 3.47$. Finally, at high frequencies, $T \to 0$ and hence the relative error can get very large. For example, at $\omega = 5$ we have $|T(i\omega)| = 0.195$, which means that the stability requirement is $|\delta| < 5.1$. The analysis clearly indicates that the system has good robustness and that the high-frequency properties of the transmission system are not important for the design of the cruise controller.

Another illustration of the robustness of the system is given in the right diagram

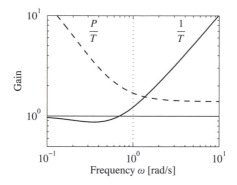
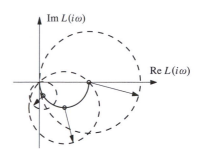

Figure 12.6: Robustness for a cruise controller. On the left the maximum relative error $1/|T|$ (solid) and the absolute error $|P|/|T|$ (dashed) for the process uncertainty Δ. The Nyquist curve is shown on the right as a solid line. The dashed circles show permissible perturbations in the process dynamics, $|\Delta| = |P|/|T|$, at the frequencies $\omega = 0, 0.0142$ and 0.05.

in Figure 12.6, which shows the Nyquist curve of the transfer function of the process and the uncertainty bounds $\Delta = |P|/|T|$ for a few frequencies. Note that the controller can tolerate large amounts of uncertainty and still maintain stability of the closed loop. ∇

The situation illustrated in the previous example is typical of many processes: moderately small uncertainties are required only around the gain crossover frequencies, but large uncertainties can be permitted at higher and lower frequencies. A consequence of this is that a simple model that describes the process dynamics well around the crossover frequency is often sufficient for design. Systems with many resonant peaks are an exception to this rule because the process transfer function for such systems may have large gains for higher frequencies also, as shown for instance in Example 9.9.

The robustness condition given by equation (12.6) can be given another interpretation by using the small gain theorem (Theorem 9.4). To apply the theorem we start with block diagrams of a closed loop system with a perturbed process and make a sequence of transformations of the block diagram that isolate the block representing the uncertainty, as shown in Figure 12.7. The result is the two-block interconnection shown in Figure 12.7c, which has the loop transfer function

$$L = \frac{PC}{1 + PC} \frac{\Delta}{P} = T\delta.$$

Equation (12.6) implies that the largest loop gain is less than 1 and hence the system is stable via the small gain theorem.

The small gain theorem can be used to check robust stability for uncertainty in a variety of other situations. Table 12.1 summarizes a few of the common cases; the proofs (all via the small gain theorem) are left as exercises.

The following example illustrates that it is possible to design systems that are robust to parameter variations.

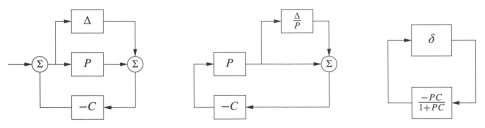

Figure 12.7: Illustration of robustness to process perturbations. A system with additive uncertainty (left) can be manipulated via block diagram algebra to one with multiplicative uncertainty $\delta = \Delta/P$ (center). Additional manipulations isolate the uncertainty in a manner that allows application of the small gain theorem (right)

Example 12.6 Bode's ideal loop transfer function

A major problem in the design of electronic amplifiers is to obtain a closed loop system that is insensitive to changes in the gain of the electronic components. Bode found that the loop transfer function $L(s) = ks^{-n}$, with $1 \le n \le 5/3$, was an ideal loop transfer function. The gain curve of the Bode plot is a straight line with slope $-n$ and the phase is constant $\arg L(i\omega) = -n\pi/2$. The phase margin is thus $\varphi_m = 90(2 - n)^\circ$ for all values of the gain k and the stability margin is $s_m = \sin \pi (1 - n/2)$. This exact transfer function cannot be realized with physical components, but it can be approximated over a given frequency range with a rational function (Exercise 12.7). An operational amplifier circuit that has the approximate transfer function $G(s) = k/(s+a)$ is a realization of Bode's ideal transfer function with $n = 1$, as described in Example 8.3. Designers of operational amplifiers go to great efforts to make the approximation valid over a wide frequency range. ∇

Youla Parameterization

Since stability is such an essential property, it is useful to characterize all controllers that stabilize a given process. Such a representation, which is called a *Youla parameterization*, is very useful when solving design problems because it makes it possible to search over all stabilizing controllers without the need to test stability explicitly.

We will first derive Youla's parameterization for a stable process with a rational transfer function P. A system with the complementary sensitivity function T can be obtained by feedforward control with the stable transfer function Q if $T = PQ$.

Table 12.1: Conditions for robust stability for different types of uncertainty

Process	Uncertainty Type	Robust Stability
$P + \Delta$	Additive	$\|CS\Delta\|_\infty < 1$
$P(1 + \delta)$	Multiplicative	$\|T\delta\|_\infty < 1$
$P/(1 + \Delta_{\mathrm{fb}} \cdot P)$	Feedback	$\|PS\Delta_{\mathrm{fb}}\|_\infty < 1$

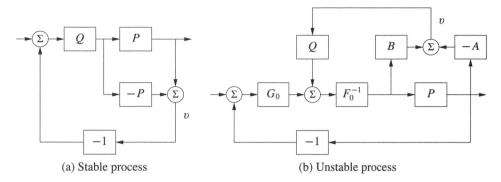

(a) Stable process   (b) Unstable process

Figure 12.8: Youla parameterization. Block diagrams of Youla parameterizations for a stable system (a) and an unstable system (b). Notice that the signal v is zero in steady state.

Notice that T must have the same right half-plane zeros as P since Q is stable. Now assume that we want to implement the complementary transfer function T by using unit feedback with the controller C. Since $T = PC/(1 + PC) = PQ$, it follows that the controller transfer function is

$$C = \frac{Q}{1 - PQ}. \tag{12.8}$$

A straightforward calculation gives

$$S = 1 - PQ, \quad PS = P(1 - PQ), \quad CS = Q, \quad T = PQ.$$

These transfer functions are all stable if P and Q are stable and the controller given by equation (12.8) is thus stabilizing. Indeed, it can be shown that all stabilizing controllers are in the form given by equation (12.8) for some choice of Q. The parameterization is illustrated by the block diagrams in Figure 12.8a.

A similar characterization can be obtained for unstable systems. Consider a process with a rational transfer function $P(s) = a(s)/b(s)$, where $a(s)$ and $b(s)$ are polynomials. By introducing a stable polynomial $c(s)$, we can write

$$P(s) = \frac{a(s)}{b(s)} = \frac{A(s)}{B(s)},$$

where $A(s) = a(s)/c(s)$ and $B(s) = b(s)/c(s)$ are stable rational functions. Similarly we introduce the controller $C_0(s) = F_0(s)/G_0(s)$, where $F_0(s)$ and $G_0(s)$ are stable rational functions. We have

$$S_0 = \frac{AF_0}{AF_0 + BG_0}, \quad PS_0 = \frac{BF_0}{AF_0 + BG_0},$$

$$C_0 S_0 = \frac{AG_0}{AF_0 + BG_0}, \quad T_0 = \frac{BG_0}{AF_0 + BG_0}.$$

The controller C_0 is stabilizing if and only if the rational function $AF_0 + BG_0$ does not have any zeros in the right half plane. Let Q be a stable rational function and

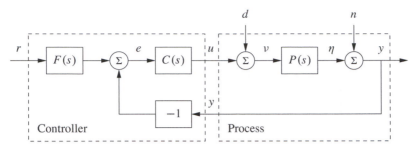

Figure 12.9: Block diagram of a basic feedback loop. The external signals are the reference signal r, the load disturbance d and the measurement noise n. The process output is y, and the control signal is u. The process P may include unmodeled dynamics, such as additive perturbations.

consider the controller

$$C = \frac{G_0 + QA}{F_0 - QB}. \tag{12.9}$$

The Gang of Four for P and C is

$$S = \frac{A(F_0 - QB)}{AF_0 + BG_0}, \qquad PS = \frac{B(F_0 - QB)}{AF_0 + BG_0},$$

$$CS = \frac{A(G_0 + QA)}{AF_0 + BG_0}, \qquad T = \frac{B(G_0 + QA)}{AF_0 + BG_0}.$$

All these transfer functions are stable if the rational function $AF_0 + BG_0$ does not have any zeros in the right half plane and the controller C given by (12.9) is therefore stabilizing for any stable Q. A block diagram of the closed loop system with the controller C is shown in Figure 12.8b. Notice that the transfer function Q appears affinely in the expressions for the Gang of Four, which is very useful if we want to determine the transfer function Q to obtain specific properties.

12.3 Performance in the Presence of Uncertainty

So far we have investigated the risk for instability and robustness to process uncertainty. We will now explore how responses to load disturbances, measurement noise and reference signals are influenced by process variations. To do this we will analyze the system in Figure 12.9, which is identical to the basic feedback loop analyzed in Chapter 11.

Disturbance Attenuation

The sensitivity function S gives a rough characterization of the effect of feedback on disturbances, as was discussed in Section 11.3. A more detailed characterization is given by the transfer function from load disturbances to process output:

$$G_{yd} = \frac{P}{1 + PC} = PS. \tag{12.10}$$

Load disturbances typically have low frequencies, and it is therefore important that the transfer function be small for low frequencies. For processes with constant low-frequency gain and a controller with integral action we have $G_{yd} \approx s/k_i$. The integral gain k_i is thus a simple measure of the attenuation of load disturbances.

To find out how the transfer function G_{yd} is influenced by small variations in the process transfer function we differentiate (12.10) with respect to P yielding

$$\frac{dG_{yd}}{dP} = \frac{1}{(1+PC)^2} = \frac{SP}{P(1+PC)} = S\frac{G_{yd}}{P},$$

and it follows that

$$\frac{dG_{yd}}{G_{yd}} = S\frac{dP}{P}. \tag{12.11}$$

The response to load disturbances is thus insensitive to process variations for frequencies where $|S(i\omega)|$ is small, i.e., for frequencies where load disturbances are important.

A drawback with feedback is that the controller feeds measurement noise into the system. In addition to the load disturbance rejection, it is thus also important that the control actions generated by measurement noise are not too large. It follows from Figure 12.9 that the transfer function G_{un} from measurement noise to controller output is given by

$$G_{un} = -\frac{C}{1+PC} = -\frac{T}{P}. \tag{12.12}$$

Since measurement noise typically has high frequencies, the transfer function G_{un} should not be too large for high frequencies. The loop transfer function PC is typically small for high frequencies, which implies that $G_{un} \approx C$ for large s. To avoid injecting too much measurement noise it is therefore important that $C(s)$ be small for large s. This property is called *high-frequency roll-off*. An example is filtering of the measured signal in a PID controller to reduce the injection of measurement noise; see Section 10.5.

To determine how the transfer function G_{un} is influenced by small variations in the process transfer, we differentiate equation (12.12):

$$\frac{dG_{un}}{dP} = \frac{d}{dP}\left(-\frac{C}{1+PC}\right) = \frac{C}{(1+PC)^2}C = T\frac{G_{un}}{P}.$$

Rearranging the terms gives

$$\frac{dG_{un}}{G_{un}} = T\frac{dP}{P}. \tag{12.13}$$

Since the complementary sensitivity function is also small for high frequencies, we find that process uncertainty has little influence on the transfer function G_{un} for frequencies where measurements are important.

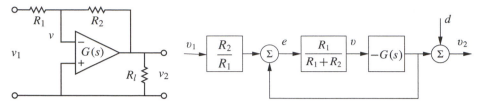

Figure 12.10: Operational amplifier with uncertain dynamics. The circuit on the left is modeled using the transfer function $G(s)$ to capture its dynamic properties and it has a load at the output. The block diagram on the right shows the input/output relationships. The load is represented as a disturbance d applied at the output of $G(s)$.

Reference Signal Tracking

The transfer function from reference to output is given by

$$G_{yr} = \frac{PCF}{1 + PC} = TF, \tag{12.14}$$

which contains the complementary sensitivity function. To see how variations in P affect the performance of the system, we differentiate equation (12.14) with respect to the process transfer function:

$$\frac{dG_{yr}}{dP} = \frac{CF}{1 + PC} - \frac{PCFC}{(1 + PC)^2} = \frac{CF}{(1 + PC)^2} = S\frac{G_{yr}}{P},$$

and it follows that

$$\frac{dG_{yr}}{G_{yr}} = S\frac{dP}{P}. \tag{12.15}$$

The relative error in the closed loop transfer function thus equals the product of the sensitivity function and the relative error in the process. In particular, it follows from equation (12.15) that the relative error in the closed loop transfer function is small when the sensitivity is small. This is one of the useful properties of feedback.

As in the last section, there are some mathematical assumptions that are required for the analysis presented here to hold. As already stated, we require that the perturbations Δ be small (as indicated by writing dP). Second, we require that the perturbations be stable, so that we do not introduce any new right half-plane poles that would require additional encirclements in the Nyquist criterion. Also, as before, this condition is conservative: it allows for any perturbation that satisfies the given bounds, while in practice the perturbations may be more restricted.

Example 12.7 Operational amplifier circuit
To illustrate the use of these tools, consider the performance of an op amp-based amplifier, as shown in Figure 12.10. We wish to analyze the performance of the amplifier in the presence of uncertainty in the dynamic response of the op amp and changes in the loading on the output. We model the system using the block diagram in Figure 12.10b, which is based on the derivation in Example 9.1.

Consider first the effect of unknown dynamics for the operational amplifier. If we model the dynamics of the op amp as $v_2 = -G(s)v$, then the transfer function for the overall circuit is given by

$$G_{v_2 v_1} = -\frac{R_2}{R_1} \frac{G(s)}{G(s) + R_2/R_1 + 1}.$$

We see that if $G(s)$ is large over the desired frequency range, then the closed loop system is very close to the ideal response $\alpha = R_2/R_1$. Assuming $G(s) = b/(s+a)$, where b is the gain-bandwidth product of the amplifier, as discussed in Example 8.3, the sensitivity function and the complementary sensitivity function become

$$S = \frac{s+a}{s+a+\alpha b}, \qquad T = \frac{\alpha b}{s+a+\alpha b}.$$

The sensitivity function around the nominal values tells us how the tracking response response varies as a function of process perturbations:

$$\frac{dG_{yr}}{G_{yr}} = S \frac{dP}{P}.$$

We see that for low frequencies, where S is small, variations in the bandwidth a or the gain-bandwidth product b will have relatively little effect on the performance of the amplifier (under the assumption that b is sufficiently large.

To model the effects of an unknown load, we consider the addition of a disturbance at the output of the system, as shown in Figure 12.10b. This disturbance represents changes in the output voltage due to loading effects. The transfer function $G_{yd} = S$ gives the response of the output to the load disturbance, and we see that if S is small, then we are able to reject such disturbances. The sensitivity of G_{yd} to perturbations in the process dynamics can be computed by taking the derivative of G_{yd} with respect to P:

$$\frac{dG_{yd}}{dP} = \frac{-C}{(1+PC)^2} = -\frac{T}{P}G_{yd} \qquad \Longrightarrow \qquad \frac{dG_{yd}}{G_{yd}} = -T\frac{dP}{P}.$$

Thus we see that the relative changes in the disturbance rejection are roughly the same as the process perturbations at low frequency (when T is approximately 1) and drop off at higher frequencies. However, it is important to remember that G_{yd} itself is small at low frequency, and so these variations in relative performance may not be an issue in many applications. ∇

12.4 Robust Pole Placement

In Chapters 6 and 7 we saw how to design controllers by setting the locations of the eigenvalues of the closed loop system. If we analyze the resulting system in the frequency domain, the closed loop eigenvalues correspond to the poles of the closed loop transfer function and hence these methods are often referred to as design by *pole placement*.

State space design methods, like many methods developed for control system design, do not explicitly take robustness into account. In such cases it is essential to

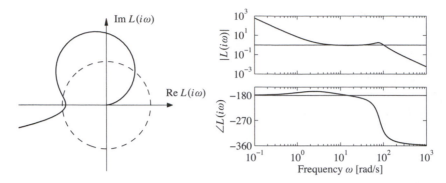

Figure 12.11: Observer-based control of steering. The Nyquist plot (left) and Bode plot (right) of the loop transfer function for vehicle steering with a controller based on state feedback and an observer. The controller provides stable operation, but with very low gain and phase margin.

always investigate the robustness because there are seemingly reasonable designs that give controllers with poor robustness. We illustrate this by analyzing controllers designed by state feedback and observers. The closed loop poles can be assigned to arbitrary locations if the system is observable and reachable. However, if we want to have a robust closed loop system, the poles and zeros of the process impose severe restrictions on the location of the closed loop poles. Some examples are first given; based on the analysis of these examples we then present design rules for robust pole (eigenvalue) placement.

Slow Stable Process Zeros

We will first explore the effects of slow stable zeros, and we begin with a simple example.

Example 12.8 Vehicle steering
Consider the linearized model for vehicle steering in Example 8.6, which has the transfer function

$$P(s) = \frac{0.5s + 1}{s^2}.$$

A controller based on state feedback was designed in Example 6.4, and state feedback was combined with an observer in Example 7.4. The system simulated in Figure 7.8 has closed loop poles specified by $\omega_c = 0.3$, $\zeta_c = 0.707$, $\omega_o = 7$ and $\zeta_o = 9$. Assume that we want a faster closed loop system and choose $\omega_c = 10$, $\zeta_c = 0.707$, $\omega_o = 20$ and $\zeta_o = 0.707$. Using the state representation in Example 7.3, a pole placement design gives state feedback gains $k_1 = 100$ and $k_2 = -35.86$ and observer gains $l_1 = 28.28$ and $l_2 = 400$. The controller transfer function is

$$C(s) = \frac{-11516s + 40000}{s^2 + 42.4s + 6657.9}.$$

Figure 12.11 shows Nyquist and Bode plots of the loop transfer function. The

Nyquist plot indicates that the robustness is poor since the loop transfer function is very close to the critical point -1. The phase margin is $7°$ and the stability margin is $s_m = 0.077$. The poor robustness shows up in the Bode plot, where the gain curve hovers around the value 1 and the phase curve is close to $-180°$ for a wide frequency range. More insight is obtained by analyzing the sensitivity functions, shown by solid lines in Figure 12.12. The maximum sensitivities are $M_s = 13$ and $M_t = 12$, indicating that the system has poor robustness.

At first sight it is surprising that a controller where the nominal closed system has well damped poles and zeros is so sensitive to process variations. We have an indication that something is unusual because the controller has a zero at $s = 3.5$ in the right half-plane. To understand what happens, we will investigate the reason for the peaks of the sensitivity functions.

Let the transfer functions of the process and the controller be

$$P(s) = \frac{n_p(s)}{d_p(s)}, \qquad C(s) = \frac{n_c(s)}{d_c(s)},$$

where $n_p(s), n_c(s), d_p(s)$ and $d_c(s)$ are the numerator and denominator polynomials. The complementary sensitivity function is

$$T(s) = \frac{PC}{1 + PC} = \frac{n_p(s)n_c(s)}{d_p(s)d_c(s) + n_p(s)n_p(s)}.$$

The poles of $T(s)$ are the poles of the closed loop system and the zeros are given by the zeros of the process and controller. Sketching the gain curve of the complementary sensitivity function we find that $T(s) = 1$ for low frequencies and that $|T(i\omega)|$ starts to increase at its first zero, which is the process zero at $s = -2$. It increases further at the controller zero at $s = 3.5$, and it does not start to decrease until the closed loop poles appear at $\omega_c = 10$ and $\omega_o = 20$. We can thus conclude that there will be a peak in the complementary sensitivity function. The magnitude of the peak depends on the ratio of the zeros and the poles of the transfer function.

The peak of the complementary sensitivity function can be avoided by assigning a closed loop pole close to the slow process zero. We can achieve this by choosing $\omega_c = 10$ and $\zeta_c = 2.6$, which gives closed loop poles at $s = -2$ and $s = -50$. The controller transfer function then becomes

$$C(s) = \frac{3628s + 40000}{s^2 + 80.28s + 156.56} = 3628 \frac{s + 11.02}{(s + 2)(s + 78.28)}.$$

The sensitivity functions are shown by dashed lines in Figure 12.12. The controller gives the maximum sensitivities $M_s = 1.34$ and $M_t = 1.41$, which give much better robustness. Notice that the controller has a pole at $s = -2$ that cancels the slow process zero. The design can also be done simply by canceling the slow stable process zero and designing the controller for the simplified system. ∇

One lesson from the example is that it is necessary to choose closed loop poles that are equal to or close to slow stable process zeros. Another lesson is that slow unstable process zeros impose limitations on the achievable bandwidth, as already

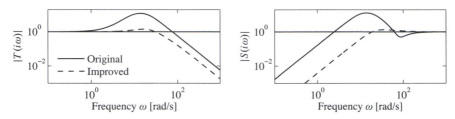

Figure 12.12: Sensitivity functions for observer-based control of vehicle steering. The complementary sensitivity function (left) and the sensitivity function (right) for the original controller with $\omega_c = 10$, $\zeta_c = 0.707$, $\omega_o = 20$, $\zeta_o = 0.707$ (solid) and the improved controller with $\omega_c = 10$, $\zeta_c = 2.6$ (dashed).

noted in Section 11.5.

Fast Stable Process Poles

The next example shows the effect of fast stable poles.

Example 12.9 Fast system poles

Consider a PI controller for a first-order system, where the process and the controller have the transfer functions $P(s) = b/(s + a)$ and $C(s) = k_p + k_i/s$. The loop transfer function is

$$L(s) = \frac{b(k_p s + k_i)}{s(s + a)},$$

and the closed loop characteristic polynomial is.

$$s(s + a) + b(k_p s + k_i) = s^2 + (a + bk_p)s + k_i b$$

If we specify the desired closed loop poles should be $-p_1$ and $-p_2$, we find that the controller parameters are given by

$$k_p = \frac{p_1 + p_2 - a}{b}, \qquad k_i = \frac{p_1 p_2}{b}.$$

The sensitivity functions are then

$$S(s) = \frac{s(s + a)}{(s + p_1)(s + p_2)}, \qquad T(s) = \frac{(p_1 + p_2 - a)s + p_1 p_2}{(s + p_1)(s + p_2)}.$$

Assume that the process pole $-a$ is much more negative than the closed loop poles $-p_1$ and $-p_2$, say, $p_1 < p_2 \ll a$. Notice that the proportional gain is negative and that the controller has a zero in the right half-plane if $a > p_1 + p_2$, an indication that the system has bad properties.

Next consider the sensitivity function, which is 1 for high frequencies. Moving from high to low frequencies, we find that the sensitivity increases at the process pole $s = -a$. The sensitivity does not decrease until the closed loop poles are reached, resulting in a large sensitivity peak that is approximately a/p_2. The magnitude of the sensitivity function is shown in Figure 12.13 for $a = b = 1$, $p_1 = 0.05$ and $p_2 = 0.2$. Notice the high-sensitivity peak. For comparison we also show the gain

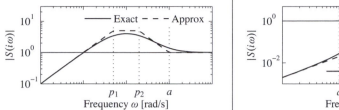
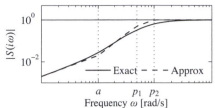

Figure 12.13: Gain curves for Bode plots of the sensitivity function S for designs with $p_1 < p_2 < a$ (left) and $a < p_1 < p_2$ (right). The solid lines are the true sensitivities, and the dashed lines are the asymptotes.

curve for the case when the closed loop poles ($p_1 = 5$, $p_2 = 20$) are faster than the process pole ($a = 1$).

The problem with poor robustness can be avoided by choosing one closed loop pole equal to the process pole, i.e., $p_2 = a$. The controller gains then become

$$k_p = \frac{p_1}{b}, \qquad k_i = \frac{ap_1}{l},$$

which means that the fast process pole is canceled by a controller zero. The loop transfer function and the sensitivity functions are

$$L(s) = \frac{bk_p}{s}, \qquad S(s) = \frac{s}{s + bk_p}, \qquad T(s) = \frac{bk_p}{s + bk_p}.$$

The maximum sensitivities are now less than 1 for all frequencies. Notice that this is possible because the process transfer function goes to zero as s^{-1}. ▽

Design Rules for Pole Placement

Based on the insight gained from the examples, it is now possible to obtain design rules that give designs with good robustness. Consider the expression (12.8) for maximum complementary sensitivity, repeated here:

$$M_t = \sup_\omega |T(i\omega)| = \left\| \frac{PC}{1 + PC} \right\|_\infty.$$

Let ω_{gc} be the desired gain crossover frequency. Assume that the process has zeros that are slower than ω_{gc}. The complementary sensitivity function is 1 for low frequencies, and it increases for frequencies close to the process zeros unless there is a closed loop pole in the neighborhood. To avoid large values of the complementary sensitivity function we find that the closed loop system should therefore have poles close to or equal to the slow stable zeros. This means that slow stable zeros should be canceled by controller poles. Since unstable zeros cannot be canceled, the presence of slow unstable zeros means that achievable gain crossover frequency must be smaller than the slowest unstable process zero.

Now consider process poles that are faster than the desired gain crossover frequency. Consider the expression for the maximum of the sensitivity function:

$$M_s = \sup_{\omega} |S(i\omega)| = \left\| \frac{1}{1 + PC} \right\|_\infty.$$

The sensitivity function is 1 for high frequencies. Moving from high to low frequencies, the sensitivity function increases at the fast process poles. Large peaks can result unless there are closed loop poles close to the fast process poles. To avoid large peaks in the sensitivity the closed loop system should therefore have poles that match the fast process poles. This means that the controller should cancel the fast process poles by controller zeros. Since unstable modes cannot be canceled, the presence of a fast unstable pole implies that the gain crossover frequency must be sufficiently large.

To summarize, we obtain the following simple rule for choosing closed loop poles: slow stable process zeros should be matched by slow closed loop poles, and fast stable process poles should be matched by fast closed loop poles. Slow unstable process zeros and fast unstable process poles impose severe limitations.

Example 12.10 Nanopositioning system for an atomic force microscope

A simple nanopositioner was explored in Example 9.9, where it was shown that the system could be controlled using an integral controller. The performance of the closed loop was poor because the gain crossover frequency was limited to $\omega_{gc} = 2\zeta\omega_0(1 - s_m)$. It can be shown that little improvement is obtained by using a PI controller. To achieve improved performance, we will therefore apply PID control. For a modest performance increase, we will use the design rule derived in Example 12.9 that fast stable process poles should be canceled by controller zeros. The controller transfer function should thus be chosen as

$$C(s) = \frac{k_d s^2 + k_p s + k_i}{s} = \frac{k_i}{s} \frac{s^2 + 2\zeta s + a^2}{a^2}, \tag{12.16}$$

which gives $k_p = 2\zeta k_i/a$ and $k_d = k_i/a^2$.

Figure 12.14 shows the gain curves for the Gang of Four for a system designed with $k_i = 0.5$. A comparison with Figure 9.12 shows that the bandwidth is increased significantly from $\omega_{gc} = 0.01$ to $\omega_{gc} = k_i = 0.5$. Since the process pole is canceled, the system will, however, still be very sensitive to load disturbances with frequencies close to the resonant frequency. The gain curve of CS has a dip or a notch at the resonant frequency, which implies that the controller gain is very low for frequencies around the resonance. The gain curve also shows that the system is very sensitive to high-frequency noise. The system will likely be unusable because the gain goes to infinity for high frequencies.

The sensitivity to high frequency noise can be remedied by modifying the controller to be

$$C(s) = \frac{k_i}{s} \frac{s^2 + 2\zeta a s + a^2}{a^2(1 + sT_f + (sT_f)^2/2)}, \tag{12.17}$$

which has high-frequency roll-off. Selection of the constant T_f for the filter is a compromise between attenuation of high-frequency measurement noise and ro-

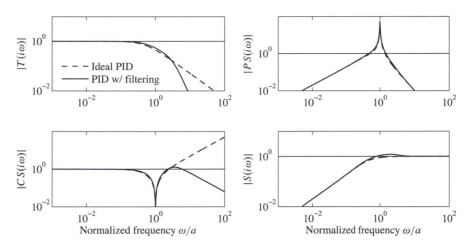

Figure 12.14: Nanopositioning system control via cancellation of the fast process pole. Gain plots for the Gang of Four for PID control with second-order filtering (12.17) are shown by solid lines, and the dashed lines show results for an ideal PID controller without filtering (12.16).

bustness. A large value of T_f reduces the effects of sensor noise significantly, but it also reduces the stability margin. Since the gain crossover frequency without filtering is k_i, a reasonable choice is $T_F = 0.2/T_f$, as shown by the solid curves in Figure 12.14. The plots of $|CS(i\omega)|$ and $|S(i\omega)|$ show that the sensitivity to high-frequency measurement noise is reduced dramatically at the cost of a marginal increase of sensitivity. Notice that the poor attenuation of disturbances with frequencies close to the resonance is not visible in the sensitivity function because of the exact cancellation of poles and zeros.

The designs thus far have the drawback that load disturbances with frequencies close to the resonance are not attenuated. We will now consider a design that actively attenuates the poorly damped modes. We start with an ideal PID controller where the design can be done analytically, and we add high-frequency roll-off. The loop transfer function obtained with this controller is

$$L(s) = \frac{k_d s^2 + k_p s + k_i}{s(s^2 + 2\zeta a s + a^2)}. \tag{12.18}$$

The closed loop system is of third order, and its characteristic polynomial is

$$s^3 + (k_d a^2 + 2\zeta a)s^2 + (k_p + 1)a^2 s + k_i a^2. \tag{12.19}$$

A general third-order polynomial can be parameterized as

$$s^3 + (\alpha_0 + 2\zeta)\omega_0 s^2 + (1 + 2\alpha_0\zeta)\omega_0^2 s + \alpha_0\omega_0^3. \tag{12.20}$$

The parameters α_0 and ζ give the relative configuration of the poles, and the parameter ω_0 gives their magnitudes, and therefore also the bandwidth of the system.

The identification of coefficients of equal powers of s with equation (12.19)

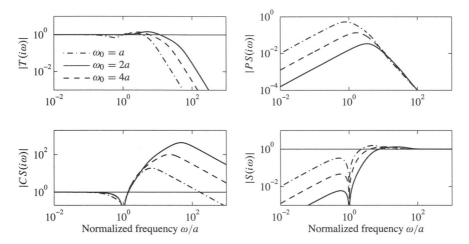

Figure 12.15: Nanopositioner control using active damping. Gain curves for the Gang of Four for PID control of the nanopositioner designed for $\omega_0 = a$ (dash-dotted), $2a$ (dashed), and $4a$ (solid). The controller has high-frequency roll-off and has been designed to give active damping of the oscillatory mode. The different curves correspond to different choices of magnitudes of the poles, parameterized by ω_0 in equation (12.19).

gives a linear equation for the controller parameters, which has the solution

$$k_p = \frac{(1 + 2\alpha_0\zeta)\omega_0^2}{a^2} - 1, \qquad k_i = \frac{\alpha_0\omega_0^3}{a^2}, \qquad k_d = \frac{(\alpha_0 + 2\zeta)\omega_0}{a^2} - 2\zeta a. \quad (12.21)$$

To obtain a design with active damping, it is necessary that the closed loop bandwidth be at least as fast as the oscillatory modes. Adding high-frequency roll-off, the controller becomes

$$C(s) = \frac{k_d s^2 + k_p s + k}{s(1 + sT_f + (sT_f)^2/2)}. \quad (12.22)$$

The value $T_f = T_d/10 = 0.1\, k_d/k$ is a good value for the filtering time constant.

Figure 12.15 shows the gain curves of the Gang of Four for designs with $\zeta = 0.707$, $\alpha_0 = 1$ and $\omega_0 = a, 2a$ and $4a$. The figure shows that the largest values of the sensitivity function and the complementary sensitivity function are small. The gain curve for PS shows that the load disturbances are now well attenuated over the whole frequency range, and attenuation increases with increasing ω_0. The gain curve for CS shows that large control signals are required to provide active damping. The high gain of CS for high frequencies also shows that low-noise sensors and actuators with a wide range are required. The largest gains for CS are 19, 103 and 434 for $\omega_0 = a, 2a$ and $4a$, respectively. There is clearly a trade-off between disturbance attenuation and controller gain. A comparison of Figures 12.14 and 12.15 illustrates the trade-offs between control action and disturbance attenuation for the designs with cancellation of the fast process pole and active damping.   ∇

12.5 Design for Robust Performance

Control design is a rich problem where many factors have to be taken into account. Typical requirements are that load disturbances should be attenuated, the controller should inject only a moderate amount of measurement noise, the output should follow variations in the command signal well and the closed loop system should be insensitive to process variations. For the system in Figure 12.9 these requirements can be captured by specifications on the sensitivity functions S and T and the transfer functions G_{yd}, G_{un}, G_{yr} and G_{ur}. Notice that it is necessary to consider at least six transfer functions, as discussed Section 11.1. The requirements are mutually conflicting, and it is necessary to make trade-offs. The attenuation of load disturbances will be improved if the bandwidth is increased, but so will the noise injection.

It is highly desirable to have design methods that can guarantee robust performance. Such design methods did not appear until the late 1980s. Many of these design methods result in controllers having the same structure as the controller based on state feedback and an observer. In this section we provide a brief review of some of the techniques as a preview for those interested in more specialized study.

Quantitative Feedback Theory

Quantitative feedback theory (QFT) is a graphical design method for robust loop shaping that was developed by I. M. Horowitz [104]. The idea is to first determine a controller that gives a complementary sensitivity that is robust to process variations and then to shape the response to reference signals by feedforward. The idea is illustrated in Figure 12.16a, which shows the level curves of the complementary sensitivity function T on a Nyquist plot. The complementary sensitivity function has unit gain on the line $\operatorname{Re} L(i\omega) = -0.5$. In the neighborhood of this line, significant variations in process dynamics only give moderate changes in the complementary transfer function. The dashed part of the figure corresponds to the region $0.9 < |T(i\omega)| < 1.1$. To use the design method, we represent the uncertainty for each frequency by a region and attempt to shape the loop transfer function so that the variation in T is as small as possible. The design is often performed using the Nichols chart shown in Figure 12.16b.

Linear Quadratic Control

One way to make the trade-off between the attenuation of load disturbances and the injection of measurement noise is to design a controller that minimizes the loss function

$$J = \frac{1}{T} \int_0^T \left(y^2(t) + \rho u^2(t) \right) dt,$$

where ρ is a weighting parameter as discussed in Section 6.3. This loss function gives a compromise between load disturbance attenuation and disturbance injection because it balances control actions against deviations in the output. If all state

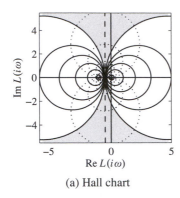

(a) Hall chart

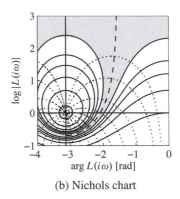

(b) Nichols chart

Figure 12.16: Hall and Nichols charts. The Hall chart is a Nyquist plot with curves for constant gain and phase of the complementary sensitivity function T. The Nichols chart is the conformal map of the Hall chart under the transformation $N = \log L$ (with the scale flipped). The dashed curve is the line where $|T(i\omega)| = 1$, and the shaded region corresponding to loop transfer functions whose complementary sensitivity changes by no more than $\pm 10\%$ is shaded.

variables are measured, the controller is a state feedback $u = -Kx$ and it has the same form as the controller obtained by eigenvalue assignment (pole placement) in Section 6.2. However, the controller gain is obtained by solving an optimization problem. It has been shown that this controller is very robust. It has a phase margin of at least $60°$ and an infinite gain margin. The controller is called a *linear quadratic control* or *LQ control* because the process model is linear and the criterion is quadratic.

When all state variables are not measured, the state can be reconstructed using an observer, as discussed in Section 7.3. It is also possible to introduce process disturbances and measurement noise explicitly in the model and to reconstruct the states using a Kalman filter, as discussed briefly in Section 7.4. The Kalman filter has the same structure as the observer designed by eigenvalue assignment in Section 7.3, but the observer gains L are now obtained by solving an optimization problem. The control law obtained by combining linear quadratic control with a Kalman filter is called *linear quadratic Gaussian control* or *LQG control*. The Kalman filter is optimal when the models for load disturbances and measurement noise are Gaussian.

It is interesting that the solution to the optimization problem leads to a controller having the structure of a state feedback and an observer. The state feedback gains depend on the parameter ρ, and the filter gains depend on the parameters in the model that characterize process noise and measurement noise (see Section 7.4). There are efficient programs to compute these feedback and observer gains.

The nice robustness properties of state feedback are unfortunately lost when the observer is added. It is possible to choose parameters that give closed loop systems with poor robustness, similar to Example 12.8. We can thus conclude that there is a

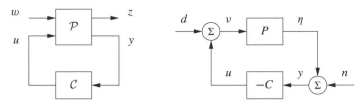

Figure 12.17: H_∞ robust control formulation. The left figure shows a general representation of a control problem used in robust control. The input u represents the control signal, the input w represents the external influences on the system, the output z is the generalized error and the output y is the measured signal. The right figure shows the special case of the basic feedback loop in Figure 12.9 where the reference signal is zero. In this case we have $w = (n, d)$ and $z = (y, -u)$.

fundamental difference between using sensors for all states and reconstructing the states using an observer.

H_∞ Control

Robust control design is often called H_∞ control for reasons that will be explained shortly. The basic ideas are simple, but the details are complicated and we will therefore just give the flavor of the results. A key idea is illustrated in Figure 12.17, where the closed loop system is represented by two blocks, the process $\mathcal{P}$ and the controller $\mathcal{C}$ as discussed in Section 11.1. The process $\mathcal{P}$ has two inputs, the control signal u, which can be manipulated by the controller, and the generalized disturbance w, which represents all external influences, e.g., command signals and disturbances. The process has two outputs, the generalized error z, which is a vector of error signals representing the deviation of signals from their desired values, and the measured signal y, which can be used by the controller to compute u. For a linear system and a linear controller the closed loop system can be represented by the linear system

$$z = H(P(s), C(s))w, \tag{12.23}$$

which tells how the generalized error z depends on the generalized disturbances w. The control design problem is to find a controller C such that the gain of the transfer function H is small even when the process has uncertainties. There are many different ways to specify uncertainty and gain, giving rise to different designs. The names H_2 and H_∞ control correspond to the norms $\|H\|_2$ and $\|H\|_\infty$.

To illustrate the ideas we will consider a regulation problem for a system where the reference signal is assumed to be zero and the external signals are the load disturbance d and the measurement noise n, as shown in Figure 12.17b. The generalized input is $w = (-n, d)$. (The negative sign of n is not essential but is chosen to obtain somewhat nicer equations.) The generalized error is chosen as $z = (\eta, \nu)$, where η is the process output and ν is the part of the load disturbance that is not compensated by the controller. The closed loop system is thus modeled by

$$z = \begin{bmatrix} y \\ -u \end{bmatrix} = \begin{bmatrix} \dfrac{1}{1+PC} & \dfrac{P}{1+PC} \\ \dfrac{C}{1+PC} & \dfrac{PC}{1+PC} \end{bmatrix} \begin{bmatrix} n \\ d \end{bmatrix} = H(P,C) \begin{bmatrix} n \\ d \end{bmatrix}, \qquad (12.24)$$

which is the same as equation (12.23). A straightforward calculation shows that

$$\|H(P,C))\|_\infty = \sup_\omega \frac{\sqrt{(1+|P(i\omega)|^2)(1+|C(i\omega)|^2)}}{|1+P(i\omega)C(i\omega)|}. \qquad (12.25)$$

There are numerical methods for finding a controller such that $\|H(P,C)\|_\infty < \gamma$, if such a controller exists. The best controller can then be found by iterating on γ. The calculations can be made by solving *algebraic Riccati* equations, e.g., by using the command `hinfsyn` in MATLAB. The controller has the same order as the process and the same structure as the controller based on state feedback and an observer; see Figure 7.7 and Theorem 7.3.

Notice that if we minimize $\|H(P,C)\|_\infty$, we make sure that the transfer functions $G_{yd} = P/(1+PC)$, representing the transmission of load disturbances to the output, and $G_{un} = -C/(1+PC)$, representing how measurement noise is transmitted to the control signal, are small. Since the sensitivity and the complementary sensitivity functions are also elements of $H(P,C)$, we have also guaranteed that the sensitivities are less than γ. The design methods thus balance performance and robustness.

There are strong robustness results associated with the H_∞ controller. It follows from equations (12.4) and (12.25) that

$$\|H(P,C)\|_\infty = \frac{1}{\delta_\nu(P,-1/C)}. \qquad (12.26)$$

The inverse of $\|H(P,C)\|_\infty$ is thus equal to the Vinnicombe distance between P and $-1/C$ and can therefore be interpreted as a *generalized stability margin*. Compare this with s_m, which we defined as the shortest distance between the Nyquist curve of the loop transfer function and the critical point -1. It also follows that if we find a controller C with $\|H(P,C)\|_\infty < \gamma$, then this controller will stabilize any process P_* such that $\delta_\nu(P,P_*) < 1/\gamma$.

Disturbance Weighting

Minimizing the gain $\|H(P,C)\|_\infty$ means that the gains of all individual signal transmissions from disturbances to outputs are less than γ for all frequencies of the input signals. The assumption that the disturbances are equally important and that all frequencies are also equally important is not very realistic; recall that load disturbances typically have low frequencies and measurement noise is typically dominated by high frequencies. It is straightforward to modify the problem so that disturbances of different frequencies are given different emphasis, by introducing

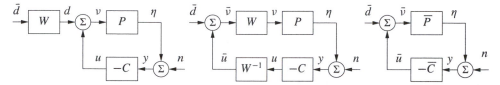

Figure 12.18: Block diagrams of a system with disturbance weighting. The left figure provides a frequency weight on processes disturbances. Through block diagram manipulation, this can be converted to the standard problem on the right.

a weighting filter on the load disturbance as shown in Figure 12.18. For example, low-frequency load disturbances will be enhanced by choosing W as a low-pass filter because the actual load disturbance is $W\bar{d}$.

By using block diagram manipulation as shown in Figure 12.18, we find that the system with frequency weighting is equivalent to the system with no frequency weighting in Figure 12.18 and the signals are related through

$$\bar{z} = \begin{bmatrix} y \\ \bar{u} \end{bmatrix} \begin{bmatrix} \dfrac{1}{1+\overline{P}\,\overline{C}} & \dfrac{\overline{P}}{1+\overline{P}\,\overline{C}} \\ \dfrac{\overline{C}}{1+\overline{P}\,\overline{C}} & \dfrac{\overline{P}\,\overline{C}}{1+\overline{P}\,\overline{C}} \end{bmatrix} \begin{bmatrix} n \\ \bar{d} \end{bmatrix} = H(\overline{P},\overline{C})\bar{w}, \qquad (12.27)$$

where $\overline{P} = PW$ and $\overline{C} = W^{-1}C$. The problem of finding a controller $\overline{C}$ that minimizes the gain of $H(\overline{P},\overline{C})$ is thus equivalent to the problem without disturbance weighting; having obtained $\overline{C}$, the controller for the original system is then $C = W\overline{C}$. Notice that if we introduce the frequency weighting $W = k/s$, we will automatically get a controller with integral action.

Limits of Robust Design

There is a limit to what can be achieved by robust design. In spite of the nice properties of feedback, there are situations where the process variations are so large that it is not possible to find a linear controller that gives a robust system with good performance. It is then necessary to use other types of controllers. In some cases it is possible to measure a variable that is well correlated with the process variations. Controllers for different parameter values can then be designed and the corresponding controller can be chosen based on the measured signal. This type of control design is called *gain scheduling*. The cruise controller is a typical example where the measured signal could be gear position and velocity. Gain scheduling is the common solution for high-performance aircraft where scheduling is done based on Mach number and dynamic pressure. When using gain scheduling, it is important to make sure that switches between the controllers do not create undesirable transients (often referred to as *bumpless transfer*).

If it is not possible to measure variables related to the parameters, *automatic tuning* and *adaptive control* can be used. In automatic tuning the process dynamics are measured by perturbing the system, and a controller is then designed automatically. Automatic tuning requires that parameters remain constant, and it has been widely applied for PID control. It is a reasonable guess that in the future many controllers will have features for automatic tuning. If parameters are changing, it is possible to use adaptive methods where process dynamics are measured online.

12.6 Further Reading

The topic of robust control is a large one, with many articles and textbooks devoted to the subject. Robustness was a central issue in classical control as described in Bode's classical book [40]. Robustness was deemphasized in the euphoria of the development of design methods based on optimization. The strong robustness of controllers based on state feedback, shown by Anderson and Moore [7], contributed to the optimism. The poor robustness of output feedback was pointed out by Rosenbrock [169], Horowitz [103] and Doyle [63] and resulted in a renewed interest in robustness. A major step forward was the development of design methods where robustness was explicitly taken into account, such as the seminal work of Zames [208]. Robust control was originally developed using powerful results from the theory of complex variables, which gave controllers of high order. A major breakthrough was made by Doyle, Glover, Khargonekar and Francis [65], who showed that the solution to the problem could be obtained using Riccati equations and that a controller of low order could be found. This paper led to an extensive treatment of H_∞ control, including books by Francis [78], McFarlane and Glover [150], Doyle, Francis and Tannenbaum [64], Green and Limebeer [90], Zhou, Doyle and Glover [209], Skogestand and Postlethwaite [181] and Vinnicombe [196]. A major advantage of the theory is that it combines much of the intuition from servomechanism theory with sound numerical algorithms based on numerical linear algebra and optimization. The results have been extended to nonlinear systems by treating the design problem as a game where the disturbances are generated by an adversary, as described in the book by Basar and Bernhard [24]. Gain scheduling and adaptation are discussed in the book by Åström and Wittenmark [19].

Exercises

12.1 Consider systems with the transfer functions $P_1 = 1/(s + 1)$ and $P_2 = 1/(s + a)$. Show that P_1 can be changed continuously to P_2 with bounded additive and multiplicative uncertainty if $a > 0$ but not if $a < 0$. Also show that no restriction on a is required for feedback uncertainty.

12.2 Consider systems with the transfer functions $P_1 = (s + 1)/(s + 1)^2$ and $P_2 = (s + a)/(s + 1)^2$. Show that P_1 can be changed continuously to P_2 with bounded feedback uncertainty if $a > 0$ but not if $a < 0$. Also show that no restriction on a is required for additive and multiplicative uncertainties.

12.3 (Difference in sensitivity functions) Let $T(P, C)$ be the complementary sensitivity function for a system with process P and controller C. Show that

$$T(P_1, C) - T(P_2, C) = \frac{(P_1 - P_2)C}{(1 + P_1 C)(1 + P_2 C)},$$

and derive a similar formula for the sensitivity function.

12.4 (The Riemann sphere) Consider systems with the transfer functions $P_1 = k/(s+1)$ and $P_2 = k/(s-1)$. Show that

$$d(P_1, P_2) = \frac{2k}{1 + k^2}, \qquad \delta_v(P_1, P_2) = \begin{cases} 1, & \text{if } k < 1 \\ \frac{2k}{1+k^2} & \text{otherwise.} \end{cases}$$

Use the Riemann sphere to show geometrically that $\delta_v(P_1, P_2) = 1$ if $k < 1$. (Hint: It is sufficient to evaluate the transfer function for $\omega = 0$.)

12.5 (Stability margins) Consider a feedback loop with a process and a controller having transfer functions P and C. Assume that the maximum sensitivity is $M_s = 2$. Show that the phase margin is at least $30°$ and that the closed loop system will be stable if the gain is changed by 50%.

12.6 (Bode's ideal loop transfer function) Make Bode and Nyquist plots of Bode's ideal loop transfer function. Show that the phase margin is $\varphi_m = 180° - 90°n$ and that the stability margin is $s_m = \arcsin \pi (1 - n/2)$.

12.7 Consider a process with the transfer function $P(s) = k/(s(s+1))$, where the gain can vary between 0.1 and 10. A controller that is robust to these gain variations can be obtained by finding a controller that gives the loop transfer function $L(s) = 1/(s\sqrt{s})$. Suggest how the transfer function can be implemented by approximating it by a rational function.

12.8 (Smith predictor) The *Smith predictor*, a controller for systems with time delays, is a special version of Figure 12.8a with $P(s) = e^{-s\tau} P_0(s)$ and $C(s) = C_0(s)/(1 + C_0(s)P(s))$. The controller $C_0(s)$ is designed to give good performance for the process $P_0(s)$. Show that the sensitivity functions are

$$S(s) = \frac{1 + (1 - e^{-s\tau})P_0(s)C_0(s)}{1 + P_0(s)C_0(s)}, \qquad T(s) = \frac{P_0(s)e^{-s\tau}C_0(s)}{1 + P_0(s)C_0(s)}.$$

12.9 (Ideal delay compensator) Consider a process whose dynamics are a pure time delay with transfer function $P(s) = e^{-s}$. The ideal delay compensator is a controller with the transfer function $C(s) = 1/(1 - e^{-s})$. Show that the sensitivity functions are $T(s) = e^{-s}$ and $S(s) = 1 - e^{-s}$ and that the closed loop system will be unstable for arbitrarily small changes in the delay.

12.10 (Vehicle steering) Consider the Nyquist curve in Figure 12.11. Explain why part of the curve is approximately a circle. Derive a formula for the center and the radius and compare with the actual Nyquist curve.

12.11 Consider a process with the transfer function

$$P(s) = \frac{(3+3)(s+200)}{(s+1)(s^2+10s+40)(s+40)}.$$

Discuss suitable choices of closed loop poles for a design that gives dominant poles with undamped natural frequency 1 and 10.

12.12 (AFM nanopositioning system) Consider the design in Example 12.10 and explore the effects of changing parameters α_0 and ζ_0.

12.13 (H_∞ control) Consider the matrix $H(P, C)$ in equation (12.24). Show that it has the singular values

$$\sigma_1 = 0, \qquad \sigma_2 = \bar{\sigma} = \sup_{\omega} \frac{\sqrt{(1+|P(i\omega)|^2)(1+|C(i\omega)|^2)}}{|1+P(i\omega)C(i\omega)|} = \|H(P, C))\|_\infty.$$

Also show that $\bar{\sigma} = 1/d_v(P, -1/C)$, which implies that $1/\bar{\sigma}$ is a generalization of the closest distance of the Nyquist plot to the critical point.

12.14 Show that

$$\delta_v(P, -1/C) = \inf_{\omega} \frac{|P(i\omega) + 1/C(i\omega)|}{\sqrt{(1+|P(i\omega)|^2)(1+1/|C(i\omega)|^2)}} = \frac{1}{\|H(P, C))\|_\infty}.$$

12.15 Consider the system

$$\frac{dx}{dt} = Ax + Bu = \begin{bmatrix} -1 & 0 \\ 1 & 0 \end{bmatrix} x + \begin{bmatrix} a-1 \\ 1 \end{bmatrix} u, \qquad y = Cx = \begin{bmatrix} 0 & 1 \end{bmatrix} y.$$

Design a state feedback that gives $\det(sI - BK) = s^2 + 2\zeta_c\omega_c s + \omega_c^2$, and an observer with $\det(sI - LC) = s^2 + 2\zeta_o\omega_o s + \omega_o^2$ and combine them using the separation principle to get an output feedback. Choose the numerical values $a = 1.5$, $\omega_c = 5$, $\zeta_c = 0.6$ and $\omega_o = 10$, $\zeta_o = 0.6$. Compute the eigenvalues of the perturbed system when the process gain is increased by 2%. Also compute the loop transfer function and the sensitivity functions. Is there a way to know beforehand that the system will be highly sensitive?

12.16 (Robustness using the Nyquist criterion) Another view of robust performance can be obtained through appeal to the Nyquist criterion. Let $S_{max}(i\omega)$ represent a desired upper bound on our sensitivity function. Show that the system provides this level of performance subject to additive uncertainty Δ if the following inequality is satisfied:

$$|1 + \tilde{L}| = |1 + L + C\Delta| > \frac{1}{|S_{max}(i\omega)|} \qquad \text{for all } \omega \geq 0. \qquad (12.28)$$

Describe how to check this condition using a Nyquist plot.

Bibliography

[1] M. A. Abkowitz. *Stability and Motion Control of Ocean Vehicles*. MIT Press, Cambridge, MA, 1969.

[2] R. H. Abraham and C. D. Shaw. *Dynamics—The Geometry of Behavior, Part 1: Periodic Behavior*. Aerial Press, Santa Cruz, CA, 1982.

[3] J. Ackermann. Der Entwurf linearer Regelungssysteme im Zustandsraum. *Regelungstechnik und Prozessdatenverarbeitung*, 7:297–300, 1972.

[4] J. Ackermann. *Sampled-Data Control Systems*. Springer, Berlin, 1985.

[5] C. E. Agnew. Dynamic modeling and control of congestion-prone systems. *Operations Research*, 24(3):400–419, 1976.

[6] L. V. Ahlfors. *Complex Analysis*. McGraw-Hill, New York, 1966.

[7] B. D. O. Anderson and J. B. Moore. *Optimal Control Linear Quadratic Methods*. Prentice Hall, Englewood Cliffs, NJ, 1990. Republished by Dover Publications, 2007.

[8] A. A. Andronov, A. A. Vitt, and S. E. Khaikin. *Theory of Oscillators*. Dover, New York, 1987.

[9] T. M. Apostol. *Calculus, Vol. II: Multi-Variable Calculus and Linear Algebra with Applications*. Wiley, New York, 1967.

[10] T. M. Apostol. *Calculus, Vol. I: One-Variable Calculus with an Introduction to Linear Algebra*. Wiley, New York, 1969.

[11] R. Aris. *Mathematical Modeling Techniques*. Dover, New York, 1994. Originally published by Pitman, 1978.

[12] V. I. Arnold. *Mathematical Methods in Classical Mechanics*. Springer, New York, 1978.

[13] V. I. Arnold. *Ordinary Differential Equations*. MIT Press, Cambridge, MA, 1987. 10th printing 1998.

[14] K. J. Åström. Limitations on control system performance. *European Journal on Control*, 6(1):2–20, 2000.

[15] K. J. Åström. *Introduction to Stochastic Control Theory*. Dover, New York, 2006. Originally published by Academic Press, New York, 1970.

[16] K. J. Åström and T. Hägglund. *Advanced PID Control*. ISA—The Instrumentation, Systems, and Automation Society, Research Triangle Park, NC, 2005.

[17] K. J. Åström, R. E. Klein, and A. Lennartsson. Bicycle dynamics and control. *IEEE Control Systems Magazine*, 25(4):26–47, 2005.

[18] K. J. Åström and B. Wittenmark. *Computer-Control Systems: Theory and Design*, 3rd ed. Prentice Hall, Englewood Cliffs, NJ, 1997.

[19] K. J. Åström and B. Wittenmark. *Adaptive Control*, 2nd ed. Dover, New York, 2008. Originally published by Addison Wesley, 1995.

[20] D. P. Atherton. *Nonlinear Control Engineering*. Van Nostrand, New York, 1975.

[21] M. Atkinson, M. Savageau, J. Myers, and A. Ninfa. Development of genetic circuitry exhibiting toggle switch or oscillatory behavior in *Escherichia coli. Cell*, 113(5):597–607, 2003.

[22] M. B. Barron and W. F. Powers. The role of electronic controls for future automotive mechatronic systems. *IEEE Transactions on Mechatronics*, 1(1):80–89, 1996.

[23] T. Basar (editor). *Control Theory: Twenty-five Seminal Papers (1932–1981)*. IEEE Press, New York, 2001.

[24] T. Basar and P. Bernhard. *H$^\infty$-Optimal Control and Related Minimax Design Problems: A Dynamic Game Approach*. Birkhauser, Boston, 1991.

[25] J. Bechhoefer. Feedback for physicists: A tutorial essay on control. *Reviews of Modern Physics*, 77:783–836, 2005.

[26] R. Bellman and K. J. Åström. On structural identifiability. *Mathematical Biosciences*, 7:329–339, 1970.

[27] R. E. Bellman and R. Kalaba. *Selected Papers on Mathematical Trends in Control Theory*. Dover, New York, 1964.

[28] S. Bennett. *A History of Control Engineering: 1800–1930*. Peter Peregrinus, Stevenage, 1979.

[29] S. Bennett. *A History of Control Engineering: 1930–1955*. Peter Peregrinus, Stevenage, 1993.

[30] L. L. Beranek. *Acoustics*. McGraw-Hill, New York, 1954.

[31] R. N. Bergman. Toward physiological understanding of glucose tolerance: Minimal model approach. *Diabetes*, 38:1512–1527, 1989.

[32] D. Bertsekas and R. Gallager. *Data Networks*. Prentice Hall, Englewood Cliffs, 1987.

[33] B. Bialkowski. Process control sample problems. In N. J. Sell (editor), *Process Control Fundamentals for the Pulp & Paper Industry*. Tappi Press, Norcross, GA, 1995.

[34] G. Binnig and H. Rohrer. Scanning tunneling microscopy. *IBM Journal of Research and Development*, 30(4):355–369, 1986.

[35] H. S. Black. Stabilized feedback amplifiers. *Bell System Technical Journal*, 13:1–2, 1934.

[36] H. S. Black. Inventing the negative feedback amplifier. *IEEE Spectrum*, pp. 55–60, 1977.

[37] J. F. Blackburn, G. Reethof, and J. L. Shearer. *Fluid Power Control*. MIT Press, Cambridge, MA, 1960.

[38] J. H. Blakelock. *Automatic Control of Aircraft and Missiles*, 2nd ed. Addison-Wesley, Cambridge, MA, 1991.

[39] G. Blickley. Modern control started with Ziegler-Nichols tuning. *Control Engineering*, 37:72–75, 1990.

[40] H. W. Bode. *Network Analaysis and Feedback Amplifier Design*. Van Nostrand, New York, 1945.

[41] H. W. Bode. Feedback—The history of an idea. *Symposium on Active Networks and Feedback Systems*. Polytechnic Institute of Brooklyn, New York, 1960. Reprinted in [27].

[42] W. E. Boyce and R. C. DiPrima. *Elementary Differential Equations*. Wiley, New York, 2004.

[43] B. Brawn and F. Gustavson. Program behavior in a paging environment. *Proceedings of the AFIPS Fall Joint Computer Conference*, pp. 1019–1032, 1968.

[44] R. W. Brockett. *Finite Dimensional Linear Systems*. Wiley, New York, 1970.

[45] R. W. Brockett. New issues in the mathematics of control. In B. Engquist and W. Schmid (editors), *Mathematics Unlimited—2001 and Beyond*, pp. 189–220. Springer-Verlag, Berlin, 2000.

[46] G. S. Brown and D. P. Campbell. *Principles of Servomechanims*. Wiley, New York, 1948.

[47] A. E. Bryson, Jr. and Y.-C. Ho. *Applied Optimal Control: Optimization, Estimation, and Control*. Wiley, New York, 1975.

[48] F. M. Callier and C. A. Desoer. *Linear System Theory*. Springer-Verlag, London, 1991.

[49] R. H. Cannon. *Dynamics of Physical Systems*. Dover, New York, 2003. Originally published by McGraw-Hill, 1967.

[50] H. S. Carslaw and J. C. Jaeger. *Conduction of Heat in Solids*, 2nd ed. Clarendon Press, Oxford, UK, 1959.

[51] H. Chestnut and R. W. Mayer. *Servomechanisms and Regulating System Design,* Vol. 1. Wiley, New York, 1951.

[52] C. Cobelli and G. Toffolo. Model of glucose kinetics and their control by insulin, compartmental and non-compartmental approaches. *Mathematical Biosciences*, 72(2):291–316, 1984.

[53] R. F. Coughlin and F. F. Driscoll. *Operational Amplifiers and Linear Integrated Circuits*, 6th ed. Prentice Hall, Englewood Cliffs, NJ, 1975.

[54] L. B. Cremean, T. B. Foote, J. H. Gillula, G. H. Hines, D. Kogan, K. L. Kriechbaum, J. C. Lamb, J. Leibs, L. Lindzey, C. E. Rasmussen, A. D. Stewart, J. W. Burdick, and R. M. Murray. Alice: An information-rich autonomous vehicle for high-speed desert navigation. *Journal of Field Robotics*, 23(9):777–810, 2006.

[55] Crocus. *Systemes d'Exploitation des Ordinateurs*. Dunod, Paris, 1975.

[56] H. de Jong. Modeling and simulation of genetic regulatory systems: A literature review. *Journal of Computational Biology*, 9:67–103, 2002.

[57] J. P. Den Hartog. *Mechanical Vibrations*. Dover, New York, 1985. Reprint of 4th ed. from 1956; 1st ed. published in 1934.

[58] L. Desborough and R. Miller. Increasing customer value of industrial control performance monitoring—Honeywell's experience. *Sixth International Conference on Chemical Process Control*. AIChE Symposium Series Number 326 (Vol. 98), 2002.

[59] Y. Diao, N. Gandhi, J. L. Hellerstein, S. Parekh, and D. M. Tilbury. Using MIMO feedback control to enforce policies for interrelated metrics with application to the Apache web server. *Proceedings of the IEEE/IFIP Network Operations and Management Symposium*, pp. 219–234, 2002.

[60] E. D. Dickmanns. *Dynamic Vision for Perception and Control of Motion*. Springer, Berlin, 2007.

[61] R. C. Dorf and R. H. Bishop. *Modern Control Systems*, 10th ed. Prentice Hall, Upper Saddle River, NJ, 2004.

[62] F. H. Dost. *Grundlagen der Pharmakokinetik*. Thieme Verlag, Stuttgart, 1968.

[63] J. C. Doyle. Guaranteed margins for LQG regulators. *IEEE Transactions on Automatic Control*, 23(4):756–757, 1978.

[64] J. C. Doyle, B. A. Francis, and A. R. Tannenbaum. *Feedback Control Theory*. Macmillan, New York, 1992.

[65] J. C. Doyle, K. Glover, P. P. Khargonekar, and B. A. Francis. State-space solutions to standard H_2 and H_∞ control problems. *IEEE Transactions on Automatic Control*, 34(8):831–847, 1989.

[66] L. E. Dubins. On curves of minimal length with a constraint on average curvature, and with prescribed initial and terminal positions and tangents. *American Journal of Mathematics*, 79:497–516, 1957.

[67] F. Dyson. A meeting with Enrico Fermi. *Nature*, 247(6972):297, 2004.

[68] H. El-Samad, J. P. Goff, and M. Khammash. Calcium homeostasis and parturient hypocalcemia: An integral feedback perspective. *Journal of Theoretical Biology*, 214:17–29, 2002.

[69] J. R. Ellis. *Vehicle Handling Dynamics*. Mechanical Engineering Publications, London, 1994.

[70] S. P. Ellner and J. Guckenheimer. *Dynamic Models in Biology*. Princeton University Press, Princeton, NJ, 2005.

[71] M. B. Elowitz and S. Leibler. A synthetic oscillatory network of transcriptional regulators. *Nature*, 403(6767):335–338, 2000.

[72] P. G. Fabietti, V. Canonico, M. O. Federici, M. Benedetti, and E. Sarti. Control oriented model of insulin and glucose dynamics in type 1 diabetes. *Medical and Biological Engineering and Computing*, 44:66–78, 2006.

[73] M. Fliess, J. Levine, P. Martin, and P. Rouchon. On differentially flat nonlinear systems. *Comptes Rendus des Séances de l'Académie des Sciences,* Serie I, 315:619–624, 1992.

[74] M. Fliess, J. Levine, P. Martin, and P. Rouchon. Flatness and defect of non-linear systems: Introductory theory and examples. *International Journal of Control*, 61(6):1327–1361, 1995.

[75] J. W. Forrester. *Industrial Dynamics*. MIT Press, Cambridge, MA, 1961.

[76] J. B. J. Fourier. On the propagation of heat in solid bodies. Memoir, read before the Class of the Instut de France, 1807.

[77] A. Fradkov. *Cybernetical Physics: From Control of Chaos to Quantum Control*. Springer, Berlin, 2007.

[78] B. A. Francis. *A Course in $\mathcal{H}_\infty$ Control*. Springer-Verlag, Berlin, 1987.

[79] G. F. Franklin, J. D. Powell, and A. Emami-Naeini. *Feedback Control of Dynamic Systems*, 5th ed. Prentice Hall, Upper Saddle River, NJ, 2005.

[80] B. Friedland. *Control System Design: An Introduction to State Space Methods*. Dover, New York, 2004.

[81] M. A. Gardner and J. L. Barnes. *Transients in Linear Systems*. Wiley, New York, 1942.

[82] E. Gilbert. Controllability and observability in multivariable control systems. *SIAM Journal of Control*, 1(1):128–151, 1963.

[83] J. C. Gille, M. J. Pelegrin, and P. Decaulne. *Feedback Control Systems; Analysis, Synthesis, and Design*. McGraw-Hill, New York, 1959.

[84] M. Giobaldi and D. Perrier. *Pharmacokinetics*, 2nd ed. Marcel Dekker, New York, 1982.

[85] K. Godfrey. *Compartment Models and Their Application*. Academic Press, New York, 1983.

[86] H. Goldstein. *Classical Mechanics*. Addison-Wesley, Cambridge, MA, 1953.

[87] S. W. Golomb. Mathematical models—Uses and limitations. *Simulation*, 4(14):197–198, 1970.

[88] G. C. Goodwin, S. F. Graebe, and M. E. Salgado. *Control System Design*. Prentice Hall, Upper Saddle River, NJ, 2001.

[89] D. Graham and D. McRuer. *Analysis of Nonlinear Control Systems*. Wiley, New York, 1961.

[90] M. Green and D. J. N. Limebeer. *Linear Robust Control*. Prentice Hall, Englewood Cliffs, NJ, 1995.

[91] J. Guckenheimer and P. Holmes. *Nonlinear Oscillations, Dynamical Systems, and Bifurcations of Vector Fields*. Springer-Verlag, Berlin, 1983.

[92] E. A. Guillemin. *Theory of Linear Physical Systems*. MIT Press, Cambridge, MA, 1963.

[93] L. Gunkel and G. F. Franklin. A general solution for linear sampled data systems. *IEEE Transactions on Automatic Control*, AC-16:767–775, 1971.

[94] W. Hahn. *Stability of Motion*. Springer, Berlin, 1967.

[95] D. Hanahan and R. A. Weinberg. The hallmarks of cancer. *Cell*, 100:57–70, 2000.

[96] J. K. Hedrick and T. Batsuen. Invariant properties of automobile suspensions. *Proceedigns of the Institution of Mechanical Engineers*, Vol. 204, pp. 21–27, London, 1990.

[97] J. L. Hellerstein, Y. Diao, S. Parekh, and D. M. Tilbury. *Feedback Control of Computing Systems*. Wiley, New York, 2004.

[98] D. V. Herlihy. *Bicycle—The History*. Yale University Press, New Haven, CT, 2004.

[99] M. B. Hoagland and B. Dodson. *The Way Life Works*. Times Books, New York, 1995.

[100] A. L. Hodgkin and A. F. Huxley. A quantitative description of membrane current and its application to conduction and excitation in nerve. *Journal of Physiology*, 117(500–544), 1952.

[101] C. V. Hollot, V. Misra, D. Towsley, and W-B. Gong. A control theoretic analysis of RED. *Proceedings of IEEE Infocom*, pp. 1510–1519, 2000.

[102] I. M. Horowitz. *Synthesis of Feedback Systems*. Academic Press, New York, 1963.

[103] I. M. Horowitz. Superiority of transfer function over state-variable methods in linear, time-invariant feedback system design. *IEEE Transactions on Automatic Control*, AC-20(1):84–97, 1975.

[104] I. M. Horowitz. Survey of quantitative feedback theory. *International Journal of Control*, 53:255291, 1991.

[105] T. P. Hughes. *Elmer Sperry: Inventor and Engineer*. John Hopkins University Press, Baltimore, MD, 1993.

[106] A. Isidori. *Nonlinear Control Systems*, 3rd ed. Springer-Verlag, Berlin, 1995.

[107] M. Ito. Neurophysiological aspects of the cerebellar motor system. *International Journal of Neurology*, 7:162178, 1970.

[108] V. Jacobson. Congestion avoidance and control. *ACM SIGCOMM Computer Communication Review*, 25:157–173, 1995.

[109] J. A. Jacquez. *Compartment Analysis in Biology and Medicine*. Elsevier, Amsterdam, 1972.

[110] H. James, N. Nichols, and R. Phillips. *Theory of Servomechanisms*. McGraw-Hill, New York, 1947.

[111] P. D. Joseph and J. T. Tou. On linear control theory. *Transactions of the AIEE*, 80(18), 1961.

[112] W. G. Jung (editor). *Op Amp Applications*. Analog Devices, Norwood, MA, 2002.

[113] R. E. Kalman. Contributions to the theory of optimal control. *Boletin de la Sociedad Matématica Mexicana*, 5:102–119, 1960.

[114] R. E. Kalman. New methods and results in linear prediction and filtering theory. Technical Report 61-1. Research Institute for Advanced Studies (RIAS), Baltimore, MD, February 1961.

[115] R. E. Kalman. On the general theory of control systems. *Proceedings of the First IFAC Congress on Automatic Control, Moscow, 1960*, Vol. 1, pp. 481–492. Butterworths, London, 1961.

[116] R. E. Kalman and R. S. Bucy. New results in linear filtering and prediction theory. *Transactions of the ASME (Journal of Basic Engineering)*, 83 D:95–108, 1961.

[117] R. E. Kalman, P. L. Falb, and M. A. Arbib. *Topics in Mathematical System Theory*. McGraw-Hill, New York, 1969.

[118] R. E. Kalman, Y. Ho, and K. S. Narendra. *Controllability of Linear Dynamical Systems*, Vol. 1 of *Contributions to Differential Equations*. Wiley, New York, 1963.

[119] J. Keener and J. Sneyd. *Mathematical Physiology*. Springer, New York, 2001.

[120] F. P. Kelly. Stochastic models of computer communication. *Journal of the Royal Statistical Society*, B47(3):379–395, 1985.

[121] K. Kelly. *Out of Control*. Addison-Wesley, Reading, MA, 1994. Available at http://www.kk. org/outofcontrol.

[122] J. M. Keynes. *The General Theory of Employment, Interest and Money*. Cambridge Universtiy Press, Cambridge, UK, 1936.

[123] H. K. Khalil. *Nonlinear Systems*, 3rd ed. Macmillan, New York, 2001.

[124] U. Kiencke and L. Nielsen. *Automotive Control Systems: For Engine, Driveline, and Vehicle*. Springer, Berlin, 2000.

[125] C. Kittel. *Introduction to Solid State Physics*. Wiley, New York, 1995.

[126] L. R. Klein and A. S. Goldberger. *An Econometric Model of the United States 1929–1952*. North Holland, Amsterdam, 1955.

[127] L. Kleinrock. *Queuing Systems,* Vols. I and II, 2nd ed. Wiley-Interscience, New York, 1975.

[128] N. N. Krasovski. *Stability of Motion*. Stanford University Press, Stanford, CA, 1963.

[129] M. Krstić, I. Kanellakopoulos, and P. Kokotović. *Nonlinear and Adaptive Control Design*. Wiley, 1995.

[130] P. R. Kumar. New technological vistas for systems and control: The example of wireless networks. *Control Systems Magazine*, 21(1):24–37, 2001.

[131] P. R. Kumar and P. Varaiya. *Stochastic Systems: Estimation, Identification, and Adaptive Control*. Prentice Hall, Englewood Cliffs, NJ, 1986.

[132] P. Kundur. *Power System Stability and Control*. McGraw-Hill, New York, 1993.

[133] B. C. Kuo and F. Golnaraghi. *Automatic Control Systems*, 8th ed. Wiley, New York, 2002.

[134] M. Kurth and E. Welfonder. Oscillation behavior of the enlarged European power system under deregulated energy market conditions. *Control Engineering Practice*, 13:1525–1536, 2005.

[135] J. P. LaSalle. Some extensions of Lyapunov's second method. *IRE Transactions on Circuit Theory*, CT-7(4):520–527, 1960.

[136] A. D. Lewis. A mathematical approach to classical control. Technical report. Queens University, Kingston, Ontario, 2003.

[137] S. H. Low, F. Paganini, and J. C. Doyle. Internet congestion control. *IEEE Control Systems Magazine*, pp. 28–43, February 2002.

[138] S. H. Low, F. Paganini, J. Wang, S. Adlakha, and J. C. Doyle. Dynamics of TCP/RED and a scalable control. *Proceedings of IEEE Infocom*, pp. 239–248, 2002.

[139] K. H. Lundberg. History of analog computing. *IEEE Control Systems Magazine*, pp. 22–28, March 2005.

[140] L.A. MacColl. *Fundamental Theory of Servomechanims*. Van Nostrand, Princeton, NJ, 1945. Dover reprint 1968.

[141] J. M. Maciejowski. *Multivariable Feedback Design*. Addison Wesley, Reading, MA, 1989.

[142] D. A. MacLulich. *Fluctuations in the Numbers of the Varying Hare (Lepus americanus)*. University of Toronto Press, 1937.

[143] A. Makroglou, J. Li, and Y. Kuang. Mathematical models and software tools for the glucose-insulin regulatory system and diabetes: An overview. *Applied Numerical Mathematics*, 56:559–573, 2006.

[144] J. G. Malkin. *Theorie der Stabilität einer Bewegung*. Oldenbourg, München, 1959.

[145] R. Mancini. *Op Amps for Everyone*. Texas Instruments, Houston. TX, 2002.

[146] J. E. Marsden and M. J. Hoffmann. *Basic Complex Analysis*. W. H. Freeman, New York, 1998.

[147] J. E. Marsden and T. S. Ratiu. *Introduction to Mechanics and Symmetry*. Springer-Verlag, New York, 1994.

[148] O. Mayr. *The Origins of Feedback Control*. MIT Press, Cambridge, MA, 1970.

[149] M. W. McFarland (editor). *The Papers of Wilbur and Orville Wright*. McGraw-Hill, New York, 1953.

[150] D. C. McFarlane and K. Glover. *Robust Controller Design Using Normalized Coprime Factor Plant Descriptions*. Springer, New York, 1990.

[151] H. T. Milhorn. *The Application of Control Theory to Physiological Systems*. Saunders, Philadelphia, 1966.

[152] D. A. Mindel. *Between Human and Machine: Feedback, Control, and Computing Before Cybernetics*. Johns Hopkins University Press, Baltimore, MD, 2002.

[153] D. Möhl, G. Petrucci, L. Thorndahl, and S. van der Meer. Physics and technique of stochastic cooling. *Physics Reports*, 58(2):73–102, 1980.

[154] J. D. Murray. *Mathematical Biology,* Vols. I and II, 3rd ed. Springer-Verlag, New York, 2004.

[155] R. M. Murray (editor). *Control in an Information Rich World: Report of the Panel on Future Directions in Control, Dynamics and Systems*. SIAM, Philadelphia, 2003.

[156] R. M. Murray, Z. Li, and S. S. Sastry. *A Mathematical Introduction to Robotic Manipulation*. CRC Press, 1994.

[157] P. J. Nahin. *Oliver Heaviside: Sage in Solitude: The Life, Work and Times of an Electrical Genius of the Victorian Age*. IEEE Press, New York, 1988.

[158] A. O. Nier. Evidence for the existence of an isotope of potassium of mass 40. *Physical Review*, 48:283–284, 1935.

[159] H. Nijmeijer and J. M. Schumacher. Four decades of mathematical system theory. In J. W. Polderman and H. L. Trentelman (editors), *The Mathematics of Systems and Control: From Intelligent Control to Behavioral Systems*, pp. 73–83. University of Groningen, 1999.

[160] H. Nyquist. Regeneration theory. *Bell System Technical Journal*, 11:126–147, 1932.

[161] H. Nyquist. The regeneration theory. In R. Oldenburger (editor), *Frequency Response*, p. 3. MacMillan, New York, 1956.

[162] K. Ogata. *Modern Control Engineering*, 4th ed. Prentice Hall, Upper Saddle River, NJ, 2001.

[163] R. Oldenburger (editor). *Frequency Response*. MacMillan, New York, 1956.

[164] G. Pacini and R. N. Bergman. A computer program to calculate insulin sensitivity and pancreatic responsivity from the frequently sampled intraveneous glucose tolerance test. *Computer Methods and Programs in Biomedicine*, 23:113–122, 1986.

[165] G. A. Philbrick. Designing industrial controllers by analog. *Electronics*, 21(6):108–111, 1948.

[166] W. F. Powers and P. R. Nicastri. Automotive vehicle control challenges in the 21st century. *Control Engineering Practice*, 8:605–618, 2000.

[167] S. Prajna, A. Papachristodoulou, and P. A. Parrilo. SOSTOOLS: Sum of squares optimization toolbox for MATLAB, 2002. Available from http://www.cds.caltech.edu/sostools.

[168] D. S. Riggs. *The Mathematical Approach to Physiological Problems*. MIT Press, Cambridge, MA, 1963.

[169] H. H. Rosenbrock and P. D. Moran. Good, bad or optimal? *IEEE Transactions on Automatic Control*, AC-16(6):552–554, 1971.

[170] F. Rowsone, Jr. What it's like to drive an auto-pilot car. *Popular Science Monthly*, April 1958. Available at http://www.imperialclub.com/ImFormativeArticles/1958AutoPilot.

[171] W. J. Rugh. *Linear System Theory*, 2nd ed. Prentice Hall, Englewood Cliffs, NJ, 1995.

[172] E. B. Saff and A. D. Snider. *Fundamentals of Complex Analysis with Applications to Engineering, Science and Mathematics*. Prentice Hall, Englewood Cliffs, NJ, 2002.

[173] D. Sarid. *Atomic Force Microscopy*. Oxford University Press, Oxford, UK, 1991.

[174] S. Sastry. *Nonlinear Systems*. Springer, New York, 1999.

[175] G. Schitter. High performance feedback for fast scanning atomic force microscopes. *Review of Scientific Instruments*, 72(8):3320–3327, 2001.

[176] G. Schitter, K. J. Åström, B. DeMartini, P. J. Thurner, K. L. Turner, and P. K. Hansma. Design and modeling of a high-speed AFM-scanner. *IEEE Transactions on Control System Technology*, 15(5):906–915, 2007.

[177] M. Schwartz. *Telecommunication Networks*. Addison Wesley, Reading, MA, 1987.

[178] D. E. Seborg, T. F. Edgar, and D. A. Mellichamp. *Process Dynamics and Control*, 2nd ed. Wiley, Hoboken, NJ, 2004.

[179] S. D. Senturia. *Microsystem Design*. Kluwer, Boston, MA, 2001.

[180] F. G. Shinskey. *Process-Control Systems. Application, Design, and Tuning*, 4th ed. McGraw-Hill, New York, 1996.

[181] S. Skogestad and I Postlethwaite. *Multivariable Feedback Control*, 2nd ed. Wiley, Hoboken, NJ, 2005.

[182] E. P. Sontag. *Mathematical Control Theory: Deterministic Finite Dimensional Systems*, 2nd ed. Springer, New York, 1998.

[183] M. W. Spong and M. Vidyasagar. *Dynamics and Control of Robot Manipulators*. John Wiley, 1989.

[184] L. Stark. *Neurological Control Systems—Studies in Bioengineering*. Plenum Press, New York, 1968.

[185] G. Stein. Respect the unstable. *Control Systems Magazine*, 23(4):12–25, 2003.

[186] J. Stewart. *Calculus: Early Transcendentals*. Brooks Cole, Pacific Grove, CA, 2002.

[187] G. Strang. *Linear Algebra and Its Applications*, 3rd ed. Harcourt Brace Jovanovich, San Diego, 1988.

[188] S. H. Strogatz. *Nonlinear Dynamics and Chaos, with Applications to Physics, Biology, Chemistry, and Engineering*. Addison-Wesley, Reading, MA, 1994.

[189] A. S. Tannenbaum. *Computer Networks*, 3rd ed. Prentice Hall, Upper Saddle River, NJ, 1996.

[190] T. Teorell. Kinetics of distribution of substances administered to the body, I and II. *Archives Internationales de Pharmacodynamie et de Therapie*, 57:205–240, 1937.

[191] G. T. Thaler. *Automatic Control Systems*. West Publishing, St. Paul, MN, 1989.

[192] M. Tiller. *Introduction to Physical Modeling with Modelica*. Springer, Berlin, 2001.

[193] D. Tipper and M. K. Sundareshan. Numerical methods for modeling computer networks under nonstationary conditions. *IEEE Journal of Selected Areas in Communications*, 8(9):1682–1695, 1990.

[194] J. G. Truxal. *Automatic Feedback Control System Synthesis*. McGraw-Hill, New York, 1955.

[195] H. S. Tsien. *Engineering Cybernetics*. McGraw-Hill, New York, 1954.

[196] G. Vinnicombe. *Uncertainty and Feedback: $\mathcal{H}_\infty$ Loop-Shaping and the ν-Gap Metric*. Imperial College Press, London, 2001.

[197] F. J. W. Whipple. The stability of the motion of a bicycle. *Quarterly Journal of Pure and Applied Mathematics*, 30:312–348, 1899.

[198] D. V. Widder. *Laplace Transforms*. Princeton University Press, Princeton, NJ, 1941.

[199] E. P. M. Widmark and J. Tandberg. Über die Bedingungen für die Akkumulation indifferenter Narkotika. *Biochemische Zeitung*, 148:358–389, 1924.

[200] N. Wiener. *Cybernetics: Or Control and Communication in the Animal and the Machine*. Wiley, 1948.

[201] S. Wiggins. *Introduction to Applied Nonlinear Dynamical Systems and Chaos*. Springer-Verlag, Berlin, 1990.

[202] D. G. Wilson. *Bicycling Science*, 3rd ed. MIT Press, Cambridge, MA, 2004. With contributions by Jim Papadopoulos.

[203] H. R. Wilson. *Spikes, Decisions, and Actions: The Dynamical Foundations of Neuroscience*. Oxford University Press, Oxford, UK, 1999.

[204] K. A. Wise. Guidance and control for military systems: Future challenges. *AIAA Conference on Guidance, Navigation, and Control*, 2007. AIAA Paper 2007-6867.

[205] S. Yamamoto and I. Hashimoto. Present status and future needs: The view from Japanese industry. In Y. Arkun and W. H. Ray (editors), *Chemical Process Control—CPC IV*, 1991.

[206] T.-M. Yi, Y. Huang, M. I. Simon, and J. Doyle. Robust perfect adaptation in bacterial chemotaxis through integral feedback control. *PNAS*, 97:4649–4653, 2000.

[207] L. A. Zadeh and C. A. Desoer. *Linear System Theory: the State Space Approach*. McGraw-Hill, New York, 1963.

[208] G. Zames. Feedback and optimal sensitivity: Model reference transformations, multiplicative seminorms, and approximative inverse. *IEEE Transactions on Automatic Control*, AC-26(2):301–320, 1981.

[209] J. C. Zhou, J. C. Doyle, and K. Glover. *Robust and Optimal Control*. Prentice Hall, Englewood Cliffs, NJ, 1996.

[210] J. G. Ziegler and N. B. Nichols. Optimum settings for automatic controllers. *Transactions of the ASME*, 64:759–768, 1942.

Index

access control, *see* admission control

acknowledgment (ack) packet, 77–79

activator, 16, 59, 129

active filter, 154, *see also* operational amplifier

actuators, 4, 31, 51, 65, 81, 178, 224, 265, 284, 311, 324, 333–335, 337
  effect on zeros, 284, 334
  in computing systems, 75
  saturation, 50, 225, 300, 306–307, 311, 324

A/D converters, *see* analog-to-digital converters

adaptation, 297

adaptive control, 20, 373, 374

additive uncertainty, 349, 353, 356, 376

admission control, 54, 63, 78, 79, 274

advertising, 15

aerospace systems, 8–9, 18, 338, *see also* vectored thrust aircraft; X-29 aircraft

aircraft, *see* flight control

alcohol, metabolism of, 93

algebraic loops, 211, 249–250

aliasing, 225

all-pass transfer function, 331

alternating current (AC), 7, 155

amplifier, *see* operational amplifier

amplitude ratio, *see* gain

analog computing, 51, 71, 250, 309

analog implementation, controllers, 74, 263, 309–311

analog-to-digital converters, 4, 82, 224, 225, 311

analytic function, 236

anticipation, in controllers, 6, 24, 296, *see also* derivative action

antiresonance, 156

anti-windup compensation, 306–307, 311, 312, 314

Apache web server, 76, *see also* web server control

apparent volume of distribution, 86, 93

Arbib, M. A., 167

argument, of a complex number, 250

arrival rate (queuing systems), 55

artificial intelligence (AI), 12, 20

asymptotes, in Bode plot, 253, 254

asymptotic stability, 42, 102–106, 112, 114, 117, 118, 120, 140
  discrete-time systems, 165

atmospheric dynamics, *see* environmental science

atomic force microscopes, 3, 51, 81–84
  contact mode, 81, 156
  horizontal positioning, 282, 366
  system identification, 257
  tapping mode, 81, 290, 299, 304, 328
  with preloading, 93

attractor (equilibrium point), 104

automatic reset, in PID control, 296

automatic tuning, 306, 373

automotive control systems, 6, 21, 51, 69, *see also* cruise control; vehicle steering

autonomous differential equation, 29, *see also* time-invariant systems

autonomous vehicles, 8, 20–21

autopilot, 6, 19

balance systems, 35–37, 49, 170, 188, 241, 334, *see also* cart-pendulum system; inverted pendulum

band-pass filter, 154, 155, 255, 256

bandwidth, 155, 186, 322, 333

Bell Labs, 18, 290

Bennett, S., 25, 290, 312

bicycle dynamics, 69–71, 91, 123, 226
  Whipple model, 71

bicycle model, for vehicle steering, 51–53

bifurcations, 121–124, 130, *see also* root locus plots

biological circuits, 16, 45, 58–60, 129, 166, 256
  genetic switch, 64, 114
  repressilator, 59–60

biological systems, 1–3, 10, 15–16, 22, 25, 58–61, 126, 293, 297, *see also* biological circuits; drug administration; neural systems; population dynamics

bistability, 22, 117

Black, H. S., 18, 20, 71, 73, 131, 267, 290, 347

block diagonal systems, 106, 129, 139, 145, 149, 212